Cosmic Discovery

COSMIC DISCOVERY

The Search, Scope, and Heritage of Astronomy

MARTIN HARWIT

Basic Books, Inc., Publishers

NEW YORK

Library of Congress Cataloging in Publication Data

Harwit, Martin, 1931–
  Cosmic discovery.

  Includes bibliographical references and index.
  1. Astronomy. I. Title.
QB43.2.H37   520   80–68172
ISBN: 0–465–01428–3   AACR2

CONTENTS

FOREWORD

When I was a student in college, just after World II, my friends and I naturally wanted to know what fields of science held the greatest excitement, the largest number of potential discoveries. We wanted to participate in the scientific enterprise and needed to know where to start. Yet no one could guide us, except in the vaguest of ways.

At roughly the same time, I imagine, planners in Washington similarly were wondering what fields of science might promise immediate advances and striking returns—where incentives might be offered to young graduate students in the form of fellowships and where new facilities might be established for novel ventures.

Such questions still remain largely unanswered. Planning commissions and advisory boards still grapple with them and seem no closer to their goal than they were three decades ago.

Cosmic Discovery is a first attempt to collect the kind of information that might be needed to answer questions on the promise of a particular science. It restricts itself to one part of one discipline and asks, How was it that we first came to discover the major phenomena we now observe in the universe? Who were the individuals responsible for the discoveries? How had they prepared for their careers? What methods led to their successes? In a different vein, the book also asks, What is the scope of future astronomical discovery? How many major cosmic phenomena remain to be found? How much remains to be done? Finally we can take all the information we can gather and ask, Are there lessons we can learn from earlier searches? Can we plan future enterprises to make them more effective than our efforts of the past? Is an imposed national plan likely to be more successful than the striving of individual, motivated scientists?

I have attempted to answer these questions by collecting the information needed to arrive at well-informed conclusions. And though these efforts must be regarded as no more than a start along a very long trek, there are new findings that clearly stand out even at this early stage of the search. I hope this approach will prove useful to others.

ACKNOWLEDGMENTS

I thank many colleagues, friends, and members of my family for their help while I was writing this book. An early draft of the manuscript was critically read by Bart J. Bok, F. Westy Dain, and Thomas A. Pauls. Later versions were read by Kenneth Brecher, Thomas F. Gieryn, Eric Harwit, Gina and Felix Haurowitz, William E. Howard III, Franklin D. Martin, Heinrich Pfeiffer, Johannes Schmid-Burgk, Alice H. Sievert, Woodruff T. Sullivan III, and Ira M. Wasserman. I thank them all for their patience, for suggestions they made, and for strongly arguing their points of view, which often differed from mine. Together they have left a strong imprint on my presentation of the subject.

Much of the history of modern astronomy has not yet been written. I thank countless colleagues in astronomy who talked to me about their personal recollections, wrote me letters, sent me material that I have quoted, and permitted extensive interviews. Without their willing collaboration the book would have lacked basic authenticity.

Successive drafts of the book were typed by Gabriele Breuer, Judith A. Marcus, Sylvia Corbin, and Barbara Davidson. Drawings were prepared by Barbara Boettcher and Walter Fusshöller. It is a pleasure to thank them for their help.

Early support for the studies that led to this book was received from the Section on the History and Philosophy of Science at the National Science Foundation. The first draft of the book was completed during the fall of 1976 while I was working at the Max-Planck Institute for Radioastronomy in Bonn, with a Senior U.S. Scientist Award from the Alexander von Humboldt Foundation of the Federal Republic of Germany. I thank Dr. Peter G. Mezger, director of the Institute, for his hospitality during that year.

The manuscript was brought to completion at Cornell University. I am grateful to students in my class, Astronomy 215, who critically read the manuscript and provided me with some of the most uninhibited commentary I received.

A book like this gives the author an opportunity to see what a wonderful institution a university can be: I thank my colleagues at Cornell, Richard N. Boyd, Terrence L. Fine, Robert McGinnis, L. Pearce Williams, and the late Raymond Bowers for elucidating conversations, respectively, on science and government, philosophy of science, viable theories of probability, the sociology of science and the history of science. I also acknowledge a most enjoyable discussion with Robert K. Merton of Columbia University and thank him for his encouraging remarks.

Finally, it is a pleasure to acknowledge the enjoyable collaboration with Martin Kessler of Basic Books and with his staff, particularly Maureen Bischoff and Ruth Gales, who brought the book into print.

PERMISSIONS

The following publishers have given their permission to reproduce copyrighted material from books and articles:

The American Association for the Advancement of Science for permission to quote from articles in the journal *Science* by J. Schmandt, M. G. Morgan, R. Orbach, A. C. Leopold, L. J. Carter, and E. B. Staats, published in 1978 and 1979.

Change Magazine Press for permission to quote from *The State of Academic Science* by Bruce L. R. Smith and Joseph Karlesky, 1977.

University of Chicago Press for permission to quote from *Personal Knowledge—Toward a Post-Critical Philosophy* by Michael Polanyi.

The Colorado Associated University Press for permission to quote from *Cosmology, Fusion and Other Matters—George Gamow Memorial Volume*, edited by Frederick Reines, Boulder Colorado, 1972.

DAEDALUS, *Journal of the American Academy of Arts and Sciences*, Boston, Massachusetts, for permission to quote from their Fall 1977 issue on *Discoveries and Interpretation: Studies in Contemporary Scholarship*, "X-Ray Astronomy" by Bruno Rossi.

The Humanities Press for permission to quote from *Boston Studies in the Philosophy of Science 2* (1965).

D. Reidel Publishing Company, for permission to quote writings of Bruno Rossi and Riccardo Giacconi from *X-Ray Astronomy*, published in 1974.

John Wiley and Sons for permission to quote from D. Edge and M. Mulkay's book *Astronomy Transformed—The Emergence of Radioastronomy in Britain*, published in 1976.

The Hale Observatory, Harvard College Observatory, Kitt Peak National Observatory, Lick Observatory, Royal Greenwich Observatory, and Yerkes Observatory are to be thanked for permission to reproduce photographs; the Royal Astronomical Society for permission to quote from an article appearing in its *Quarterly Journal;* and the journals *Astronomy and Astrophysics, Astronomical Journal, Astrophysical Journal, Nature,* and *Publications of the Astronomical Society of Japan* for permission to reproduce pictures and quotations. Finally, I particularly wish to thank colleagues who permitted me to use photographs or drawings they had produced or to quote their writings.

Preface

Cosmic Discovery is an investigation into the complexity of the universe. It is addressed to a wide range of readers interested in astronomical discovery from the astronomer's, the historian's, and the policymaker's points of view. I have tried to avoid the specialized jargon of each of these three fields and have appended a glossary* to explain those technical and lesser-known terms and abbreviations that had to be included.

The book contains five chapters. The first summarizes the most important findings and conclusions of the study. Readers largely interested in ideas and results, rather than substantiation and evidence, may find themselves satisfied by this chapter-length essay. Others, particularly professionals more interested in a thorough examination of the subject, will find full documentation in the remaining four chapters—and may, in fact, prefer to read chapter 1 only after reading these later chapters. Extensive bibliographic notes facilitate access to original sources. Two technical appendices containing tables and background material complete the text.

The universe contains stars that shine steadily like the sun, variable stars that pulsate regularly, and eruptive variables that periodically flare. There are supernovae, pulsars, and X-ray stars. Clouds of luminous gas permeate the spaces around bright blue stars, while dark clouds of dust linger just beyond. Faint red stars in the hundreds of thousands aggregate in globular clusters. Galaxies that rival or exceed the Milky Way in size populate the universe out to all distances our telescopes can reach. Here and there galaxies emit their energy, not in visible

* See the Glossary/Index beginning on page 315 for these explanations.

light but predominantly as X rays or infrared radiation. Clusters of galaxies abound. Quasars are interspersed, some seemingly ejecting mass at velocities exceeding the speed of light.

These phenomena and the circumstances of their discovery are described in chapter 2, which is meant to provide not only factual information but also a sense for the flavor of cosmic discovery: We encounter immense variations in scale—size, luminosity, variability, energy of emitted particles and waves—that differentiate some forty-three phenomena. We see the role that theory and ideas play in the discovery of each new phenomenon. We become aware that many of the discoverers come to astronomy from other disciplines, bringing with them new tools with which they stumble on the unexpected.

Many cosmic phenomena have only come to be recognized in the past thirty-five years, largely through the introduction into astronomy of radio, X-ray, infrared, and gamma-ray techniques. None of the new phenomena had been anticipated before World War II, and it is natural to wonder how many more remain unrecognized even today, how rich and complex the universe might be. Further, if technological advances already have helped us uncover so many new cosmic features, how many more innovations of similar kinds could we put to use in future cosmic searches?

These are some of the questions I will seek to answer in chapter 3. And while this attempt may appear brash or even presumptuous, there are good reasons why success may be expected in astronomy, though it eludes us in other sciences.

Astronomy is largely an observational science, and for at least the next century our technology will be insufficiently advanced to permit exploration of the universe beyond the solar system. The distances simply are too great. So enormous is the distance to the nearest stars, so overwhelming our separation from the nearest galaxies, that these journeys might never be tried even in remote future aeons.

Because astronomy is so dependent on observations, it is relatively simple to assess the impact that further technological advances are likely to make. In the experimental sciences such an assessment would be far more complex. The experimentalist studies a system by imposing constraints and observing the system's response to a controlled stimulus. The variety of these constraints and of stimuli may be extended at will, and experiments can become arbitrarily complex (figure P.1).

Astronomy is different. The observer has only two choices. He can seek to detect and analyze signals incident from the sky, or he may choose to ignore them. But he has no way of stimulating a cosmic source to alter its emission. He can only observe what is offered. He is entirely dependent on the carriers of information that transmit to him all he may learn about the universe.

Information carriers, however, are not infinite in their variety. All the information we currently have about the universe beyond the solar system has been transmitted to us by means of electromagnetic radiation (light, radio waves, X rays, infrared radiation) or by cosmic-ray particles (electrons and atomic nuclei). Two other carriers are known in physics,

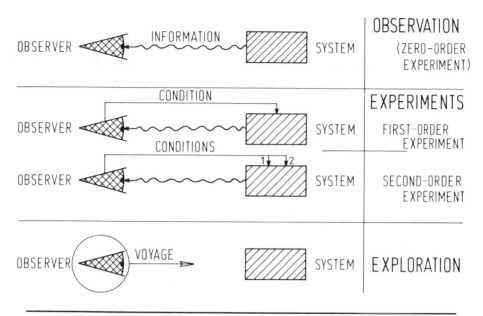

Figure P.1 *Observation, Experiments, and Exploration*

Observation is the most passive means for gathering data. The observer receives and analyzes information transmitted naturally by the system he is studying. The experimenter, in contrast, stimulates the system under controlled conditions to evoke responses in some observable fashion.

Exploration is an attempt to gather increasing amounts of information by means of a voyage which brings the experimenter or observer closer to the system to be studied.

When an experimenter is permitted to change no more than a single imposed condition—such as temperature—the system's range of responses is relatively confined. We may call such an experiment a first order experiment. When two experimental parameters—perhaps temperature and ambient magnetic field—are varied, the system's potential responses become more complex. Such an experiment can be considered a second order experiment. Correspondingly, an experiment in which some ten parameters are varied at will becomes a tenth order experiment. And a set of observations, in which no conditions at all are imposed can be considered a zero-order experiment.

Observation is the simplest form of experimentation. Because of this simplicity the scope of a purely observational discipline, such as the study of the universe beyond the solar system, should be simpler to analyze than the potential wealth and complexity of an experimental science. That is the premise of this book.

neutrinos and gravitational waves. Both are difficult to detect; both have eluded the scrutiny of astronomers.

Even among electromagnetic waves and cosmic-ray particles there are many classes that can never reach us; and others arrive so transmuted and jumbled that all traces of origin are lost: Powerful cosmic radio beacons emitting their energy at wavelengths of 100 kilometers could never be directly observed. We might infer their existence from other observations, but the emitted waves would be totally lost—absorbed by traces of gas between Earth and even the nearest neighboring stars.

No magic of technology, no inventiveness of man could help us detect these or other waves that never reach Earth. Technology can only help the astronomer reach the natural boundaries imposed by the universe itself.

* * *

When we discover a new cosmic phenomenon, how do we know that we have found something new, rather than just a variant of an already recognized species? Chapter 4 examines this question in order to arrive at a quantitative estimate of the number of phenomena we actually have discovered to date.

If we know the ultimate limits of astronomical observation, we can also attempt to estimate the number of phenomena that remain undiscovered. Such estimates usually encounter incredulity and arouse controversy if not outright hostility. Nevertheless, I cannot see how a study of cosmic discoveries can be complete or how such a study can help us to improve methods or policies for future searches unless we provide at least a tentative sense of scale; and that scale is determined by the number of phenomena we have discovered and the number remaining unrecognized.

The number of cosmic phenomena I estimate to exist can be verified long before our knowledge of astronomy becomes complete. The estimate itself takes the form of a procedure that can be applied by anyone at any future stage in the development of astronomy to obtain either the same number or perhaps a quite different one. If the two numbers differ, my assessment may prove to have been wrong; or the recipe—the formula telling us what to do—may need to be revised in view of developments that I had not foreseen.

What is important here is not so much whether my appraisal of cosmic complexity is correct; rather, it is that I provide a prescription so anyone can make that judgment himself—perhaps with astronomical data better than those available to me—in order to arrive at a result which he himself can trust. It is a way of making the best informed estimate of the scope of astronomy, though that estimate may still have shortcomings.

That is the novelty of the approach.

Chapter 5 discusses ways in which we might best continue cosmic searches in the future, the directions we will need to follow for rapid progress, the main technological gaps that will have to be spanned, and the manpower needed to accomplish all this.

Most human enterprises involve planning. The larger the venture, the more detailed are our plans. The more expensive the project, the more scientific are the analyses. Only in planning science itself do we often lack scientific insight. I know of no systematic studies that attempt to predict the rate of progress or the ultimate scope of even one of the many branches of science. In fact, there have been few attempts at systematic examination of how the scope of a scientific discipline might be correctly assessed. The only astronomer to have recognized the need for such an examination appears to have been Fritz Zwicky, who twenty years ago discussed his ideas in his book *Morphological Astronomy*.[1]

Yet there is a pressing need for clear analysis. In a thoughtful editorial written for the interdisciplinary journal *Science,* Jurgen Schmandt of the Lyndon B. Johnson School of Public Affairs at the University of Texas in Austin has summarized this need.

> Difficult and controversial policy decisions often need a factual base that can only be provided by careful scientific investigation. . . . Without extensive research, embodied in numerous individual studies, such policy decisions would be blind. However, the results of scientific research do not enter the decision-making process in an automatic fashion, nor should they be allowed to be used in a haphazard way. To be used responsibly, scientific data must first be summarized, evaluated, and interpreted. What does the evidence add up to? How solid is it? Are the results tentative or final? Is there consensus or disagreement among the experts about the significance and meaning of the data? What is suggested by contradictory evidence? What is needed to fill gaps in available knowledge? . . .
>
> Policy analysis is in heavy demand in government. . . . While the level of activity is increasing, little is known about the quality and impact of its results. . . .[2]

In a subsequent editorial M. Granger Morgan of the Carnegie-Mellon University elaborates.

> Good policy analysis recognizes that physical truth may be poorly or incompletely known. Its objective is to evaluate, order, and structure incomplete knowledge so as to allow decisions to be made with as complete an understanding as possible of the current state of knowledge, its limitations, and its implications. Like good science, good policy analysis does not draw hard conclusions unless they are warranted by unambiguous data or well-founded theoretical insight. Unlike good science, good policy analysis must deal with opinions, preferences, and values, but it does so in ways that are open and explicit and that allow different people, with different opinions and values, to use the same analysis as an aid in making their own decisions. . . . . Scientists who find policy analysis alien must strive to understand its value and importance, even if they cannot bring themselves to engage in its practice.[3]

Even though most scientists would agree that systematic studies of scientific planning could yield an improvement on the intuitive approaches we normally take, two attitudes have seemed to prevail. First, scientists tend to be skeptical about the value of any predictions concerning the future of science; and second, they worry about the potential abuses of centralized planning, no matter how accurate the predictions on which the plans are based.

An uneasy feeling persists that long-term predictions on the progress of science are doomed to fail. Among the many anecdotes concerning great scientists of the nineteenth and early twentieth centuries who bungled their predictions on the future of physics, here are two frequently recalled stories.

In 1902, only five years before he was to become the first American scientist to win the Nobel Prize, Albert A. Michelson was able to write:

> The more important fundamental laws and facts of physical science have all been discovered, and these are now so firmly established that the possibil-

ity of their ever being supplanted in consequence of new discoveries is ex-
ceedingly remote.[4]

Three years later Albert Einstein announced his startling new principle
of relativity which found convincing support in measurements Michelson himself had carried out some two decades earlier.

In a similar vein, Walter Meissner, a colleague of Max Planck for
thirty years, recalls the young Planck's choice of a career after matricu-
lating in preparation for entry to the University of Munich at the age
of seventeen.

> At first he was uncertain whether to select classical philology, music, or
> physics, but he finally decided on physics in spite of the fact that [Philipp
> von] Jolly, then professor of physics at the University of Munich, advised
> him against it, since in the field of physics there was nothing new to be
> discovered.[5]

A quarter of century later Max Planck was to lay the foundations for
the quantum theory of physics, an approach to prove vital for progress
in the investigation of atomic, nuclear, and subnuclear structures.

Whether stories concerning men like Michelson or Jolly are representative of nineteenth-century thinking is not clear. Stephen Brush
and Lawrence Badash have debated this question.[6] Certainly men like
James Clerk Maxwell and William Thomson (Lord Kelvin) were con-
stantly finding imaginative ways to probe the wonders of Nature and
were corresponding with each other about their latest discoveries.[7]
Theirs was nothing like an attitude of complacency.

Nevertheless, the anecdotes most frequently recalled today portray
the late nineteenth-century scientist as confident in his own understanding of Nature, unwilling to grant the possibility of further revolutionary
discoveries.

I believe it is this caricature that has left most of us reluctant to
venture serious predictions about the future course of science. What if
we should fail just as dismally as Michelson or Jolly? Worse yet, what
if the predictions were to be taken seriously? Many scientists are concerned that their disciplines might be threatened through centralized
management if detailed scientific or social scientific studies were to
err in prescribing just how best to proceed in our scientific ventures.

David Edge, writing on the sociology of innovation in British astronomy, has summarized this attitude:

> In my experience . . . scientists tend to think that sociologists are trying to
> discover the "one true theory" of how science should organize itself and
> proceed, if it is to advance more efficiently and effectively. Once that theory
> is established, our lords and masters . . . will then attempt to beat scientists
> into appropriate conformity. In other words, the first sense of threat stems
> from the idea that the sociology of science is normative.[8]

Edge's article is based on a talk delivered in Edinburgh at the April,
1977, meeting of the Royal Astronomical Society, and he tries to reassure
his audience that sociological analyses cannot be *normative*—cannot
set up new procedural standards. He writes, "The aim of sociology is

to explain and understand, not to evaluate or judge."[9] This last statement, while true, need not really lessen the potential normative impact of sociological and other analytical studies on how a science like astronomy should progress.

Scientists are quick to pick up and put to good use any successful new research tool. Any social or procedural strategy shown to be effective would quickly become assimilated into plans for the future. Were this not so, whole classes of potentially effective approaches would be permitted to go unused—a waste quite uncharacteristic of most scientific efforts.

We should therefore acknowledge that any reliable study concerned with progress in science—with procedures, attitudes, or working conditions under which great advances are made—may be useful in bringing about further advances and might ultimately influence how we actually conduct science. Changes in the conduct of science, however, tend to involve centralized planning. We see that most clearly whenever and wherever large expenditures are required for steps likely to lead to particularly useful advances.

Our dilemma, then, is this: Communally reached decisions seldom provide the flexibility that appears to have been an essential ingredient of the most startling astronomical discoveries of recent decades. And yet centrally imposed decisions seem unavoidable, especially where costly, highly promising investigations are to be initiated. We must, therefore, worry about regulating our major plans so they will not inadvertently choke scientific progress.

There are clear grounds for concern about the ways in which a grand scientific strategy might be implemented. The scientific spirit firmly believes in challenging dogma through confrontation with new facts. How can this confrontation continue to succeed if a bureaucracy is to prescribe specific areas a scientist should investigate and others that are to be left untouched?

This is a reasonable and important fear. The scientific method has led to great discoveries, primarily when freedom to investigate new paths has not been curtailed. If centralized planning is to play an important role, as it now does in most fields that require massive funding, then ways must be found to assure freedom of objective investigation no matter where it leads. How this freedom is to be made compatible with the security of society is a difficult question. Most recently it has been raised in discussions on studies of recombinant DNA.

Complex issues concerning science and its service to society will no doubt continue to require complex solutions. Responsible government can, nevertheless, encourage daring science: Steps to implement innovative research in astronomy are not difficult to find once we have thoroughly understood measures that have succeeded in the past and attempts that have led nowhere. An analysis of these successes and failures leads directly to a set of specific recommendations that occupy the final portions of the book. Some of these recommendations involve communal endeavor; others depend on the imagination and enterprise of the individual. Together they are meant to provide incisive approaches to astronomical ventures and promise a rich and exciting era of cosmic discovery.

CHAPTER

1

The Search

Five Turning Points

In September of 1608 a Dutchman at the annual fair in Frankfurt offered for sale "an instrument by means of which the most distant objects might be seen as though quite near."[1] On October 2, not a month later, the States-General of the Netherlands recorded in The Hague the patent application for a similar instrument by Hans Lipperhey, a spectacle maker from Middelburg in Zeeland. But there were several competing claims and no patent was awarded. By April, 1609, spyglasses could be bought in the shops of spectacle makers in Paris, and by that summer Thomas Harriot in England was using a spyglass to make observations of the moon and was drawing maps of its surface.*

In Italy, Galileo Galilei, an experimenter of the greatest skill who had previously invented an improved military compass, also went to work. He constructed a spyglass which he presented to the Venetian senate in August of that year, pointing out that its ninefold magnification could prove of utmost importance in war. Galileo then undertook the construction of an even more powerful instrument that magnified some 20 times and incorporated a number of improvements in design. Although he had regularly lectured on the Ptolemaic system at the University of Padua and given public lectures on the supernova of 1604, Galileo's work up to that time had largely been concerned with the laws of moving bodies and with practical inventions. Now, however, he entered observational astronomy with exuberant energy. He aimed his spyglass—the word *telescope* was not to be coined until the following

* Technical terms used in this book and abbreviations that may be foreign to readers are explained in the glossary at the end of the book.

year—at the moon, stars, and planets and quickly discovered four of
the moons orbiting Jupiter, found the surface of our own moon to be
covered with mountains and valleys much like Earth's, and in these
ways immediately elucidated the general nature of moons. Turning to
the Milky Way, Galileo noted that there were vastly larger numbers of
stars than ever seen before: Viewed through his spyglass the milky
patches condensed into swarms of stars.

These results Galileo described in his *Sidereus Nuncius*—roughly
translated as *The Sidereal Messenger,* or *The Starry Messenger.*[2] This
booklet, published in March, 1610, is so short it can be read in an hour.
But its reports of Galileo's findings caused an immediate furor among
the learned of the day and started a revolutionary era of astronomical
discovery.

At six o'clock on the morning of August 7, 1912, the Austrian physicist Viktor Hess and two companions climbed into a balloon gondola
for the last of a series of seven launches. The flight, which had started
at Aussig on the Elbe, was under the command of Captain W. Hoffory.
The meteorological observer was W. Wolf, and Hess listed himself as
"observer for atmospheric electricity." Over the next three or four hours
the balloon rose to an altitude above 5 kilometers, and by noon the group
was landing at Pieskow, some 50 kilometers from Berlin. During the
six hours of flight Hess had carefully recorded the readings of three
electroscopes he used to measure the intensity of radiation and had
noted a rise in the radiation level as the balloon rose in altitude.

In the *Physikalische Zeitschrift* of November 1 that year Hess wrote,
"The results of these observations seem best explained by a radiation
of great penetrating power entering our atmosphere from above. . . ."[3]
This was the beginning of cosmic-ray astronomy. Twenty-four years
later Hess shared the Nobel Prize in physics for his discovery.

Late in 1931 Karl Jansky, a radio engineer at Bell Telephone Laboratory, set up an observing station at Holmdel, New Jersey, to track down
the source of static noise that interfered with transoceanic radio telephone reception. He was particularly interested in short radio waves
and built an antenna tuned to a wavelength of 14.6 meters. In records
kept day after day, three sources of noise soon became apparent. Local
thunderstorms were the most obvious. More distant thunderstorms
seemed to provide a second source. A third source gave a steady hiss
that varied with the time of day. At first Jansky thought he was seeing
radio emission associated with the sun. In a 1932 paper in the *Proceedings of the Institute of Radio Engineers* he wrote:

> During the latter part of December and the first part of January the direction
> of arrival of this static coincided, for most of the daylight hours, with the
> direction of the sun from the receiver. However, during January and Febru-
> ary the direction has gradually shifted so that now (March 1) it precedes
> in time the direction of the sun by as much as an hour.[4]

In a footnote inserted in proof Jansky adds, "Since this paper was written the curve has shifted much further to the left. Now (May 25) it crosses south at 4:30 A.M."

A year later Jansky was almost sure that this radiation was arriving from the central portions of the Milky Way. The directional discrimination of his equipment, however, was not sufficiently accurate, and he noted in a second paper in the *Proceedings* that the direction of peak radio emission "is also very near the point in space toward which the solar system is moving with respect to other stars."[5] Jansky's surmise about the Milky Way was, however, to prove correct, and he is credited with the initiation of radio astronomy. Subsequent technical strides have permitted the discovery of radio galaxies that are far more powerful emitters of radio waves than the Milky Way. These techniques have also led to the disclosure of quasars, pulsars, interstellar masers, and many other surprises.

In 1948 a group of researchers at the U.S. Naval Research Laboratories (NRL) began to place X-ray detectors aboard German V-2 rockets captured at the end of World War II. On the earliest flights, the physicist T. R. Burnight used simple detectors consisting of photographic films covered by thin metal plates. With a rocket launched at White Sands, New Mexico, on August 5, 1948, he observed solar X rays that could penetrate through beryllium plates three-quarters of a millimeter thick.[6] This flight provided evidence for X-radiation in unexpected intensities. A few months later Burnight's colleagues, Richard Tousey, J. D. Purcell, and K. Watanabe, confirmed these observations.[7] And on September 29, 1949, Herbert Friedman, S. W. Lichtman, and E. T. Byram, also of NRL, were able to launch a V-2 payload containing modern photon counters that would usher in a whole new era of X-ray astronomy.[8]

The sun is a weak X-ray source. It would not have been detected had it been at the distance of other stars; and for a decade no one expected to find X-ray emission from stars other than the sun. By 1960, however, theoretical predictions made the detection of X rays from supernova remnants, from flare stars, or from a hot extragalactic plasma distinct possibilities.

With a payload designed mainly to search for X rays generated at the surface of the moon by energetic radiation arriving from the sun, Riccardo Giacconi, Herbert Gursky, Frank Paolini, and Bruno Rossi of the American Science and Engineering Corporation, in 1962, detected the first signals from an X-ray source outside the solar system. Soon the NRL team, led by Herbert Friedman, accurately located the position of this source. It lies in the constellation Scorpius.[9] Like other X-ray stars it is now known to be a member of a close binary. Later, improvements in resolving power led to the identification of X-ray emission from galaxies and clusters of galaxies and showed the existence of rapidly pulsing emission from one variety of X-ray stars.

In an age of mistrust among nations, extreme measures are taken to keep track of weapons tests carried out by other countries. During the 1960s the United States placed the *Vela* series of spacecraft in distant orbits around Earth. They were deployed to leave no part of our planet

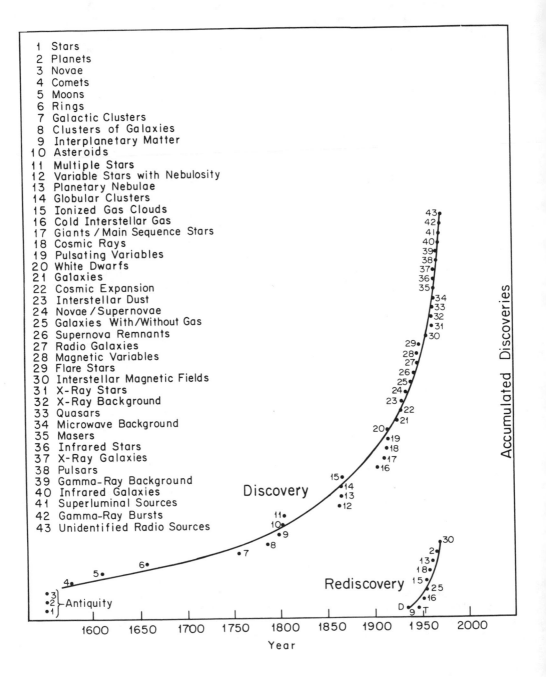

1 Stars
2 Planets
3 Novae
4 Comets
5 Moons
6 Rings
7 Galactic Clusters
8 Clusters of Galaxies
9 Interplanetary Matter
10 Asteroids
11 Multiple Stars
12 Variable Stars with Nebulosity
13 Planetary Nebulae
14 Globular Clusters
15 Ionized Gas Clouds
16 Cold Interstellar Gas
17 Giants / Main Sequence Stars
18 Cosmic Rays
19 Pulsating Variables
20 White Dwarfs
21 Galaxies
22 Cosmic Expansion
23 Interstellar Dust
24 Novae / Supernovae
25 Galaxies With/Without Gas
26 Supernova Remnants
27 Radio Galaxies
28 Magnetic Variables
29 Flare Stars
30 Interstellar Magnetic Fields
31 X-Ray Stars
32 X-Ray Background
33 Quasars
34 Microwave Background
35 Masers
36 Infrared Stars
37 X-Ray Galaxies
38 Pulsars
39 Gamma-Ray Background
40 Infrared Galaxies
41 Superluminal Sources
42 Gamma-Ray Bursts
43 Unidentified Radio Sources

Discovery

Rediscovery

Antiquity

Accumulated Discoveries

Year

unsurveyed at any time. Each spacecraft had the capability of detecting bursts of gamma rays—energetic X rays—generated in nuclear explosions in space.

In mid-1973 Ray W. Klebesadel, Ian B. Strong, and R. A. Olson at the Los Alamos Scientific Laboratory, New Mexico, surprised astronomers everywhere by announcing that short bursts of gamma rays lasting only a few seconds had been observed by several of the spacecraft and appeared to arrive from beyond the solar system.[10] Bursts of considerable strength appear to occur four or five times a year, but we know little else of this phenomenon.

Questions

The search for cosmic phenomena is one of mankind's grandest adventures and one of the most ambitious enterprises of modern science. To understand its conduct we will need to discern the nature of the phenomena, identify the men and women caught up in the search, describe the skills these scientists possess, analyze the plans they follow to their goals, and identify the tools required for their quest. We will need to know how the most significant discoveries are being made, whether

---

Figure 1.1 *Discovery and Rediscovery Dates for 43 Cosmic Phenomena, Showing Progressive Accumulation*

Though there is no unique definition of a cosmic phenomenon, most astronomers would compile a list much like the one shown here if asked to name the principal phenomena characterizing the universe. The 43 phenomena named are given prominence in most astronomical texts and standard reference works. Conferences and symposia concern themselves with individual phenomena on the list, and books or review articles frequently focus on one or another of these entries.

The discovery dates for the phenomena shown in the top curve cannot always be precisely pinpointed because the realization of a discovery sometimes dawns slowly. Where possible, however, the year of discovery is taken to be the year in which the first unambiguous report of the discovery is published.

At times a discovery involves the recognition that a previously known phenomenon actually comprises two quite distinct sources or classes of events—giant and main sequence stars, novae and supernovae, galaxies that contain gas—while others do not. Such discoveries are indicated by a slash (/) in the designation of the two phenomena that become resolved.

The lower curve shows phenomena that are redundantly recognized through two totally independent techniques. Thus, the existence of planets in the solar system would by now have been discovered even if optical telescopes had never been invented. An astronomer on an ever-cloudy body, such as Venus, would by now also have discovered the system of planets by virtue of planetary radio emission alone.

One phenomenon, interplanetary matter, is known not only in doubly (D) but in triply (T) independent ways. We see the faint zodiacal light in the night sky; we observe meteors and meteorites burning as they enter the atmosphere and can collect meteorites that hit the ground; finally, we obtain radar reflections from fine dust grains that burn on entering the upper atmosphere.

The most striking aspect of the two curves is the increasingly accelerated rate of discovery shown by the rapid rise in the top curve. Simultaneously with this rise comes an increasing recognition of previously known phenomena, now rediscovered by means of radio telescopes. Soon we may recognize many of these phenomena independently, in yet other ways—by means of the X-ray, infrared, or perhaps neutrino signals they emit.

---

past successes can guide us to further discoveries, whether there are ways to gauge the future scope of the search—of deciding whether our inventory of cosmic phenomena is nearly complete, whether we are close to being the last generation of astronomers needed to unravel the complexities of the universe, or whether there will be an endless cadre of cosmic researchers stretching into an uncertain future.

There is only one way to approach this study: We must look at the conduct of past searches in order to discern trends that can lead to assessments of the future.

Later on we will examine many other astronomical searches. The five turning points just presented, however, illustrate a number of common trends.

Trends

Seven traits common to many discoveries, particularly to those just cited and to others listed in figure 1.1, and more carefully analyzed in table 5.1 of chapter 5, are these:

1. *The most important observational discoveries result from substantial technological innovation in observational astronomy.*

Galileo's spyglass enabled him to resolve features on the moon considerably finer than any that can be distinguished with the unaided eye. He could also see stars several magnitudes fainter. Viktor Hess's discovery of cosmic rays led to an increasing awareness that electromagnetic radiation is not the only carrier of astronomical information reaching Earth; there also exist energetic subatomic particles that inform us about catastrophic events far away in the cosmos.

Karl Jansky looked out into the universe and saw signals from our galaxy at wavelengths 20 million times longer than anything the eye can see; the scientists at the Naval Research Laboratory and at the American Science and Engineering Corporation were able to detect signals at wavelengths 1,000 times shorter than visible light, and these techniques led to the discovery of X-ray stars and galaxies. The Vela military satellites could detect brief bursts of gamma rays and found just such bursts arriving from unknown parts in the sky; and short pulses detected at radio wavelengths similarly had led Anthony Hewish and Jocelyn Bell to the discovery of pulsars in 1968.[11]

Technological innovations that led to discovery have usually involved completely new wavelength ranges never used in astronomy before, or have made use of instruments with unprecedented precision for resolving sources to exhibit structural features, time variations or spectral features never seen before. Mere telescope size—light-collecting area alone—appears not to have been crucial to discoveries. Relatively few discoveries have been made with the largest telescope in existence at the time.

2. *Once a powerful new technique is applied in astronomy, the most profound discoveries follow with little delay.*

Many discoveries are made within weeks or months of the introduc-

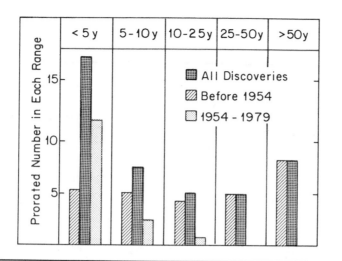

Figure 1.2   *Age of Required Technology at Time of Discovery for 43*
Cosmic Phenomena

It is not surprising that a cosmic phenomenon cannot be discovered until technical means
exist that make the discovery possible. An X-ray star cannot be discovered unless we construct
telescopes sensitive to X-rays; bursts of gamma rays reaching us from the universe
will not be noticed unless apparatus capable of sensing these rays is taken above the atmosphere to perform the observation. What is surprising, however, is the speed with which
major discoveries follow technical innovation. This is particularly striking in recent times.

A large fraction of the 43 phenomena listed in figure 1.1 was discovered less than five
years (<5y) after the introduction into astronomy of those techniques essential to the discovery.
Prior to 1954 the time lag between innovation and discovery tended to be longer and the
rate of discovery shown in figure 1.1 was slower as well.

For many of the discoveries the recognition of the new phenomenon, or its rediscovery
by independent means, occurred in a series of discrete steps. For these the age of the contributing technology was separately assessed for each step and prorated to keep the total at 43.

---

tion of new observing equipment. In the past twenty-five years there
have been no discoveries that could have been made with instruments
available a quarter-century earlier; and only a few of these recent findings could have been established as little as ten years earlier (figure
1.2). Occasionally a discovery is verified by an observer who can point
to records he had obtained some two or three years before; on these
the new phenomenon may already be discerned, though not perhaps
convincingly enough to stand out.

3. *A novel instrument soon exhausts its capacity for discovery.*

This corollary to the speed with which new discoveries follow technical innovations does not imply that new apparatus quickly becomes
useless. It just means that the instrument's function changes. It joins
an existing array of tools available to the astronomer for analytical work,
rather than for discovery, and continues useful service in that capacity.

To revitalize a technique for further searches for new phenomena,
its sensitivity or resolving power must be substantially increased, often
by as much as several factors of 10. Thus the discovery of quasars became
possible only after radio astronomical techniques had advanced to a

stage at which sources subtending angles no larger than 1 second of arc could be identified and accurately located in the sky.[12] The radio equipment available earlier to Jansky could simply not have coped with such observations.

4. *New cosmic phenomena frequently are discovered by physicists and engineers or by other researchers originally trained outside astronomy.*

In a wide-ranging study of the emergence of radio astronomy in Great Britain, the sociologists David Edge and Michael Mulkay have noted that all of the early British radio astronomers were physicists or electrical engineers and that radio astronomy, worldwide, was originally staffed largely by workers trained in physics or electrical engineering.[13] The same trait can be discerned all across modern astronomy. We speak of "cosmic-ray physicists," and, in fact, cosmic-ray research is largely carried out in the physics departments of universities. Similarly, none of the early X-ray or gamma-ray astronomers appear to have had advanced degrees in astronomy. Most of them had been trained as physicists.

This trend is not just confined to modern times. We need only think of the work of Galileo, William Herschel, Joseph Fraunhofer, and E. C. Pickering, who brought new methods and techniques to astronomy after working in other fields (see figure 1.3). Many of these pioneers initially worked at what professional astronomers would have considered the outskirts of astronomy. They could at best be marginally considered astronomers; and Edge and Mulkay have termed them *marginal workers.*[14] The rapid expansion of astronomy since World War II by now has greatly extended earlier boundaries so that radio observers, as well as infrared, X-ray, and gamma-ray astronomers, now find themselves at the center of current astronomical activity, though twenty years ago they all were marginal.

5. *Many of the discoveries of new phenomena involved use of equipment originally designed for military use.*

The rapid development of radio astronomy after World War II was made possible by existing radar equipment developed for the war effort.[15] Even before the war's end, however, James Stanley Hey, at the time a young troubleshooter for the British radar network, had noted occasional strong radio emission from the sun[16] and had discovered radar reflections from meteor trails.[17] Subsequently both solar radio astronomy and radar meteor observations developed into subdisciplines with their own practitioners. In 1946, shortly after the end of the war, Hey and his collaborators, still using wartime equipment available to them, also discovered a curiously undulating cosmic radio signal that proved to be the first extragalactic radio source, Cygnus A.[18]

In the United States the earliest postwar solar X-ray and far-ultraviolet observations were carried out with German V-2 rockets. Advances in infrared astronomy were similarly based on progress in the construction of military detectors, but the most sensitive far-infrared detectors were not developed until some time after the war and did not become available for astronomical work until the late 1950s and early 1960s.

This relationship between the tools of surveillance in war and in

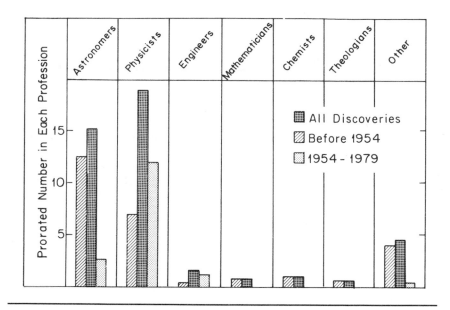

Figure 1.3   *Career Background for the Discoverers of 43 Cosmic Phenomena*

Most of the major cosmic phenomena were discovered by individuals prepared for careers other than astronomy—instrumentalists with novel techniques for looking at the sky. This educational background and early work experience outside astronomy is particularly apparent for the twenty-five-year period between 1954 and 1979. Even prior to 1954, however, half of all discoveries were made by outsiders. Both original discoveries and rediscoveries of phenomena listed in figure 1.1 are included in this compilation. Where several discoverers were involved, the contributions are prorated so that the number of careers also totals 43.

astronomy is not new: Galileo had pointed out the military value of the spyglass to the Venetian senate. His gift of the spyglass so pleased the senators that they at once doubled his salary and made his professorship at the University of Padua a lifetime appointment.[19]

The same interdependence of scientific and military progress can be found in a variety of fields. Many of the urgent problems in navigation, communication, detection, logistics, and medicine, as well as in willful destruction, have occupied military minds throughout the centuries, and it is not surprising that pure science and constructive technology often inherit some gains. As early as 1938, the sociologist Robert K. Merton noted this longstanding relationship in his study *Science, Technology and Society in Seventeenth-Century England.*[20] Astronomy has benefited most directly from attempts to advance the arts of navigation, communication, and detection.

6. *The instruments used in the discovery of new phenomena often have been constructed by the observer and used exclusively by him.*

The instruments used by Galileo, Hess, Jansky, and the early X-ray astronomers were built by them and were under their sole control. The Los Alamos group that discovered gamma-ray bursts had access to all the data gathered by the passive Vela satellites. Men like Christian Huygens, James Bradley, William Herschel, Friedrich Wilhelm Bessel, William Huggins, and William Parsons, third earl of Rosse, responsible for prime discoveries in the seventeenth, eighteenth and nineteenth centuries, also had instruments that they alone might use: When faced with a surprising result, they could check their apparatus at leisure, repeat the observations, and cross-check their results until they were certain that a finding was genuine and not due to some instrumental artifact. More than half of the phenomena discovered in historical times were made with instruments under exclusive control of the astronomers responsible for the discovery.

At national centers, equipment is widely shared among both resident and visiting astronomers. These centers were first established in the United States in the late 1950s, and by now require two-thirds of the National Science Foundation's astronomy budget for their support. None of the forty-three cosmic phenomena listed in figure 1.1—including fourteen phenomena discovered since 1960—were originally discovered at a national center.

7. *Observational discoveries of new phenomena frequently occur by chance—they combine a measure of luck with the will to pursue and understand an unexpected finding.*

Before Viktor Hess set out to make his balloon observations, the slow discharge that always takes place in electroscopes had generally been attributed to radioactivity in the earth. If that were true, Hess argued, absorption of the radiation in 5 kilometers of air should shield his instruments and cause them to discharge more slowly. Instead, Hess noted an increased discharging rate at higher altitudes and concluded that the rays were cosmic.

Karl Jansky, in 1931, was concerned with the mundane problem of finding ways to lower interference in radio telephone transmission. Instead he is now remembered for finding radio emission from the Milky Way.

The Los Alamos gamma-ray physicists were searching for surreptitious nuclear bomb tests and found cosmic gamma-ray bursts instead.[21] The first X-ray star was discovered through a search for lunar X-ray fluorescence.[22] Cosmic microwave background radiation was found when Arno Penzias and Robert Wilson at Bell Laboratories found they could not reduce below a certain level the noise registered by a new sensitive radiometer they had developed. Pulsars forced attention upon themselves when Anthony Hewish and Jocelyn Bell were actually set to measure scintillation—twinkling—of radio sources.[23] About half the cosmic phenomena now known were chance discoveries. The number of these surprises is a measure of how little we know our universe.

Surprises

We can argue all we wish that theoretical astrophysicists should be able to protect us against surprises—that they should be able to calculate, define, and predict what the observer should expect and not expect, but the possibilities in astronomy are too numerous for accurate prediction. This feature of the discipline perhaps should be further emphasized.

Our theorists know very well how to draw correct inferences and conclusions from astronomical observations, provided there are no unanticipated, as yet unobserved, phenomena to invalidate their predictions. But what if such phenomena do exist? The possible outcomes of calculations that attempt to anticipate unobserved phenomena in different combinations then become too numerous, and predictions can be wildly wrong. Hence, when entirely new observations are contemplated, most theorists prefer to admit the uncertainties and make no predictions at all. This reluctance is a measure of how much remains undiscovered.

To make no prediction is not the same as predicting that nothing new will be found. But we tend to forget this distinction and frequently are surprised and startled when we discover a genuinely new phenomenon by means of an innovative observing technique. Perhaps we should be no more surprised at finding a new phenomenon than at discovering nothing at all. When we lack predictions, there is nothing that favors a new discovery, and nothing against it. Our prejudice, however, is to think that "absence of prediction" implies "prediction of absence." Consequently, we are surprised by each unpredicted discovery.

Astrophysical Theory

Theoretical astrophysics relates wide varieties of different measurements and on this basis predicts results of new, previously untried observations. When these predictions prove to be correct, we gain assurance that our theoretical approach—our perception of the main physical processes underlying a phenomenon—is sound.

Astrophysical theory is a remarkably useful tool for dealing with a well-studied phenomenon. It guides us toward key analytical observations whose results help us decide between alternative models and thereby remove ambiguities and uncertainties in our understanding. But these theoretical steps have to be founded on observational data. Where no data base exists, the logic of theory alone provides no help. Rarely can we use theory to leap from a thorough understanding of one phenomenon to a prediction of the existence of a totally different, never-before-observed phenomenon. The span involved is too wide, and theory normally fails in the attempt.

For this reason, an observer given a powerful new tool rarely restricts himself to an examination of sources that theorists consider important. He knows that no amount of calculation can be as useful to him now as a thorough search of the sky to see what the universe will reveal, what new phenomena might await discovery. This is the thrill of the search: The excitement and anticipation of that first look.

Analysis Versus Discovery

We have alluded several times to two different kinds of astronomical observations, those aimed at discovery, and more routinely those meant to analyze. The distinction is seldom emphasized, but it is important, particularly because the tools needed for analysis differ from those required for discovery.

Analytical observations comprise compilations or catalogues of various astronomical objects, searches for possible regularities in the behavior of stars and nebulae, tests of theoretical models designed to explain astronomical phenomena—in short, data and correlations that are needed if we are to understand the basic processes underlying a set of cosmic events. A complete set of analytical measurements may entail all the observational techniques at our disposal.

Analytic work presupposes that we know what the universe contains and that we ask only, How does it work? In contrast, searches for new cosmic phenomena are aimed at determining the wealth of the universe. We want to answer questions, such as, How rich in phenomena is our cosmos? What are the phenomena we do not yet recognize?

A balance between programs of analysis and aspirations for further discovery is complicated by the conflicting requirements of these two modes of observational work. Two of the more important differences are these:

Discovery requires a new observational approach independent of, and substantially different from, previously used technique.

Analysis requires few extraordinary tools, though standardization of equipment for cross-reference between studies conducted by different groups is desirable. Analysis, therefore, can be relatively inexpensive.

But if analytical methods fail to recognize new phenomena, they will not properly advance astronomy. For this reason careful judgments are required to decide on the respective cost effectiveness of plans for analysis and for discovery. We will return to a discussion of proper balance between analysis and discovery later on.*

Messengers

We have walked the surface of the moon, sent spacecraft across the solar system to Jupiter and beyond, and achieved feats of exploration beyond our own imaginations only twenty-five years earlier. When can we start exploring the stars, the galaxy and the universe beyond?

Let us start with the most ambitious of these ventures, and then pare back. To reach the Andromeda Nebula, the nearest galaxy compa-

* See page 248 ff.

rable in stature to the Milky Way, within two million years, we would have to devise a spacecraft that travels at the speed of light and start our journey at once. The nebula lies 2 million light-years away, and our laws of physics tell us that we cannot build spacecraft that travel faster than the speed of light. To reach the center of our own galaxy, at the speed of light, only 30,000 years would be required—still a major investment of time.

If we were willing to wait a century to be informed, how far could we travel? What could we achieve? At best we could build a spacecraft that traveled a distance of 50 light-years and returned or sent back a message, all at the speed of light. It could visit any one of a thousand different stars in our nearest neighborhood. But we would have to launch the spacecraft today to receive the information within the allotted 100 years.

At present we have nothing like the means to start such a venture. Our spacecraft to the outer planets amble along at one ten-thousandth the speed of light. They would require ten millenia to reach even the nearest stars. The best we might hope to achieve within a century might be to build a spacecraft within fifty years that travelled at a tenth the speed of light, reached the very closest of stars, Proxima Centauri, within another half century, and radioed its findings back to Earth. Even then, the energy we would have to invest for every gram of spacecraft weight to accelerate the craft to this high speed would be a million times the energy we have had to expend on planetary spacecraft to date and would require capital investments and technologies of quite different orders. And when we were all done, the information returned might tell us not much more than we already knew through observations from a distance.

For the next few decades if not longer, we will, therefore, have to depend entirely on information carriers such as light or cosmic-ray particles to portray the universe beyond the planets. Given that, we will find that the detail and complexity we can ultimately discern in the cosmos depends not so much on the innate variety of cosmic phenomena as on the limited capacities of the carriers of information. These messengers cannot always reach us without hindrance. Information on cosmic events that occurred billions of years ago in distant parts of the universe may simply be absorbed by cosmic matter—never to reach us, never to be registered. The paths of the message-carrying particles or waves also may be so distorted that bits and pieces of information reach us at haphazard times, from arbitrary directions, scrambling the message.

The scope of astronomical observations, then, is limited, in part by cosmic interference with the transmission of carriers. Some of these limitations may be overcome with technical skill; others are fundamental and cannot be surmounted. If we analyze those carriers we understand best, we find there is a limited range of transmission that passes unhindered. Correspondingly, there is a limited range of observations through which we can learn more about the universe. To see how these limits arise we must enumerate the channels through which cosmic information can be conveyed and consider the restrictions the carriers encounter.

Channels of Information

Before World War II practically all our information about the universe was brought to us by visible light. Stars, novae, and interstellar dust clouds—all were known through visual or photographic observations. Whatever information we had about the galaxies also was borne across the universe by visible light.

There were minor exceptions: Karl Jansky in 1932 had found radio waves emitted by the Milky Way. Some astronomical clues at that time also were available from observations based on cosmic rays, meteors, and meteorites. But that was the extent of our grasp on the cosmos.

Today we have a firmer grip on incoming information, but our control still is far from complete. We can enumerate five separate channels through which information from the universe reaches us. Each is distinguished by a different kind of carrier (table 1.1). The electromagnetic channel still provides us with the bulk of information we receive. Cosmic-ray particles and meteorites add valuable complementing data, and someday neutrinos and gravitational waves may become important reporters of cosmic events.

TABLE 1.1

The Five Known Types of Carrier that Transport
Information Across the Universe

---

1. Electromagnetic radiation: Gamma rays, X rays, ultraviolet radiation, visible light, infrared radiation and radio waves—quanta of radiation, or photons, that differ from each other only in wavelength and energy. Gamma-ray photons have the shortest wavelength and are most energetic; radio waves have the longest wavelength and are least energetic.

2. Cosmic-ray particles: Highly energetic electrons, protons, and heavier nuclei, as well as neutrons and mesons which cannot cross large cosmic distances because they are unstable and short-lived. Some cosmic-ray particles consist of antimatter.

3. Solid bodies: Meteors and meteorites. Those we detect appear to originate in the solar system, but sooner or later we may identify chunks of matter from interstellar space as well.

4. Neutrinos and antineutrinos: No neutrinos from celestial sources have been reliably identified, to date. Theoretical predictions suggest that solar neutrinos should by now have been observed with available equipment.

5. Gravitational waves: Theoretically predicted radiation from rapidly accelerating massive bodies such as exploding stars. So far gravitational radiation has not been directly observed, and considerable effort may be required to construct adequately sensitive instruments.

Restrictions on Carriers of Information

We might at first think that there is an infinite variety of carriers available to us in each of these five channels of communication. In

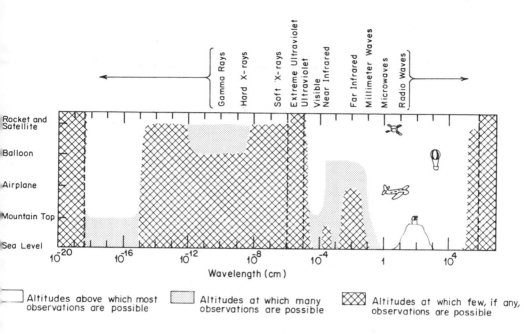

Figure 1.4 *Astronomical and Atmospheric Limitations on Observations Made Across the Electromagnetic Spectrum*

Astronomical observations are possible with ground-based telescopes in only a few portions of the electromagnetic spectrum. In most parts of the spectral range the radiation is absorbed in the upper layers of Earth's atmosphere and cannot penetrate to the level of mountaintop observatories, let alone down to sea level. At these wavelengths astronomers are forced to take apparatus to airplane altitudes, 12 to 15 kilometers above sea level, balloon altitudes, ranging up to 40 kilometers, or rocket and satellite altitudes, above 100 and 300 kilometers, respectively.

Sea level observations tend to be satisfactory only at radio wavelengths. Mountaintop observations are satisfactory in the visible part of the spectrum and for the highest energy gamma rays. In the far-infrared, at wavelengths of 1/100 of a centimeter, aircraft observations are satisfactory for some observations, but balloon observations are better, and rocket or satellite observations permit most measurements to be carried out. Finally, as explained in the text, cosmically determined limitations confine the observable part of the electromagnetic spectrum to a finite range bracketed by the extreme gamma-ray region at short wavelengths, and the extreme radio region at long wavelengths. These extreme waves and also extreme ultra-violet radiation are blocked even at rocket and satellite altitudes.

fact, the range is restricted. We cannot, for example, expect to make observations with electromagnetic radiation at very short or very long wavelengths—very high or very low frequencies (see figure 1.4)

At radio frequencies as low as 1 MHz—wavelengths of 300 meters*—the earth's ionosphere rarely permits transmission of any waves, because these low frequencies are strongly absorbed. While we now have radio telescopes that can be taken above the ionosphere in space vehicles, the extension to longer wavelengths that can be achieved in this way is not great. We soon encounter frequencies that fall below the plasma

---

* For an explanation of these units, see the glossary and especially tables G.1 and G.3.

frequency of the interstellar medium—the limiting frequency below which radiation is rapidly absorbed in its attempt to cross the space between stars. At wavelengths above 3 kilometers (frequencies below ~ 100 kHz) astronomical observations therefore become excluded.

At the high frequency end of the spectrum there is another barrier. Gamma rays with energies above 5×10^{14} eV (that is, at wavelengths shorter than 3×10^{-19} centimeters) readily collide with photons that make up the all-pervasive cosmic microwave radiation. In this process electron-positron pairs are produced and the gamma photons are destroyed. So large is the probability for this that gamma rays at this high energy typically cannot traverse a distance greater than some 30 thousand light-years. We might therefore just barely expect to see as far as the center of our Milky Way at these wavelengths, but galaxies beyond our own are forever obscured.

Another, intermediate wavelength interval comprises low frequency X rays and ultraviolet radiation barely energetic enough to ionize hydrogen atoms. In this wavelength range little information is likely to reach us even from the very nearest stars, and across larger distances the radiation is totally absorbed by traces of interstellar gas.

A similar set of restrictions affects cosmic-ray particles. At energies below some 10^{11} eV, cosmic rays cannot readily enter the solar system because a solar wind that carries outflowing magnetic fields enmeshes the particles and blows them back into the interstellar void. At the opposite energy extreme, particles having energies above 10^{21} eV can cross distances no greater than 30 million light-years before being destroyed in a collision with a microwave photon. At this energy protons could reach us from the nearest galaxies but not from beyond.

Neutrinos and gravitational waves are likely to have corresponding limitations. But we know too little about interactions with these carriers to specify their limitations as information carriers with any accuracy.

Restrictions on Observations

In astronomy we generally deal not with single quanta of radiation but with whole streams of photons that enter our telescope. And we classify the observations we make in terms of particular sets of photons that are singled out by the instrument we use: The telescope selects only those photons arriving from a narrow portion of the sky; a color filter defines the wavelength range we choose to view; the instrument's quickness of response determines the intensity variations we are equipped to discern; an added polarization analyzer permits us to search for polarization in the incoming beam. Table 1.2 lists seven traits that fully describe an elementary observation. Any larger observing program is just a collection of such elements for each of which we can specify a seven-parameter classification.

TABLE 1.2

Parameters Characterizing an Elementary Astronomical Observation

1. Type of carrier to which the observing instrument responds—photon, cosmic-ray particle, neutrino, and so on
2. Wavelength or energy of the carrier that can be detected
3. Angular resolution of the observing instrument
4. Spectral resolution of the apparatus
5. Time resolution of the instrument
6. Polarization sensitivity—if any
7. Time and date and direction observed

This classification only describes the conditions under which we observe. Nothing has been said about a result. The actual result of an elementary observation is given by a single number, the intensity of radiation measured under the specified observing conditions.

Let us see how this works.

1. We point the 5-meter telescope on Mount Palomar, in California, in the direction of the quasar 3C 273, employ a blue filter, and limit our efforts to a search for light variations on a time scale of minutes. This specifies the observations we undertake.

2. Our apparatus then registers the intensity of these time variations. That intensity is *the* result of the observation described.

The distinction between the description of an observation and its result is important, but not particularly puzzling. The description merely lays down the conditions under which we search for astronomical information but does not in any way fortell the result of the observation or by itself convey any information. Only when we have the result of the observation—the intensity registered—are we informed of cosmic events.

Even though we cannot foretell the results of observations that have never been undertaken, we can learn a great deal about future astronomical searches merely by enumerating all the observations that ultimately might prove possible in our universe. This enumeration permits us to estimate the total number of observational results we might ever hope to obtain, quite independent of what those results would be. The range and variety of possible observations is a measure of the universe's capacity to provide us—perhaps surprise us—with information.

Finite Limits

Since there are seven parameters listed in table 1.2, we must now investigate whether any of them become infinite. If all the seven parameters are constrained to finite values will we be facing a finite future of astronomical observations.

Table 1.1 lists the five known classes of astronomical carriers. We know of no other stable carriers of information—carriers that can survive the long trek through interstellar space unaffected by radioactive decay.

As we just saw, the energy or wavelength range of electromagnetic radiation and cosmic-ray particles transmitted large distances through the universe is finite.

In chapter 3 we will see that the ranges of angular spectral and time resolution, as well as the span of polarization sensitivities, are likely to be finite for all carriers of information. Some of these limitations arise from uncertainties inherent in all physical measurements, constraints imposed by the Heisenberg uncertainty principle. Others arise through confusion produced by the variety of existing sources that populate our galaxy and the universe beyond. Further limitations still arise from the interference of interstellar matter and interstellar magnetic fields with carriers bringing information from distant regions by altering their directions of propagation, their sense of polarization, their energy, or some other trait.

Finally, the range of positions in the sky, at which we may direct our telescopes, and the range of times and dates on which we may observe also are finite over any finite observing interval. Because our observing techniques are limited to resolving some smallest angle in the sky and some smallest time element, they also define a smallest patch of the sky and a smallest time interval we can actually examine. For these reasons there is only a finite number of these independent elements we can view in the sky.

The variety of observations we can undertake in a limited epoch, perhaps lasting a few millennia, is therefore finite. In other conceivable universes this might not be so. There the variety of potential observations could well be infinite. In our universe, however, the laws of physics that govern the behavior of the carriers of information, as well as the contents of the universe itself, limit the types of carriers transmitted across space and conspire to keep the range of observations finite.

Homogeneity

Despite restrictions already cited, we would be feverishly active if we had to carry out each and every one of the many enumerable observations. In practice our work is far more limited because experience has shown that the universe, at least statistically, is homogeneous and isotropic. This means that events observed in our neighborhood of the galaxy and in our neighborhood of the universe, by and large, mimic events occurring elsewhere. There will be rare events that occur only here and there, perhaps with long intervening periods of inactivity. Since these events occur so seldom they may have to be sought in nearby galaxies, even though they will occasionally also be found in our own. All the supernovae observed in the present century have exploded in other galaxies. Not one has been seen in our own galaxy during this time.

Statistical homogeneity implies that we need not observe every event that occurs in the universe: If we miss one particularly important occurrence, there will be others like it later. And conversely, if we study in detail the behavior of the thousand variable stars that lie closest to our solar system, we will perhaps detect all the major features characterizing another billion variables just like them at larger distances.

Time and place of observation therefore represent a subordinate class of parameters. What we can observe at a particular time viewing a particular direction in the sky is less important than what we observe with a particular carrier and at a particular resolving power. The discovery of quasars illustrates this point.

Recognition

In a note published in 1965, Halton Arp of the Hale Observatories pointed to a distinctive restriction on visual observations of extended sources in the sky.[24] Such measurements are peculiarly limited in that they permit discovery only of certain diffuse objects that have a severely constrained combination of brightness and diameter and lie along a narrowly defined strip (figure 1.5). If the celestial source is too large and faint, it merges into the brightness of the night sky and its discovery is thwarted. If it is too bright or compact, it appears stellar and may be mistaken for any normal star. In order to be noticed as an extended source, its brightness and dimensions must be confined to the diagonal strip on the diagram.

The quasars 3C 48 and 3C 273 shown in the figure do not lie on the strip and, correspondingly, remained unrecognized for many decades. We can go to the collection of old astronomical photographs stored in the Harvard College Observatory and find these two quasars detected and recorded on photographic plates nearly a century before their unusual radio emission finally allowed us to recognize them as a new phenomenon. Before that, their photographic appearance alone (figure 1.6) had led us to mistake them for quite ordinary stars of which hundreds of millions are within observational reach.

The distinction between the mere detection and actual recognition of a new phenomenon is therefore crucial. We may now be detecting many phenomena that we do not recognize because we have not made the crucial observation that will make them stand out. For quasars the striking earmark was their high radio brightness and compact size at radio as well as optical wavelengths. The narrowness of the strip in Arp's diagram suggests that many objects lying off this band may have escaped discovery. How many are there? Can we simply compare the area of the strip to the total available area in the plot and arrive at some statistical estimate?

These questions are somewhat oversimplified. Arp's plot, for one, is restricted to the visual domain. We need to rephrase the problem to

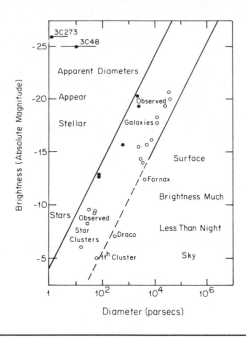

Figure 1.5   *Arp's Diagram*

In 1965 Halton Arp published this figure showing that all extended visual sources in the sky fall along a narrow diagonal strip on a plot of the absolute brightness of a source, measured in absolute magnitudes, against the diameter of the source measured in parsecs. One parsec is roughly 3 light-years.

Arp explained that if extended sources did exist beyond the confines of this strip, we could not yet have seen them. To the right of the strip and below it would be highly extended, faint sources that could not be seen through the normal atmospheric glow at night. To the left of the strip and above it the sources would be bright and compact and, like the quasars 3C 273 and 3C 48, could at first be mistaken for like ordinary stars.

Arp's diagram emphasizes three important points: First, there is a great difference between mere detection and actual recognition of a new phenomenon. Second, many diffuse visual sources might be recognized if we took appropriate apparatus above the atmosphere, or beyond the region of interplanetary scattered radiation. Such observations would widen the strip toward the right. Third, new diffuse visual sources might be found through use of instruments with higher angular resolving power that could widen Arp's strip toward the left.

Diagram courtesy of Dr. H. Arp and the *Astrophysical Journal*

Figure 1.6   *The First Photograph of Quasar 3C 273*

Though this plate was obtained on April 17, 1887, at Harvard's Boyden Station at Arequipa, Peru, quasars were to remain undiscovered for another seven decades.[25]

Quasar 3C 273 is an inconspicuous speck at the center of the picture. The accompanying sketch of the star field points to its precise location. It is easy to see that the quasar could not have been discovered by photographic means alone. There simply are too many stars in the sky to scrutinize each individually. A different characteristic—high radio brightness—was needed to bring the quasar to our attention.

How many more cosmic phenomena do we now detect, but fail to recognize?

Photograph courtesy of the Harvard College Observatory

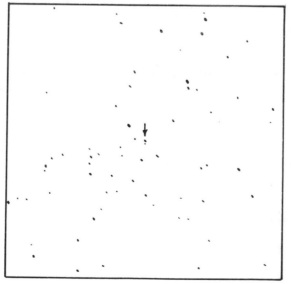

include all possible observations and then pose our query again: What fraction of all possible kinds of observations can now be undertaken? What fraction of the phenomena ultimately destined for discovery have we already found?

We have already taken the first step toward an answer by describing all possible types of observations in terms of the properties of the available carriers of information—electromagnetic waves, neutrinos, cosmic rays, gravitational waves and meteorites. We now must look at the fraction of possible observations already undertaken and the extent to which increased capabilities have led to new discoveries.

Discovery

Just as high sensitivity to cosmic X-rays was the key to the discovery of X-ray stars, both high spectral and high angular resolution at radio wavelengths was requisite for the discovery of cosmic masers, and high time resolution was crucial for the discovery of pulsars.

All the discoveries made in recent decades have come about because of new subchannels of information that made available a flow of information to which we had no access before. We may expect that these discoveries will continue, and on this basis we can make rough estimates of the number of discoveries we might expect in the future.

We now possess about 5 percent of all the electromagnetic observing capabilities that could ultimately come into use in astronomy.* If we similarly consider the other channels of information to which we now have access—the meteors and the cosmic-ray particles—and make a guess at the range of gravitational-wave and neutrino observations that ultimately will become possible, we may estimate that we have by now developed 1 percent of all the observing techniques we will eventually need to fully survey the universe.

Does that mean that we have only made 1 percent of all the discoveries ultimately possible? Probably not. The evidence we have amassed strongly suggests that we have come to recognize a far larger fraction, perhaps one-tenth or even one-third, of all the phenomena that can be found in the cosmos.

Before turning to the evidence for this conclusion, we should provide a definition of what we mean by a cosmic phenomenon. The word phenomenon can have somewhat different meanings for different people. Corresponding to any given definition, a particular count can be expected to emerge, both for the number of phenomena we already know and for the number remaining to be discovered. The important point then is to have a procedure that permits quantitative estimates for a variety of definitions to see whether these numbers are more or less independent

* See page 190 ff.

of how we choose to define the word phenomenon. The next three sections show how that is done.

Phenomena

Most astronomers have an intuitive grasp of what is meant by a new phenomenon and are quick to recognize it when it is first discovered. We are able to detect far more sources than we can discuss in detail, but readily recognize those that seem distinct and therefore of particular interest in some sense or other. Asked to construct a list of some fifty important phenomena, most of us would include galaxies, quasars, pulsars, and supernovae. There might be an even split on whether regular spiral galaxies and barred spirals represent different phenomena or are part of a single class, but few would include, say, all five subtypes of supernovae.

Despite this possible consensus we nevertheless need an objective specification for the designation *cosmic phenomenon*. This specification must correspond quite closely to our intuitive sense of a phenomenon. Otherwise we will not be able to answer satisfactorily the imprecise question, How many different phenomena characterize our universe?

This intuitive concept of a phenomenon, moreover, is slightly altered and enriched each time a new discovery is made, and it may therefore be necessary to also change our definition from time to time. Linguists and philosophers are not always comfortable with this shifting frame of reference; but our attitude here is similar to one that has gained the acceptance at least of philosophers of science like Hilary Putnam and Michael Polanyi.

Putnam points out some of the difficulties we face in describing the meaning of any given word—in our case the word phenomenon. The theory of meaning in Putnam's view depends on the idea that every language has rules of which we are largely unaware.

> The unconscious character of linguistic rules is important in understanding what happens when someone asks for the "meaning" of a word. What the inquirer wishes to gain is a knowledge of the rules governing the employment of the word—so that *he* will be able to employ it too. But he does not wish for an explicit statement of these rules, but for [a] kind of implicit knowledge. . . . Thus it is the respondent's task, in such a situation, to say something from which the inquirer, employing his considerable [implicit] linguistic knowledge, can "pick up" the information he wishes—"information" which neither the inquirer nor the respondent can verbalize, and "pick up" by a process which *no one* today understands.[26]

For example, we all have an intuitive feeling of what we mean by the word temperature. A cup of tea that feels warm when touched generally has a higher temperature than one that feels cold, though our senses can sometimes be fooled. A more precise definition of temperature can be given in terms of the mean kinetic energy of the molecules that constitute the cup of tea. But even without the kinetic theory of heat, the word temperature has a well-defined meaning which would still be correct even if we were to find some day that the kinetic theory had in some sense to be amended or modified.

Michael Polanyi has explained that the meanings of words are continuously changing in response to new observations. All of us have a personal sense of changes in meaning that are acceptable and can distinguish them from those that are not. Polanyi is concerned with much the same difficulties of scientific classification that we will need to face here. He views the problem in these terms:

> Language is continuously re-interpreted in its everyday use without the sharp spur of any acute problem, and some kindred questions of nomenclature are usually settled in a similarly smooth fashion in science. . . . In this changing world our anticipatory powers have always to deal with a somewhat unprecedented situation, and they can do so in general only by undergoing some measure of adaptation. More particularly: since every occasion on which a word is used is in some degree different from every previous occasion, we should expect that the meaning of the word will be modified in some degree on every such occasion. For example, since no owl is exactly like any other, to say "This is an owl," a statement which ostensibly says something about the bird in front of us, also says something about the term "owl," that is about owls in general. . . . If we can say of an unprecedented owl, belonging perhaps to a new species: "This an owl," using this designation in an appropriately modified sense, why should we not equally well say of an owl: "This is a sparrow," meaning a new kind of sparrow, not known so far by that name? . . .
> . . . We call a new kind of owl an owl, rather than a sparrow, because the modification of the conception of owls by which we include the bird in question as an instance of "owls" makes sense; while a modification of our conception of sparrows, by which we would include this bird as an instance of "sparrows," makes nonsense.[27]

Some scientists have a special gift for distinguishing different species and are particularly adept at this kind of classification: This expertise, Polanyi points out, should not be dismissed lightly simply because we cannot articulate absolutely objective criteria by means of which membership in a class is recognized. He recognizes the role in any kind of classification of an expertise or connoisseurship based on familiarity and experience and cites this facility as displayed by the great naturalist Sir Joseph Hooker;

> In 1859 he brought together and published evidence of nearly 8000 species of flowering plants in Australia, more than 7000 of which he himself had collected, seen and catalogued. The 8000 generic entities which Hooker derived from the individual specimens coming under his notice, have been recognized as valid in the vast majority of cases by subsequent observations of botanists.[28]

Someday the classification of species may be based solely on the molecular structure of genes, a procedure that could lead to greater objectivity. However, connoisseurship, while more subjective, does play an important and useful role not only in botany but in all developing sciences. And modern astronomy is in every sense a developing science.

Appearance and Reality

We should now think of ways in which questions about the number of cosmic phenomena can be answered at least somewhat objectively. Later, in chapter 4, we will become more precise. For now, however, let us restrict ourselves to the simple premise that cosmic features—objects, events—that appear different, actually are different. And cosmic features that appear vastly different represent different cosmic phenomena.

Such a designation may appear superficial. For one, no attempt is made to understand the nature or meaning of the phenomena, nor do we ask about interrelations among them. A second difficulty is that differing phenomena specified in this way do not necessarily represent events produced by vastly different physical processes. According to our premise, phenomena are classified simply by noting their gross formal appearance.

Nevertheless, form and function do often go hand in hand in nature. The ant and the elephant differ not only in size. Their alimentary, circulatory, and visual processes—to name just a few—differ just as radically. An elephant just could not be built to the size of an ant merely by scaling down all parts. The viscosity of the blood would be so great that it would refuse to circulate through the tiny capillaries. The animal's surface area in comparison to its volume would be too large to permit maintenance of the body at constant temperature, and so on. For an animal to be as small as an ant, its construction must radically differ from that of an elephant.

We will therefore assume that any two cosmic sources that transmit radically different-appearing signals also represent different phenomena. This should not disturb us.

Pulsars were given a name and considered a new phenomenon well before we had any consensus on what kind of entity might be blinking out there. We had never before seen anything in the sky that pulsed regularly, emitting a sharp radio pulse on a time scale of seconds or fractions of seconds. This was a new phenomenon! Quasars were recognized initially because they had a high surface brightness in radio waves and a stellar appearance at optical wavelengths. These traits were unmatched by any other sources in the sky. Gamma-ray bursts were discovered because nothing else we had ever seen gave off a single sharp pulse of gamma rays for an interval of a few seconds with no repeat pulse to follow for at least months thereafter, perhaps forever.

We do not truly understand the nature of any of these three phenomenon. Could the gamma bursts just be a release of energy at the surface of a neutron star? Are quasars evolutionary phases through which normal galaxies occasionally pass? Eventually, we should be able to sort all this out and come to understand the nature of these processes. For now, however, we are satisfied that we at least are aware of these events—that we have recognized their existence in the universe.

To be more specific, let us call two sets of cosmic events different phenomena if their appearance differs by a factor of 1,000 or more in at least one observational trait. In particular, let us consider building an instrument designed to respond optimally to a particular signal characterizing a given cosmic phenomenon. For such an instrument, the first six parameters listed in table 1.2 will take on quite specific values. For pulsars, for example, we would build an instrument sensitive at radio wavelengths with the capability of viewing compact sources and pulses that have a duration of only fractions of seconds. For gamma-ray bursts, on the other hand, we would build an instrument designed to detect gamma rays and to sense pulses lasting a few seconds. These two instruments differ in the wavelength to which they respond by a factor of a billion. We would therefore say that the two phenomena differ sufficiently in their observational traits to qualify as different phenomena—a factor of 1,000 would already have sufficed.

The factor of 1,000 is somewhat arbitrary but brings the definition into rather good agreement with lists of cosmic phenomena on which most astronomers might agree.

A scale factor of 1,000 distinguishes the two main types of star clusters. Galactic clusters generally contain somewhere between a hundred and a thousand stars. Globular clusters contain a hundred thousand to a million stars. The distinction is clear.

A factor of 1,000 permits differentiation between planetary nebulae that emit almost all their light in two or three spectral lines of great color purity and globular clusters that emit their light in a continuous spectrum comprising all colors. It permits the identification of cosmic masers through their pointlike radio appearance, their high spectral purity, and their high polarization. X-ray stars are recognized because of their high ratio of X-ray emission to visible light, and so on.

A factor of 1,000 also permits distinction between novae, stars that suddenly flare in brightness by a factor of 10,000 from supernovae explosions that are another factor of ten-to-a-hundred thousand brighter. But detailed differences between different classes of supernovae are not noted on this scale of coarseness, nor are all the minor differences between all kinds of peculiar variable stars emphasized. And this is what we would wish to see in a system that classifies distinct phenomena. It should emphasize clearly important factors and neglect small variations between subspecies.

Rediscovery

In addition to providing new discoveries, novel astronomical techniques have also provided another, almost equally important datum, the rediscovery of phenomena recognized from earlier observations. Frequently even this rediscovery occurs unexpectedly. The detection of strong radio emission from Jupiter by Bernard Burke and Kenneth Franklin in 1955 represented the first, unexpected measure of radio emission from a planet. Previously, planets had been considered to be undetectable with available radio telescopes because their theoretically predicted flux was too low to permit detection.

We say that a phenomenon is recognized in two completely independent ways if it could be discovered equally well with separate instruments that differ by at least a factor of a 1,000 in one of their observing capacities—in one of the first six traits listed in table 1.2. Spiral galaxies, for example, can be recognized equally well through radio observations and through optical studies at wavelengths a million times shorter.

This independent recognition of phenomena through widely differing channels of information provides a statistical key to the total number of observational discoveries we might ultimately hope to make.

A simple example illustrates the idea: Many of us have children who collect baseball or football cards. At the beginning of each season when the collection is still small, all the cards differ from each other. But as the collection grows, an increasingly large fraction of the cards becomes duplicated. Assume that all the athletes' faces are equally represented—rather than having one face appear far more often than others— and assume further that the cards are obtained in some statistically random order. *Then the very first duplicate obtained tells us an important characteristic of the set—namely, that it is finite.* In a collection containing an infinite number of different cards randomly distributed, no duplicates would ever be found unless an infinite number of cards were examined.

As the collection of cards grows and the number of duplicates increases, the relationship between the number of single cards in the collection and the number of duplicates can yield an increasingly accurate estimate of the total number of different cards in a complete set. Well-known statistics apply to this type of sampling, and the calculation is straightforward.*

The statistics governing our search for cosmic phenomena through different channels of observation is similar to the statistics of baseball cards. Each newly opened channel corresponds to a newly opened package of cards. Initially, as our technical expertise grows and the number of available channels increases, we discover an increasing number of new phenomena. But as the number of our discoveries grows, a survey carried out through a completely new channel will mainly uncover an increasing number of duplicates—phenomena already known from earlier discovery through previously established channels. We currently recognize quite a number of duplicates—the rediscoveries of figure 1.1— such as the planets and the spiral, gas-containing galaxies. Thus the number of cosmic phenomena is finite and can be estimated.

An Estimate of the Number of Cosmic Phenomena

Let us return to the example of baseball cards and denote the number of single cards by the letter A, the number of duplicates by B, and the total number of cards constituting a complete set by n. When both A

---

* For an explanation of this statistical argument see page 219 ff. and appendix A.

and *B* are small compared to *n*, we can write an equation for our best estimate of *n* as:

$$n = \frac{A}{2B}[A + 2B]$$

And knowing how many baseball cards, *n*, are contained in a complete set allows us to calculate how many cards, namely *n* − *(A + B)* are still missing from our collection. While this number is informative, it does not in any way provide any insights into the appearance of the missing cards. Similarly, knowing how to estimate the number of cosmic phenomena that remain undiscovered is a far cry from knowing what these phenomena will turn out to be (figure 1.7).

Understanding how many phenomena the universe contains, nevertheless, is important in its own right since it provides a sense of the complexity of the cosmos, a measure of the scope of work lying ahead.

When we tally the number of singly recognized astronomical phe-

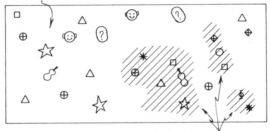

Entire volume of phase space

Subvolumes observed to date

Independently observed phenomena ✳ and ▫

Figure 1.7   *Distribution of Observed Phenomena in Phase Space*

The phase space of observations is a multidimensional space, each point of which corresponds to a different basic observing capability. The rectangular border represents a region of this space containing all conceivable astronomical observations that can be carried out in our universe. This diagram anticipates a more realistic depiction in figures 3.6 to 3.9 which show projections of such a space onto the two-dimensional page of a book. The regions of this space in which a given phenomenon appears, are determined by its observable traits and not by its position in the sky. The phenomenon's location in the phase space will therefore not change appreciably from one phase in the evolution of our galaxy to the next, though the constellations of stars seen in the sky would change appreciably during that time.

The framed portion of the diagram represents the entire accessible multidimensional space. The shaded portions are regions in which observations have already been undertaken. The cosmic phenomena are represented by a variety of familiar symbols. Some of these symbols appear in several portions of the space. Violins appear both in a region to which we already have access and in a portion of the space in which we currently lack the instrumental capabilities to observe. Some phenomena, such as the one represented by the Greek letter ϕ appear only once. Some, like the question marks or the little men—which might represent life in the universe—appear outside current instrumental reach. These phenomena remain to be discovered. Others, still, have been observed in two widely separated shaded regions. These are exemplified by the asterisks and open squares and represent phenomena independently observed in two different ways.

nomena and compare them to the number recognized in two independent ways, the result we find is this: We recognize some 43 different cosmic phenomena today; some $A = 35$, are recognized just one way, and some $B = 7$, are recognized in two independent ways. Using our formula for n we now can estimate the total number of phenomena that we may ultimately hope to recognize—roughly 123. Only one-third of these are known today.

Verification

The number n is a property of the universe, and our estimate of n should therefore be constant, independent of the epoch in which the estimate is made. With the data shown in figure 1.1, we can obtain estimates for n, based on values for the number of singly and doubly recognized phenomena known in 1959, 1969, and 1979. Use of our equation relating n to A and B then permits us to calculate the best estimate for n that we would have obtained in each of these years. Table 1.3

TABLE 1.3
*The Number of Phenomena Estimated
in Different Years*

	Year of Tally		
Cosmic Phenomena	1959	1969	1979
Total discovered $(A + B + C)$	30	39	43
Singly recognized A	25	31	35
Doubly recognized B	4	7	7
Triply recognized C	1	1	1
Best estimate for total in universe, n	103	99	123
Recognized fraction of total	29%	39%	35%

shows that these estimates only differ by about 20 percent. With the limited available data we can expect no greater constancy than that.

In chapter 4 we will examine a variety of ways of defining and enumerating cosmic phenomena and will find that n generally assumes a value of approximately 130.

Unimodal and Multimodal Phenomena

The estimate just made of the wealth of cosmic phenomena, n, was based on the assumption that each of the phenomena could potentially be found in several different locations in the volume represented by figure 1.7. However, we have no assurance that this assumption is valid. A substantial number of phenomena could exist that revealed themselves only through a unique astronomical observing mode and were otherwise beyond reach of observational discovery.

How many of these unimodal phenomena does the universe contain? Are they more numerous than the multimodal phenomena we estimate to number $n \sim 130$?* How can we tell?

We can estimate the maximum number of unimodal phenomena in this way: We examine all the phenomena A that have thus far been singly recognized. Among these there will be a subgroup numbering A', for which we cannot be certain whether a redundant recognition is bound to be possible. For the remaining $(A - A')$ phenomena, astrophysical theory will permit the prediction of an alternate mode of observation, based on our more intimate understanding of these phenomena. Right now, cosmic-gamma bursts represent a phenomenon that could be unimodal. We know too little about these bursts to tell whether they will reveal themselves in any other way than just through a burst of gamma rays. At present there may be as many as four phenomena of this kind, and we can therefore set $A' = 4$ for the current epoch of astronomy. We may divide A', by a factor r, representing current competence for making observations. That competence is just the fraction of the observing space that technical developments have permitted us to search—the shaded fraction of the plot in figure 1.7. The ratio A'/r then provides an upper limit to the number of phenomena that ultimately would remain singly recognized. In chapter 3 we will show that $r \sim 0.01$ at present, and therefore $A'/r \sim 400$. We therefore expect a maximum number of about 400 unimodal phenomena. These may quadruple the total number of phenomena ultimately to be found, making it ~ 500 rather than ~ 130, if many cosmic phenomena are unimodal.

At our present stage of development in astronomy, this is a satisfactorily close estimate which should continually improve as we learn more about the phenomena we uncover. We can see at least that we are not dealing with a list of phenomena numbering in the thousands or the millions, an estimate that could have been quite conceivable without the approach developed here.

If this way of viewing the universe as a system of finite variety seems slightly improbable right now, we must only remember that geography, too, was once an unimaginably variegated field. During the fifteenth and sixteenth centuries, discovery followed great discovery. But then it all came to an end. Today there are few surface features on Earth that remain unknown. The probability of discovering a new ocean or finding a mountain higher than Mount Everest is nil. Uncharted features on the earth's surface are now measured in meters, not kilometers or miles. In a similar way we may expect the undiscovered cosmic phenomena to shrivel in number until only minor features remain unnoticed.

The End of the Search

The history of most efforts at discovery follows a common pattern, whether we consider the discovery of varieties of insects, the exploration

---

* The symbol $\sim$ denotes approximate equality.

of the oceans for continents and islands, or the search for oil reserves in the ground. There is an initial accelerating rise in the discovery rate as increasing numbers of explorers become attracted. New ideas and new tools are brought to bear on the search, and the pace of discovery quickens. Soon, however, the number of discoveries remaining to be made dwindles, and the rate of discovery declines despite the high efficiency of the methods developed. The search is approaching an end. An occasional, previously overlooked feature can be found or a particularly rare species encountered; but the rate of discovery begins to decline quickly and then diminishes to a trickle. Interest drops, researchers leave the field, and there is virtually no further activity.

The discovery rate in this course of events follows a bell-shaped curve, roughly as drawn in figure 1.8. Correspondingly, there also is a steep S-shaped curve which represents the accumulation of discoveries.

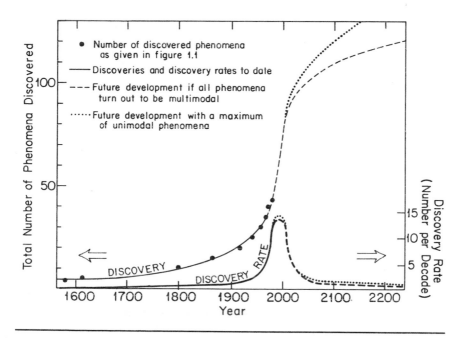

Figure 1.8 *Discovery and Discovery Rate Projected into the Future*

The discovery rate for cosmic phenomena, averaged over early decades when discoveries were sparse, is shown in the bottom, bell-shaped curve. The dashed portion is an extension into the future, based on the rates measured in the past. It is computed for a total wealth of multimodal phenomena estimated at 130, as explained in the text. The tall S-shaped curve clearly exhibits the accelerating rate of discovery in recent decades and represents data shown in figure 1.1. The dashed and dotted lines extend the curve into the future, taking into account the general rise and fall of discovery rates seen in other ventures, as well as the total estimated number of phenomena still awaiting discovery. The dashed line assumes that there are no unimodal phenomena and levels off at $n \sim 130$ phenomena. The dotted curve assumes a maximum number of unimodal phenomena and can reach a level as high as 500. By the year 2200 we should have become aware of some 90 percent of all multimodal phenomena characterizing the universe.

Each point on that curve represents the cumulative area under the bell-shaped curve at a given stage in the history of the search.

Projections into the future are based on two factors, a belief that a smooth bell-shaped curve, symmetric about its peak, well represents the rate of discovery throughout the search and a belief that the total number of multimodal phenomena to be found will equal our best estimate of roughly 130.

If this curve accurately corresponds to future developments, we should have found some 90 percent of all the multimodal phenomena by the year 2200 (figure 1.9). Thereafter, however, it might take several millennia to find the remaining few percent. Just as astronomical discovery started a few thousand years ago with an awareness of planets that regularly move through the starry sky, so too, the search may continue thousands of years into the future. This is particularly true if many cosmic phenomena are unimodal. Discoveries will then take place as long as new observational techniques can be introduced into astronomy—as long as any portions of the volume drawn in figure 1.7 remain unshaded. The phenomena to be discovered last might be those that occur with great rarity and emit too little energy to be found at great distances across the universe. Alternatively, the last few phenomena might be those that emit carriers for which observing apparatus is developed very late in the history of technological progress. The rate of discovery in that case would be delayed by technical factors rather than by limitations intrinsic to the nature of the undisclosed phenomena.

The smooth curves extending into the future assume that there will be no sudden changes in our way of conducting the search. Such changes, however, could be legion. Political factors might dictate that astronomy receive less support in the future. A war might slow the search to a virtual halt, though the postwar era, if there was one, could provide astronomers with discarded military equipment that would again accelerate the discovery rate.

On a less drastic scale, the greatest threat to astronomical progress might come from strict, centralized control of the search.

How Much Control?

Of the cosmic phenomena we now recognize about half were discovered unexpectedly. Is there some way of eliminating or at least reducing this element of chance? Can we systematize our search to make ourselves less dependent on luck? It seems we can, because the traits shared by many astronomical discoveries tell us what to emphasize.

Discoveries of novel phenomena generally follow shortly after the introduction into astronomy of powerful new observing techniques. Some of the instrumentation is inherited from the military or from the communications industry. The new techniques frequently are transplanted into astronomy by technically gifted outsiders who not only in-

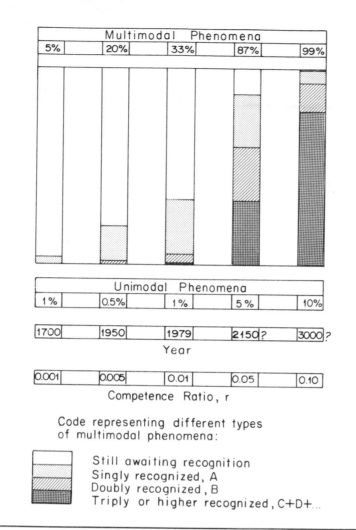

Figure 1.9 *Approach to Completion*

As the search for cosmic phenomena approaches completion, we find that the number of unrecognized phenomena dwindles. Multimodal phenomena become redundantly recognized through a variety of independent techniques; the fraction of singly and multiply recognized phenomena is readily calculated and is shown here for different stages of completion. When the search for multimodal phenomena is only 5 percent complete, none of the phenomena is doubly recognized.

If our estimate for the total number of multimodal phenomena is correct and equals ~ 130, this state of affairs corresponds to the year 1700. With 20 percent of all phenomena recognized, roughly 1 phenomenon in 10 becomes doubly recognized. Triply recognized phenomena start to appear at a level of completeness, amounting to 33 percent. This is the level attained around 1979. Eighty-seven percent and 99 percent of these phenomena, respectively, should have been observed by the years 2150 and 3000.

The competence ratios shown represent the fraction of all physically permitted observations that can be undertaken with techniques available to astronomy at each epoch. With only 10 percent of all observations completed, 99 percent of all multimodal phenomena should already be recognized. For unimodal phenomena the approach to completion is separately shown and is numerically equal to the competence ratio r. The rise in discovery of unimodal phenomena is currently much slower but will also persist longer than the discovery of multimodal phenomena.*

* See text and appendix A for an explanation of these statistical arguments.

troduce the new instrumentation but also make the important discoveries.[29] Seldom is the required apparatus purchased ready-made by an astronomer and effectively put to use in a discovery. Large telescopes, by themselves, rarely are crucial, and national facilities shared among many astronomers may lack the ingredients demanded for discovery. Theoretical predictions also fall short; otherwise, discoveries would not continually surprise us. Evidently our powers of prediction are too weak.

It might be argued that major discoveries will come along anyway, with no need for change, just as they always have in the past, largely by chance. But clearly the approach we have taken in the past has been quite inefficient.

Were we to enter an entirely new universe today, with all conceivable observational equipment at our disposal, we would no doubt quickly put to use our whole kit of tools to study the major features characterizing this new cosmos. We would certainly not repeat history, spending centuries first concentrating on visual observations alone before ever proceeding to radio, infrared, and X-ray observations.

To facilitate future discoveries we can take the following steps:

1. We can emphasize the introduction into astronomy of new techniques to expand our observational skills. In terms of long-range progress this means developing techniques that ultimately will permit the detection of cosmic neutrinos and gravitational waves, carriers of information to which we now are totally blind. In the electromagnetic domain the required directions of progress are easily specified: We must introduce instruments with sensitivity and resolving power currently lacking in different wavelength ranges. Similarly the lack of capabilities for cosmic-ray observations and meteorite studies point up the need for technological progress to fully utilize these final two channels of communication.

2. We must, however, keep in mind that there are two distinct strategies to be followed in observational astronomy. Only the first is aimed at discovering the contents of the universe, the phenomena that characterize it. The other is analytic and is aimed at understanding those phenomena we already recognize. The instrumentation needed for these two efforts differs. Discovery requires looking at the universe through new instruments—new eyes. Analysis needs a systematic approach, a variety of cross-checks, frequently aided by large telescopes able to gather more routine information quickly.

Discovery implies pioneering new areas; but with normal fiscal constraints the funding available for new ventures is limited. In times of constant budgets, at best some 20 percent of a national astronomy budget can be dedicated to new ventures.* A corresponding decrease is simultaneously required in ongoing projects to accommodate those new ventures that prove most promising. Such cutbacks always are painful. But without a policy aimed at phasing out less productive programs, new starts cannot be undertaken. A lack of determination in eliminating traditional subdisciplines of astronomy limits the rate at which the field can pro-

* See page 249 ff.

gress in its search for unrecognized phenomena. The 20 percent figure is approximate but is based on typical growth rates of projects, retirement or transfer of researchers, and obsolescence rates on equipment.

3. Incorporation of new techniques for our searches frequently requires the talents of physicists, chemists, and engineers who have pioneered new instrumentation in their own fields. While skilled scientists from other fields have often been transplanted into astronomy, we should install systematic means for making such a transition easy and attractive. At the same time, students planning astronomical careers should be better trained in technical skills so they may more readily adapt themselves to unexpected directions in astronomy. Grants that help established researchers learn new techniques for astronomy in mid-career could also help, both in phasing out older projects that have run their course and in providing the skills and tools required for new ventures.

4. Our inability to rely on astrophysical theory for guidance to new discoveries has several consequences that may require changes in deeply rooted astronomical institutions. The first of these is the peer review—

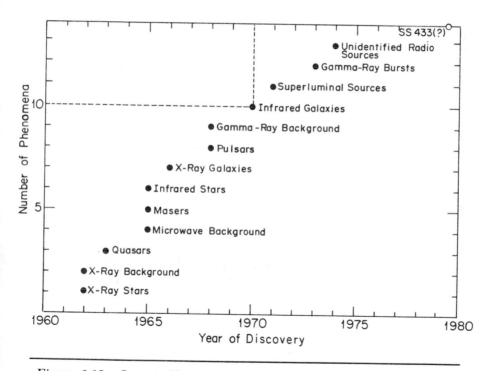

Figure 1.10 *Cosmic Phenomena Discovered Between 1960 and 1979*

The decade between 1960 and 1970 brought about the discovery of nine new cosmic phenomena. Seven of these were discovered by American researchers. Between 1970 and 1979 only four further discoveries were made, but as 1979 drew to a close, a fifth apparently new phenomenon, known only by the catalogue number SS 433 assigned to the observed source, appeared.[31]

the review of proposed new research by a group of peers. Traditionally, a new venture must be justified to other astronomers through its utility in clarifying conflicting astronomical views or through its promise for improved observations of known phenomena.

This justification is generally phrased in terms of established astrophysical theory. But since major discoveries rarely are predicted by theory, it becomes almost impossible to justify a search for new phenomena. To counter this trend, we must insist on the importance, in their own rights, of new techniques capable of providing novel observations to which we have had no access to date. To justify this recommendation, we need only note that the use of new observational techniques is the only demonstrated road to success in our search for new phenomena, a success that neither theory nor planning have matched. The lack of peer review may account for some of the success of astronomical programs carried out with military support. These programs were not much concerned with astrophysical theory but could well judge the merits of new techniques that might help in detection, surveillance, communication, navigation, and other martial needs. Many of these needs involved tools imaginative astronomers were eager to put to use in their searches.

5. Past plans have dictated the conduct of astronomy ten years in the future, while major discoveries were being made with new apparatus less than five years old. The planners therefore had no knowledge of the exciting steps that might be taken toward the end of any planned decade. Plans for astronomy would probably be more successful if they were updated at shorter intervals and showed a more flexible format. Astronomers in 1969 constructed a ten-year plan for the 1970s, but we may be fortunate that no such plan had been established in 1959 for the 1960s. Between 1960 and 1969, we discovered X-ray stars and galaxies, quasars, pulsars, the microwave background radiation, infrared stars and galaxies, and a variety of other new phenomena. By the end of the decade, the 1959 plan would have been hopelessly out of date—and if faithfully followed might have constrained astronomical progress dur-

---

Figure 1.11   *Federal Funding for Astronomy, NASA Funding for Basic Research, and U.S. National Research and Development Expenditures, By Year, in Constant 1972 Dollars.*[32]

Astronomy has benefited a great deal from applied research, from the testing of new materials and devices, and from the development of new instruments or new technical skills. Expenditures specifically designated for astronomical purposes, therefore, are not the sole measure of U.S. support of astronomy. Research and development efforts, both federally and industrially funded, can also aid astronomical efforts. The level of research and development in the U.S. between 1963 and 1976 is shown by the curve labeled "National R & D." Basic research conducted by the National Aeronautics and Space Administration (NASA) also can benefit astronomy. Direct expenditures on astronomy are presented showing both total funding and the fraction of the funds provided through the National Science Foundation (NSF). Dashed lines represent estimates. The top three curves show a peak in expenditure around 1969 but roughly comparable overall funding in the two decades immediately before and after 1970. NSF support for astronomy rose slightly in the 1970s. These data are of interest in comparing the number of new cosmic phenomena discovered in the 1960s and the 1970s.

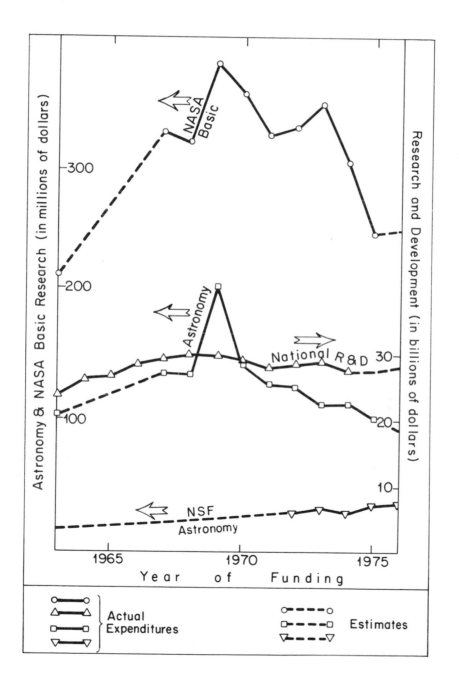

ing the 1960s to the point where some of that decade's most striking discoveries might never have been made.

Figure 1.10 shows the rapid discovery rate of the 1960s. Some of the most imaginative work ever seen in the history of astronomy was going on, much of it with military support or with the use of discarded military equipment. No master plan existed. Individual applications for research support were handled by the individual government agencies as proposals were submitted. With the Mansfield Amendment, prohibiting military support of basic research, the 1970s, in contrast, became an era governed by a detailed astronomical plan set up by the National Academy of Sciences through a committee headed by Jesse L. Greenstein, professor of astronomy at the California Institute of Technology.[30] In the judgment of most astronomers, the Greenstein committee did an excellent job. It is doubtful that any other committee working at that time would have come up with a vastly superior plan. But despite almost equal funding in the 1960s and 1970s, the rate of discovery, if anything, declined in the last decade. (Figure 1.11). It is therefore not clear to what extent the plan for the 1970s worked to the benefit or to the detriment of searches for new phenomena.

Perhaps the cornerstone for astronomical progress still is the individual researcher, not the large committee. We may need to give serious thought to deregulation of much of astronomy, making firm long-range plans only for clearly needed major facilities—facilities that may be vital to the large-scale analytic efforts that constitute much of the routine but important work astronomers carry out to deepen their understanding of cosmic events. Progress in the search for major cosmic phenomena, however, would best be served by a flexible policy that assured substantial funding for the rapid implementation of powerful new techniques as they become available without too much worry about just what these new tools might reveal about the universe.

6. Shared, multipurpose facilities at national centers, though again potentially important for analytic work, appear not to have been successful in aiding the search for new phenomena.* Discoveries largely have been made elsewhere with instruments dedicated to special tasks and under the sole control of individuals or groups of astronomers working on a specific project, sometimes unrelated to astronomy. If national centers are to prove useful in promoting discoveries, they may have to be operated differently to permit at least occasional concentrated searches by small teams or even by individuals who need uninterrupted use of a facility for several months or years at a time. The centers may also need to incorporate greater flexibility, sometimes entering altogether new areas of research or revising the scope of their missions.

* See page 259 ff.

Enlightened Planning

To date our theories still are too poorly developed to predict where novel phenomena are to be found. Half of all cosmic phenomena are discovered by individuals who come into the field with no formal astronomical training, though most practicing astronomers in fact do have that formal training.* This suggests that we do not know how to train astronomers properly to make these major discoveries themselves. Current plans look ten years into the future, but many of the prime discoveries are made with tools unknown to astronomers five years earlier. This says that even the timing of our plans ill fits the current needs of the field. Under such conditions centralized planning for the future of astronomy could easily choke progress and restrict the rate of discovery.

Our best hope for the future is to study those factors that most help in the search for cosmic phenomena, incorporate these ideas into our planning, and then keep our plans as flexible as possible, so that we can change direction in response to new findings and improved understanding. With this outlook it should not be long before we approach the end of mankind's most exhilarating era—the epoch of *Cosmic Discovery.*

* See page 246.

CHAPTER

2

Discoveries

Two Descriptions

An astronomical phenomenon may be described in two essentially quite independent ways.

The first method consists of a listing of factual statements. Each statement combines two parts, a description of observing conditions, and a report on the observational results. In simplified form such a two-part statement might read like this:

> In July 1967 a large radio telescope operating at a frequency of 81.5 MHz was brought into use at the Mullard Radio Astronomy Observatory [Cambridge, England]. This instrument was designed to investigate the angular structure of compact radio sources by observing the scintillation caused by the irregular structure of the interplanetary medium. The initial survey includes the whole sky in the declination range $-08° < δ < 44°$ and this area is scanned once a week. A large fraction of the sky is thus under regular surveillance. Soon after the instrument was brought into operation it was noticed that signals which appeared at first to be weak sporadic interference were repeatedly observed at a fixed declination and right ascension; this result showed that the source could not be terrestrial in origin.
>
> Systematic investigations were started in November and high speed records showed that the signals, when present, consisted of a series of pulses each lasting ~ 0.3 s and with a repetition period of about 1.337 s which was soon found to be maintained with extreme accuracy.

Thus starts the paper by the Cambridge University group led by Anthony Hewish and Jocelyn Bell reporting the discovery of the first pulsar.[1]

In specifying the radio frequency 81.5 MHz and the capability for measuring pulse durations of 0.3 seconds and pulse repeat rates of 1.337 seconds, the authors of the paper are describing some of the parameters of a complex filter through which they were looking at the sky. A complete description of this filter would require the added specification of the spectral resolution, the angular resolution, and the polarization sensitivity of the equipment. In adding a statement that lists limitations on the directions in the sky from which the instrument was able to receive signals at the time the pulses were observed, the authors also specify a set of blinders that restricted and further defined the characteristics of the received radiation.

Presented in this fashion, a report on an observed phenomenon amounts to the specification of the intensity of radiation received through a complex combination of filters and blinders. Most of the observations we carry out permit us to see only a highly selected set of rays taken from a far richer matrix of radiation that pounds the earth and is incident on our apparatus from all sides. The rejection of so much of the available radiation may look like a waste, but if we permitted larger numbers of quanta to interact with our equipment, we would be lowering our selectivity. Sources in the sky would now appear blurred, and we would lose much of the specific information we need to recognize a new phenomenon and comprehend its nature.

Let us next look at an alternate type of description. Typically it will have this form:

> The case that neutron stars are responsible for the recently discovered pulsating radio sources appears to be a strong one. No other theoretically known astronomical object would possess such short and accurate periodicities as those observed, ranging from 1.33 to 0.25 [seconds]. . . .
>
> Since the distances are known approximately from interstellar dispersion of the different radio frequencies, it is clear that the emission per unit emitting volume must be very high; the size of the region emitting any one pulse can . . . not be much larger than the distance light travels in the few milliseconds that represent the lengths of the individual pulses. No such concentrations of energy can be visualized except in the presence of an intense gravitational field.
>
> The great precision of the constancy of the intrinsic period also suggests that we are dealing with a massive object. . . . Accuracies of one part in 10^8 belong to the realm of celestial mechanics of massive objects.
>
> There are as yet not really enough clues to identify the mechanism of radio emission. It could be a process deriving its energy from some source of internal energy of the star, and thus as difficult to analyse as solar activity. But there is another possibility, namely, that the emission derives its energy from the rotational energy of the star. . . , and is a result of relativistic effects in a co-rotating magnetosphere.
>
> . . . A magnetic field of a neutron star may well have a strength of 10^{12} gauss at the surface of the 10 km object. At the "velocity of light circle," the circumference of which for the observed periods would range from 4×10^{10} to 0.75×10^{10} cm, such a field will be down to values of the order of 10^3–10^4 gauss. . . .
>
> If this basic picture is the correct one it may be possible to find a slight, but steady, slowing down of the observed repetition frequencies. Also, one would then suspect that more sources exist with higher rather than lower

repetition frequency, because the rotation rates of neutron stars are capable of going up to more than 100/s, and the observed periods would seem to represent the slow end of the distribution.

These words are taken from the first essentially correct description of a pulsar published by Thomas Gold of Cornell University in 1968, just a few months after the initial pulsar discovery.[2]

This second type of description has quite different qualities. Our first means of describing a pulsar had provided observational data and nothing else. The aim of the second is to mold isolated data into a conceptual model of what actually is occurring in the scene observed. It attempts to present a coherent view of the structure and composition of the observed sources, displays the driving forces, and tries to extrapolate both into the past and into the future to show how the phenomenon originated and how it might evolve.

When we talk about a cosmic phenomenon, muse about it, or speculate about its significance, most of us depend entirely on this second, more abstract model, not on the raw data. This has many advantages. The model extracts, from a morass of thousands or millions of individual observations, a conceptual picture that permits us to understand the nature of the phenomenon without the need for recalling every known data fragment. The model also allows us to make predictions on what results to expect in novel types of observations. The ability to predict, in fact, is the main test of the model. A good model predicts well; a model that predicts poorly is correspondingly useless and is soon discarded. The slowdown of pulsars predicted in Gold's theoretical picture was in fact soon observed and helped to establish the rotating neutron star hypothesis as a useful model.

In physics theoretical predictions often are tested in the laboratory. The information gathered in experiments is then fed back into the theoretical models, and new experiments are tried to verify the reformed theory. Occasionally an impasse between theory and experiment is reached, and then the difficulty may be settled either experimentally or through a major revision of the theory or by means of both. The historian of science Thomas Kuhn has analyzed how mature sciences deal with this kind of impasse and has illustrated his findings with examples drawn from history.[3]

In astronomy the major discoveries seem to differ qualitatively from the revolutions that Kuhn describes. Despite its tradition, which stretches back many millennia, astronomy does not appear to qualify as a mature science in Kuhn's sense of the word—a science with an established framework of theory and understanding. The observational discovery of each new major phenomenon does provide us with a revolution; but there is no impasse, no conflict with theory, largely because theorists seldom worry ahead of time about phenomena not yet discovered. Thus the new discoveries pose no challenge to defensible preconceptions. Instead, there is a complex interplay between ideas and observations, encountered in many different forms. We shall see this, next, in looking at the observational appearance, physical structure, and original discovery of the known cosmic phenomena.

Known Cosmic Phenomena

The accounts of discoveries presented here are usually based on the writings of the participants, though, in describing more recently discovered phenomena, I have often included new, personal recollections conveyed to me by the discoverers, particularly where previous accounts were not generally available. The prime aim, however, of all these historical sketches, whether based on older material or new, is to present a picture of how observational discoveries in astronomy come about and to sort out elements that appear to favor discovery.

1. Dust and Stones Between the Planets: Meteorites, Zodiacal Dust

If we stand at equatorial latitudes on a clear dry night not long after darkness has set in, we can see a faint tongue of light reaching upward from where the sun has passed below the western horizon (figure 2.1). If we wait through the night, this zodiacal light reappears in the eastern sky, some two hours before dawn. As day breaks, the faint glow submerges in a brightening sky and disappears.

The zodiacal glow is sunlight scattered by countless fine dust grains that float in space between the planets. Individual grains can be registered by means of special sensors mounted on space probes that journey through the solar system.

Occasionally, a larger chunk of matter from the zodiacal cloud enters the earth's atmosphere. Such meteors orbit the sun along elliptical paths, much like the smaller dust grains or the far larger planets. As a meteor enters the atmosphere, air drag heats it intensely, and it emits visible light. Often we see a meteor's bright trail, but only for an instant, and we have to be prepared to make all our observations in a flash. The larger meteors, the meteorites, survive this fierce passage and can be dug up where they hit ground (figure 2.2).

That meteors could be hitting the earth from space was not established until late in the eighteenth century. The name meteor already tells us that these shooting stars formerly were considered to be meteorological phenomena, much like lightning. But in 1798, two twenty-two-year-old students at Göttingen, H. W. Brandes, who later was to become professor of physics at Leipzig, and J. F. Benzenberg, destined to become professor of physics in Düsseldorf, decided to observe meteors from two different locations. By ordinary methods of triangulation, they concluded that the meteors were at heights far greater than the atmosphere had been thought to extend.[4] The velocity of the meteors can be determined by observing the rate at which meteors cross the sky. This velocity is high, and it soon became clear that meteors must be originating in space and falling onto Earth.

In the early hours of November 12, 1799, Alexander von Humboldt, on his South American expedition in Cumaná, had observed a meteor

Figure 2.1 *The Zodiacal Light*

The zodiacal light is the glow of sunlight scattered from tiny dust grains that orbit the solar system like myriad microscopic planets. A shower of this dust continually strikes Earth's upper atmosphere, and the larger particles produce a visible flash as they are heated by atmospheric friction. These are the shooting stars, or meteors. Meteorites are the larger chunks which survive the intense heat of passage through the atmosphere and hit Earth. They have survived almost 5 billion years without the melting and mixing that has taken place on Earth and bring a record of chemical processes that took place during the formation of the solar system.

Photograph courtesy of NASA

shower and noticed that the meteors all appeared to be arriving from a single direction. This became widely appreciated thirty-four years later, on November 12, 1833, when fantastic meteor showers were observed in Europe and in America. Again, all the meteors appeared to come from one portion of the sky. It did not take long to establish that meteor showers correspond to the earth's passage through portions of space previously crossed by comets, and that the in-falling matter represents cometary debris orbiting the solar system.

A cosmic origin for meteorites had become apparent around 1803, when the young French physicist Jean Baptiste Biot subjected to a thorough study the fall of a set of stones at L'Aigle in France.[5] Thereafter

Figure 2.2 *The Barringer Crater near Winslow, Arizona*

A continuous hierarchy of matter orbits the sun. It ranges in size from tiny grains of dust to Jupiter, the largest planet. Occasionally a huge interplanetary boulder strikes the earth's surface leaving a crater that can survive for millions of years until erosion finally erases all memory of the impact. On the moon, where weathering is incomparably slower, these craters survive for billions of years and account for the pockmarked surface seen today. The largest craters seen on the moon were left by masses that rival the smaller asteroids in size. The Barringer Crater is many thousand years old. Its enormous size is made evident by the buildings just barely seen on the lower right of the photograph, straddling the crater's rim.

Photograph courtesy of Yerkes Observatory

chemical analyses, as well as dynamic investigations, of these forms of extraterrestrial matter became possible.

Finally, the interplanetary origin for the zodiacal light itself became clear around 1934. Up to that time it was possible to argue that the zodiacal glow was just an extension of the solar corona. But on September 10, 1923, the German astronomer H. Ludendorff had obtained a set of coronal spectra in the course of a solar eclipse. These spectra were analyzed eleven years later by W. Grotrian of the Potsdam Observatory in Germany, who noticed that close to the sun the spectrum appeared washed out, while at a larger distance it exhibited sharp, dark Fraunhofer lines, named for Joseph Fraunhofer who first had seen the lines in the solar spectrum in 1817. Grotrian correctly concluded that near the sun the coronal light is produced through scattering by fast-moving electrons which Doppler shift the light and thereby smooth the spectrum. Further away, he decided, the sun must be surrounded by dust grains that move slowly and scatter sunlight without altering its absorption line spectrum.[6]

The sun's radiation causes dust grains to slowly spiral into the sun. The zodiacal dust cloud must therefore be replenished, aeon after aeon. Where does this dust originate? Some of it is cometary debris; some of it originates in the collisions of asteroids. But we do not know if these two sources account for all of the dust we see. There may be other components originating in ways still unknown.

Visual observations of the zodiacal glow and direct observations of individual meteors and meteorites provide us with two independent means of studying interplanetary matter. A third powerful technique was accidentally discovered toward the end of World War II, when radar sets of the British Anti-Aircraft Command were used to search for approaching German V-2 rockets. While they did this efficiently, they also registered frequent false alarms. At the time, the physicist James Stanley Hey and his colleagues at the Army Operational Research Group (AORG) had chief responsibility for tracing and identifying any sources of radar interference. Hey investigated these false alarms and suggested that many of the recorded radar echoes probably originated in the earth's ionosphere and might be associated with the passage of meteors.[7] After the war this was verified, and radar studies of meteors have now become one of our prime sources of information about interplanetary grains. Hey, however, was not the first to note reflection of radio waves by meteors. J. P. Schafer and W. M. Goodall had already noted this effect during the Leonid meteor shower of November 1931, although their conclusions were made somewhat uncertain by magnetic disturbances present at the time.[8] Schafer and Goodall, both researchers at Bell Telephone Laboratories, had undertaken these studies at the suggestion of their colleague, A. M. Skellett,[9] who proposed that meteors could account for significant ionization in the Kennelly-Heavyside regions of the upper atmosphere from which radio waves were known to be reflected. Interestingly, the work of Skellett, Schafer, and Goodall appears in the *Proceedings of the Institute of Radio Engineers* right after Karl Jansky's first article on atmospherics in which he tentatively identified an extraterrestrial radio noise. A year later, when Jansky had definitely established the astronomical character of this noise, he concluded his celebrated paper[10] by acknowledging "the help of Mr. A. M. Skellett, also of the Bell Telephone Laboratories, in making some of the astronomical interpretations of the data." Skellett, who then was studying for his Ph.D. in astronomy at Princeton while working as a radio engineer at Bell Laboratories, was a friend and bridge partner of Jansky. He introduced Jansky to basic astronomy and celestial coordinates and evidently played an important role in the early radio astronomical developments at Bell.[11]

2. Planets

Planets initially were detected in antiquity as visual objects that wander across the night sky in a time of months or years. These driftings were known to the Assyrians many centuries before the birth of Christ. They knew all the naked-eye planets: Mercury, Venus, Mars, Jupiter, and Saturn. But it took two millennia and the invention of the telescope before the three remaining planets, Uranus, Neptune, and Pluto, were discovered in 1781, 1846, and 1930. Pluto, the most remote planet, orbits the sun at a distance roughly 40 times further than Earth and takes two and a half centuries to complete its circuit.

The elliptical orbits that the planets and Earth describe around the sun were first recognized by Johannes Kepler in the early years of the seventeenth century. The data on which he based the calculations, which after many years enabled him to stumble on the elliptical form of the orbit of Mars, were the measurements compiled by Tycho Brahe during the course of twenty years of work on the Danish island of Ven. Arthur Koestler has traced out the tortuous path followed by Kepler in his search.[12] It is clear how much Kepler owes to Tycho since Kepler himself was not an observer. The discovery of the elliptical orbits could not have taken place without precise measurements, and only Tycho possessed these.

In the last quarter of the sixteenth century, Tycho had set a new course for astronomy, founded on precise measurements carried out over a succession of many years. Kepler had soon recognized the importance of these observations, and at an early opportunity had joined Tycho, then exiled in Prague. Following Tycho's death late in 1601, Kepler became the inheritor of Tycho's treasury of observation data and set to work to unravel the orbits of the planets.

While the Polish astronomer Nicolaus Copernicus had first presented strong arguments for believing the planets to orbit the sun, Kepler was able to establish the shape these orbits apparently assumed and the rate at which the planets sped along their way.

Still, the ideas of Copernicus and Kepler were widely held to be debatable, and it took two quite different observations to establish their heliocentric view beyond a shadow of a doubt. The first of these came in 1610, a few months after Galileo constructed his telescope. He pointed it at Venus and found that the planet presented the shape of a crescent. The phase of the crescent changed with time, and Venus clearly had to be orbiting the sun in order to exhibit these variations in illumination. Copernicus himself had well understood this consequence of his theory, but with the naked eye the Venus crescent cannot be resolved.[13]

While Galileo's observation proved that Venus moves about the sun, the earth's own motion still remained a controversial question. The theory of Copernicus required that an annual parallax—an apparent displacement of the positions of stars in the sky—result from the earth's orbital motion about the sun. The lack of an obvious parallax, however, implied that the stars are immensely distant. After the death of Copernicus in 1543, countless observers looked for the small parallactic displacement in vain. This endeavor continued without success for nearly three centuries.

One of the unsuccessful aspirants was James Bradley, born in England in 1693. He and a companion, James Molyneux, set out to determine the parallax of the bright star Gamma Draconis, which passes almost directly overhead at the latitude of London. They used a 24-foot-long telescope firmly attached to a stack of brick chimneys at Kew where Molyneux was living. This vertically mounted telescope could be moved slightly by a screw, so as to tilt it accurately toward the star. Bradley expected the tilt angle to reach extreme values in December and June, as shown in figure 2.3. But instead, he found that the tilt was most extreme in March and September, and was far larger than the expected parallax. This presented a puzzle.

Bradley decided to build a second instrument with which he could observe not just Gamma Draconis and the one other star visible through the original telescope but some fifty to seventy additional stars, twelve of which he could observe the year round because they were bright enough to be seen in the daytime. He installed this device in the house of his aunt who kindly permitted him to cut holes in her floor and roof. The objective lens was out on the roof and the occular twelve and a half feet down at a lower level. With this instrument, Bradley confirmed the earlier results. (Molyneux had died in the meantime, and Bradley had continued alone.)[14]

The theoretical explanation for these results occurred to Bradley during a pleasure cruise on the Thames. Every time the boat changed course, a weather vane, mounted on the boat's mast, shifted a little as though there were a slight change in the wind. This apparent wind, Bradley realized, reflected the motion of the boat itself. It occurred to him that the displacement of the stars he was observing similarly reflected the changing motion of the earth. Just as rain or snow always appears to be coming from the direction in which an automobile is traveling, so starlight appears to shift slightly, according to the direction of the earth's motion.[15] This effect on the rays is now known as the aberration of starlight.

Bradley published his findings in 1728. He had proved that the earth moves about the sun, though his dream of discovering the stellar parallax was to remain unfulfilled.

The effect Bradley had found was not small by the instrumental standards of his time. It amounted to a total annual excursion of some 40 seconds of arc. Indeed, occasional anomalies of 10 to 15 seconds of arc in the declination of stars had been noted as early as 1667 by Jean Picard, who was primarily interested in using positions of stars for purposes of geodesy. Picard used a zenith sector with a 5-second reproducibility in measurements, and he might have been able to discover the aberration had he persisted. Others had also been puzzled by these anomalies.[16] It is to Bradley's credit that he was able to eliminate a variety of potentially confusing errors—atmospheric refraction, nutation of the earth's axis of rotation, and so on—to establish a clear annual pattern and that he then was able to head directly toward the correct explanation of his observations.

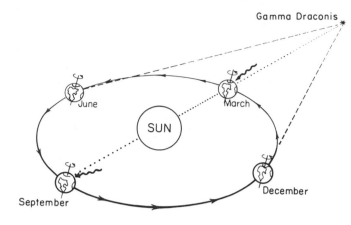

ABERRATION OF LIGHT

————Expected direction of arrival of light from Gamma Draconis
if parallax due to Earth's changing position along its
orbit had been dominant.

∿∿➤Actual direction of arrival of light produced by Earth's
appreciable orbital velocity.

Figure 2.3   *Aberration of Light—Earth as a Planet*

In the year 1728, the English astronomer James Bradley was able to show convincingly
that the earth moves about the sun and that the Copernican view of Earth as a planet is
correct. Late in 1725 Bradley and a friend, James Molyneux, had ventured to observe the
parallax of the bright star Gamma Draconis.

They expected that the apparent displacement of the star viewed through their telescope
would be greatest in December and June, when the earth reaches extreme points in its
orbit about the sun, as seen from the position of Gamma Draconis. In fact, the star's apparent
displacement was greatest in March and September; and at these times the star appeared
displaced toward the direction in which the earth was moving—by about 20 seconds of
arc. Bradley realized that this effect must be due to Earth's orbital velocity about the sun,
which, though small compared to the speed of light, cannot be entirely neglected.

Though he had discovered the motion of Earth, Bradley never did manage to measure
the parallax of a star, nor did William Herschel who set out to find parallax and instead
discovered the existence of double stars. Eventually, in 1838, three centuries after Copernicus
had predicted its existence, Bessel, the German mathematician-astronomer, obtained a measurable parallax for the star 61 Cygni. The search for the annual parallax shows how refined
astronomical observations that stretch technological capabilities, often lead to unexpected
discoveries though they may fail to approach the ends originally sought.

The discovery of Uranus by William Herschel in 1781 was the first
discovery of a planet since antiquity and a matter of great excitement
that catapulted him into the spheres of royalty.[17] Herschel, who had
been born in Hannover in 1738, went to England when he was nineteen
and quickly gained renown as a musician. A man of great energy, he
taught himself Italian, Greek, mathematics, optics, and astronomy in
spare hours and at the age of thirty-four, rented a small telescope to
look at the sky. He knew at once that he needed a better instrument,
but having too little money to buy one, he acquired the necessary appara-

tus secondhand and set to grinding and polishing his own speculum mirrors. Two years later, in 1774, he succeeded in constructing his first telescope of quality and started serious observations of the Orion Nebula. His first scientific paper was written after the age of forty, but his next forty years of indefatigable activity served to make him one of the foremost astronomers of all time.

Many of the hundreds of parabolic specula he constructed served other astronomers all over Europe. The king of England ordered four 10-foot-long telescopes for himself and had Herschel construct other instruments as gifts to European royalty. For Herschel, the sale of these was a significant source of support for his work, particularly in his early years as an astronomer. He invented the Herschelian mounting which made use of but a single mirror and thus minimized the reflection losses of the speculum metal—the best material available in his time. He constructed the most powerful telescope that had ever been built, a 40-foot-long instrument with a 4-foot-diameter reflector. He invented grinding and polishing machines, and the instruments he constructed were unsurpassed in quality.[18] Where Galileo's spyglass had magnified just about 20 times, Herschel, 170 years later, was using magnifications ranging from 227 to 2,010, and his resolution was roughly 1 second of arc.

On first seeing Uranus, Herschel thought he had discovered a new comet. He could partly resolve its disk, whereas stars remained unresolved points of light at all the different magnifications available to him. He also noted the small motion of the source relative to the fixed stars. This clearly showed it to be a member of the solar system, and before long astronomers accepted the new object as a planet.

Charles Messier, the great French astronomer, wrote Herschel expressing surprise at Herschel's ability to discern a body that looked so nearly starlike and moved so slowly across the sky. Just how challenging this observation was can be gauged by the fact that Uranus had been observed and mistakenly recorded as a star on at least twenty occasions between 1690 and its final recognition as a planet in 1781.[19] But the observers, men like John Flamsteed and Bradley, who had recorded its position earlier, had evidently not been able to resolve Uranus to indicate a nonstellar appearance, nor had they apparently repeated their observations after a few days to note the small displacement of the planet against the background of fixed stars.

Still, these early observations were not wasted but led to the discovery of yet another planet—Neptune. For when the orbit of Uranus was plotted from 1690 on, it showed small deviations from a path predicted by Isaac Newton's theory of gravitation. In turn, this led first to the conjecture that an unknown planet might be perturbing the orbit of Uranus and then brought about the actual discovery of that other planet.

Many of the planets can now be studied almost as easily through the faint radio signals they emit as through their visual reflection of sunlight. But this technique has only been known since the mid-1950s. The radio discovery of Jupiter was quite unexpected and unpremeditated. Two young radio astronomers at the Department of Terrestrial Magnetism (DTM) of the Carnegie Institution of Washington, Bernard Burke and Kenneth Franklin, were carrying out observations on the

Crab Nebula during the winter of 1955. Writing about these observations some four years later, Franklin recalled:

The records themselves showed the characteristic hump as the Crab Nebula passed through the pencil beam. This was followed by a smaller hump, lasting the same 15 [minutes], attributed to [the source] IC 443. At times the records exhibited a feature characteristic of interference, occurring some time later than the passage of the two known sources. This intermittent feature was curious, and I recall saying once that we would have to investigate the origin of that interference some day. We joked that it was probably due to the faulty ignition of some farm hand returning from a date.

We decided to present the material we had to the Princeton meetings of the [American Astronomical Society] in April 1955. Accordingly, Burke assembled all the records of the Taurus region for the first three months of 1955 in preparation for the reductions. . . . He was then startled to find that the interference always occurred at almost the same sidereal time. A strange rural romance this was turning out to be! As spring drew nearer, our swain was returning home earlier and earlier, each evening.

Since the source of the "interference" was clearly attached to the sky, we immediately went to an atlas to find anything that might be obvious. A peculiar galactic cluster and an interesting planetary nebula were candidates, but they were ruled out when we noticed that this strange source was not always at the same right ascension. It appeared to drift westward, slightly, over the three-month interval, so that in March the two interesting objects were not in the beam at the time of the recorded event. The late Howard Tatel had looked at Jupiter a few nights before. . . . Having this in mind, he somewhat facetiously suggested to Burke and me that our source might be Jupiter. We were amused at the preposterous nature of this remark, and for an argument against it I looked up Jupiter's position in the American Ephemeris and Nautical Almanac. I was surprised to find that Jupiter was just about in the right place, and so was Uranus. Here was something which needed clearing up. . . .

The next morning, I plotted the right ascensions as a function of the date for the points of beginning and ending of each recorded event, and drew a smooth curve through the points locating each phenomenon. . . . At this point in the growth of the diagram, I began to plot the right ascensions of Jupiter. As I plotted each point, Burke, who was watching over my left shoulder, would utter a gasp of amazement. Each point appeared right between the boundary lines representing the beginning and end of each event! The meaning was exquisitely clear: these events were recorded only when the planet Jupiter was in the confines of the narrow principal beam of the Mills Cross. . . .

Naturally, we wrote to our colleagues in other parts of the world. C. A. Shain, in Australia, immediately began observations which confirmed our identification, and he searched his old records for possible prediscovery observations. It turned out that he had actually received noise from Jupiter in 1950, but had attributed it to interference. Those prediscovery records have proved of great value as early-epoch data, and have been discussed in the literature.

Our identification of Jupiter as a radio source is not based directly on reasoning, but more on luck. . . . We were led into it by the nature of our equipment: a very narrow pencil beam. Shain had a broad beam which was suited to his needs, but which enabled him to overlook the celestial source of "interference" appearing on his records.[20]

I have quoted Franklin's recollections so extensively because the events he recounts bear a remarkable resemblance to many other astronomical discoveries—particularly to the discovery of pulsars by Anthony Hewish and Jocelyn Bell.

Within a decade or two following the radio discovery of Jupiter at least some of the other planets, as well as their motions around the sun, would probably have been similarly discovered even in the total absence of any visual data. Radio astronomers are not bothered much by cloudy skies, and planetary radio astronomy has by now provided us with significant information independent of earlier visual work. Independent techniques such as these are valuable because they can provide a completely fresh, unbiased perspective of our universe, free of prejudices that might have escaped earlier notice.

3. Asteroids

Asteroids, sometimes called planetoids because they appear to be very small planets, were unknown before 1800. Then on New Year's Eve of that year, Father Giuseppe Piazzi of Palermo saw a small, unfamiliar light in the constellation Taurus. The following night its position had shifted. Another man might not have noticed the small shift, but Piazzi was a meticulous worker with an intimate knowledge of the sky. Three years later he was to publish the first of two star catalogues that contained 7,600 stars and were models of precision for their time.

Piazzi could not resolve the image of the new body at all. It had all the appearances of a star, except each night it moved a tiny amount relative to other stars in the sky. He carefully recorded this motion and on January 23, 1801, wrote two other astronomers, Barnaba Oriani of Milan and Johann Elert Bode in Berlin, that he had found a comet.[21]

Bode had predicted the possible existence of a new planet at a distance from the sun of 2.8 earth-orbits, in the unusually wide gap between the orbits of Mars and Jupiter. In fact, he had felt so sure of this that he had organized a search for such a planet, and Piazzi's new object seemed to have all the right traits. However, by the time the news reached Bode on March 20, the new source had shifted into the daytime sky, and there was fear that it might never be found again because its displacement from its original position would be too great when it reappeared in the night sky some months later. During the observations Piazzi had made in January and February, the asteroid had moved across the sky by no more than 3 degrees, which was not much data for computing an extended orbit. Yet predictions of its further motion, worked out with a method specifically invented for the purpose by the mathematician Karl Friedrich Gauss, led to its rediscovery a year later by F. X. von Zach at Gotha and, independently, on the following night, the anniversary of the original discovery, by H. W. M. Olbers at Bremen.[22]

These were the first sightings of Ceres, the largest of the asteroids; and it does in fact orbit the sun at a distance of 2.8 earth-orbits, just as Bode had predicted. Its radius of about 500 kilometers is some 5 times smaller than that of the smallest planet, Mercury. Its mass probably is a 100 times less than Mercury's. But compared to the smallest known asteroids, whose radii are only of the order of 1 kilometer, Ceres is enormous.

By the time Olbers had found Ceres, he had become so familiar with that particular stretch of sky that he also noted a second, apparently new, starlike body. To the surprise of everyone, this turned out to be a second small planet. Several more planetoids were to be found in following years.

Asteroids are distinguished by their orbital motions about the sun, by their rapid—approximately two-hour—rotation periods, by a reflection spectrum that mimics the sun's emission, and by an infrared flux peaking at wavelengths around 10 microns. There is some question about the difference between a planet and an asteroid, and size certainly is the most striking difference. The asteroids are not visually resolved at present, and we have little information about surface features.

4. Moons

A child discovers the moon at age three or four. Primitive man must therefore have been aware of our satellite early in prehistoric times. Yet the concept of moons in general became established no earlier than 1610 when Galileo constructed his spyglass and found four specks of light that slowly drifted around Jupiter. In relation to their parent planet, their orbits were like the moon's trajectory around Earth. Galileo's spyglass also enabled him to see mountains and valleys on the moon similar to those on Earth (figure 2.4) and to determine the height of the highest mountain, which exceeds 4 miles. Rather than consisting of some completely unknown form of matter reserved for celestial objects, the moon appeared to have a structure similar to that of Earth; and moon matter presumably also obeyed terrestrial laws of physics and chemistry.

Galileo was not the first scientist to turn a spyglass toward the moon, nor was he the first person to use a telescope for astronomical purposes. In July 1609, more than four months before Galileo started scanning the skies, Thomas Harriot in England had used a simple telescope to sketch the moon.[23] But Harriot did not pursue these efforts with comparable instruments or competing vigor, and it was left to Galileo to announce the new and amazing discoveries that the spyglass had made possible.

We take such observations for granted today, but in Galileo's time his scrutiny of these natural phenomena represented an arrogant intrusion of man into the affairs of the heavens; and Galileo had to suffer the Church's punishment for this arrogance.

Moons exhibit two periodicities: an orbital period about the sun and an orbital period about the parent planet. The main time scales are years for orbits about the sun and days or months for orbits about the planet. While angular resolutions of the order of 1 minute of arc are needed to observe the motion of satellites about the planets, the orbital

Figure 2.4 *The Earth Seen from the Moon*

The crescent Earth seen from the spacecraft *Apollo 17* on the last of the manned lunar flights in 1972. The mountains and valleys of the moon, which Harriot and Galileo had first vaguely discerned in 1609, are clearly seen in the foreground. Galileo first saw a similar crescent Venus in 1609. The shape of the crescent changed, and Galileo correctly argued that that could only happen if Venus orbited the sun as required by Copernican theory.

motion about the sun occurs, for a terrestrial observer, over angles large enough to be measured in radians. The main difference between moons and asteroids or even small planets is that asteroids and planets merely move in simple ellipses about the sun, while moons orbit around their parent planets, as well as around the sun.

5. Rings

Christian Huygens was one of the most versatile scientists and inventors of the seventeenth century. Physicists know him for his ideas on the wave propagation of light. As an inventor, he is best known for the development of the pendulum clock, which revolutionized positional astronomy, and for his design of a spring balance regulator for chronometers that could be taken to sea. The pendulum clock, used in conjunction with a transit instrument, permitted astronomers to determine the right ascension of stars with great precision; chronometers were needed in navigation to fix a ship's longitude by observing the meridian transit of those stars whose positions had already been accurately determined.[24]

In 1655, when he was twenty-six, Christian Huygens and his brother Constantin constructed a 12-foot focal-length telescope with a lens they had ground themselves. Its aperture was 57 millimeters, about 2½ inches. When the telescope had been built, Huygens pointed it toward Saturn and immediately discovered Titan, the planet's first-known satellite. He also was able to solve the puzzle of what had appeared to Galileo to be two ear-shaped appendages sticking out on opposite sides of Saturn. To Galileo's surprise, these ears vanished in time, and he did not know what to make of it. Huygens, with the greater resolving power of his new instrument, was able to observe the same phenomenon more clearly and noted that Saturn appeared to be surrounded by a thin flat ring which did not touch the planet anywhere (figure 2.5). He explained the apparent disappearance of the rings as a projection effect—a periodic occurrence which took place whenever the ring was seen edge on from Earth's position in space.

Huygens's discovery was made at a time when the rings were close to disappearance so that he never actually saw the ring shape until much later. Instead, the projection of the rings presented the aspect of a thin line piercing the planet. Because of his superior instrument Huygens was able to note that this thin line did not shrink in length as it got thinner and ultimately disappeared. This observation, together with sketches others had made of Saturn's appendages, led Huygens to conclude the existence of a ring around the planet, a conclusion that was long challenged by his contemporaries. Increasingly improved telescopes, however, removed all doubt.[25,26]

We now know that the belt around Saturn actually consists of several concentric rings separated by gaps, but these could not be resolved at that time. Huygens's telescope was probably the finest in existence in 1655, though the glass available to him never was of very good quality. Later, he and his brother built a number of telescopes having enormous

Figure 2.5 *Saturn and Its Rings*

In 1655, Christian Huygens and his brother Constantin constructed a 12-foot-long telescope having a resolving power appreciably greater than Galileo's smaller spyglasses. On pointing the telescope at Saturn, Huygens discovered its largest satellite, Titan. Others later reported that they too had seen this speck of light but had mistaken it for a background star. Huygens, however, maintained that he had had the superior telescope and that this had also permitted him to decide that Saturn was encircled by a ring, where others only saw puzzling append-ages. Twenty years later, in 1675, J. D. Cassini's still more powerful telescope showed that Huygens's ring appeared divided, and we now know that the planet actually is surrounded by a disk that consists of myriad rings separated by gaps.[26]

As Saturn and Earth orbit the sun, Earth occasionally passes through the plane of these fine flat rings, and the projection of the rings becomes so thin when seen edge on that the rings appear to vanish. Huygens explained this projection effect when he published his findings.

Photograph courtesy of Hale Observatories

focal lengths. The Royal Society of London has three of their lenses, ranging up to 23.2 centimeters in diameter and reaching focal lengths up to 64 meters.[27] Telescopes having such great focal lengths at the time provided the only means for obtaining highly magnified images without chromatic aberration—the fracture of the images into different colors. Besides building some of the most powerful telescopes, Huygens also devised an improved eyepiece which became widely adopted.

More than three hundred years had to pass before a second planet was found to have rings. The discovery was made by James L. Elliot of Cornell University in 1976. Elliot and his collaborators, E. Dunham and D. Mink, had planned to observe the occultation of the ninth-magnitude star SAO 158687 by the planet Uranus. The observations were carried out at an altitude of 41,000 feet, aboard NASA's Kuiper Airborne Observatory—a converted C-141 military cargo plane, having a 91 centimeter aperture telescope mounted to look out the side of the plane. Another similar set of observations was being conducted simultaneously from the ground by Robert L. Millis of the Lowell Observatory in Flagstaff, Arizona.[28]

Elliot had expected the light from the star to be extinguished just once as it passed behind the planet's disk, but instead he found that it repeatedly was extinguished for short durations, ranging from one second to several seconds—both before and again after the occultation by Uranus. The star's light appeared to be turned off at least five times before occultation and almost symmetrically again five times after occultation. The only sensible interpretation appeared to involve the star's passage behind a number of thin rings, some no more than $\sim$ 12 kilometers wide, surrounding Uranus. Measurement of the thickness of these rings would have required an angular resolution better than $\frac{1}{1,000}$ of an arc second. In Elliot's observations, this resolution was achieved through careful timing of the occultations with a resolution of the order of 10 milliseconds. The shadow of Uranus was moving at a speed of $\sim$ 12 kilometers per second at the time, and that velocity together with the occultation duration provided a measure of the width of the rings.

Early in March 1979, a ring was also discovered around Jupiter.[29] Observations by the NASA spacecraft *Voyager,* made from within Jupiter's system of satellites, registered a streak of light when *Voyager* passed through the planet's equatorial plane. Within three days of NASA's announcement, these findings were confirmed through telescopic observations from the ground—observations that could have been carried out many months earlier had the idea occurred to anyone.[30]

6. Comets

Comets have been known since antiquity and were thought to presage calamities. Together with novae, comets were classed as "ominous stars" or "guest stars" in the ancient Chinese dynastic histories, official histories compiled following the termination of individual dynasties.[31] In Western civilization, comets had a bad reputation as well.

Comets often are spectacular visual objects extending many degrees across the sky. They move through the solar system, first approaching the sun and then receding over a period of months; they brighten and decay on a time scale of a month, but hour-to-hour variations also occur. They exhibit visual spectral lines, a significant infrared flux, and weak spectral features in the radio wavelength range.

In the latter part of the sixteenth century, Tycho Brahe had compiled precise data and had constructed reliable instruments that provided minute-of-arc accuracy for the measurements of planetary and stellar positions. By thorough observations, Tycho was able to show that the comet of 1577 pursued an orbit far more distant than that of the moon. He based his conclusion on a careful search for the comet's diurnal parallax which he found too small to measure, and certainly far smaller than the moon's. He then went on to establish the path of the comet and found that it moved around the sun in an orbit beyond that of Venus, an orbit that was not circular but rather oblong or oval.

Tycho, incidentally, had not been the first to suggest the use of parallax to establish the distance of a comet. Somewhat earlier, in 1550, the Italian physician and mathematician Girolamo Cardano had also used the method to show that comets could not be atmospheric phenomena. But Cardano apparently had not been able to demonstrate his results convincingly.[32] At any rate, the precise determination of the comet's parallax clearly placed it beyond the realm of atmospheric phenomena and led to the recognition that comets are cosmic bodies like stars and planets, rather than atmospheric features as claimed by the Aristotelians.

The best-known comet is named after the English astronomer Edmund Halley, who used Newton's then new principles to compute an orbit for the comet of 1682 (figure 2.6). Halley noted in 1705 that this orbit was nearly identical with those of the comets of 1607 and 1531 and concluded that he was dealing with three successive apparitions of the same comet. In fact, he predicted the comet would return again no later than 1759. The comet actually did reappear in that year and became the first periodic comet to be discovered. Periodic appearances of Halley's comet can now be traced through historical records back as far as 1057 B.C.; and after 240 B.C. each reappearance, except perhaps for one that must have occurred around 163 B.C., has been recorded either in China or Europe.[33]

Through his discovery, Halley succeeded in destroying the ancient superstitions about calamities presaged by comets. When his comet reappeared as predicted in 1759, it became demonstrably clear that the comets were simply obeying Newton's laws of gravitation in following their orbits about the sun. This was no small matter when one considers that even great men like Tycho Brahe and Johannes Kepler, both still alive early in the seventeenth century, cast horoscopes and, as far as we know, were sincere in their astrological beliefs.[34]

Comets often are long-tailed objects that contain a solid nucleus consisting largely of frozen materials—ices—that can be volatilized through heating on approach to the sun. As the material—water, carbon dioxide, and perhaps ammonia—evaporates, inclusions of solid, possibly sandy, materials become loosened and float into the surrounding space. The solids are repelled by sunlight and form a curved tail generally pointing away from the sun; a solar wind blows away the ions, which can reach speeds in excess of a 100 kilometers a second and form a straight tail pointing away from the sun.

Figure 2.6   *Halley's Comet on Its Passage Through the Solar System in 1910*

Comets are the largest bodies in the solar system. But they are so tenuous that a billion comets would still fall far short of rivaling the mass of Earth. In this photograph, obtained on May 12, 1910, the comet's tail appeared 30° long and could have spanned the distance between Venus and Earth.

7. Main Sequence Stars

The sun's yellowish color, its spectrum, and its total light output are typical of a large proportion of the stars we see in the sky. Only our proximity to the sun deludes us. For us, the sun is incomparably brighter than any other celestial source. It is this very brightness which for millennia prevented man's recognition of the sun as simply a star.

A first comparison of the sun's brightness to that of the nearer stars became possible in the seventeenth century when the size of planetary orbits in the solar system had become established. The invention of the telescope had permitted an approximate size determination of the planets themselves. In this way, it was already known in Newton's time that Saturn intercepted about one part in a thousand million of the sun's light and that Saturn appears as bright as a first-magnitude star. Newton guessed that Saturn might reflect roughly a quarter of the sun's incident radiation. This meant that the light received from a first magnitude star amounted to no more than one part in four thousand million of the light from the sun. If its luminosity—its absolute light output— were the same as the sun's, it would have to be nearly a hundred thousand times further away.[35]

Unfortunately, at that time there was no way to confirm this distance. Nearly two centuries had to pass before Friedrich Wilhelm Bessel first correctly measured the distance to a star. He sought to observe the annual parallax—the position relative to fainter and therefore probably more distant sources—of a particular star in the constellation of the Swan. This star is catalogued as 61 Cygni. By observing it at different times of the year, as the earth moved in its orbit around the sun, Bessel was able to measure the apparent periodic displacement of 61 Cygni relative to the stellar background. Knowing the distance Earth had traveled in producing this parallactic displacement, Bessel was able to determine the remoteness of the star, much as a surveyor measures distances. His report on his first year's work, published in 1838, and a report two years later, based on added measurements, gave a distance to 61 Cygni amounting to some 11 light-years, a distance six hundred thousand times greater than our distance from the sun.[36]

Within a year, two other astronomers, Thomas Henderson and Wilhelm Struve, respectively, had published distances to Alpha Centauri and Vega.[37] They were of the same order as the distance to 61 Cygni and could be taken as typical of distances between independent stars in the Milky Way. The partial success of these observations followed a long list of failures and, worse, false claims. The idea of observing stellar parallaxes had suggested itself to astronomers as soon as the Copernican theory was taken seriously. Copernicus recognized the importance of these measurements, but his instruments could hardly measure angles of 10 minutes of arc, let alone fractions of seconds.

Thinking that the brightest stars must also be the nearest, astronomers had long sought in vain to measure the parallaxes of these stars. Many of the brightest stars in the sky, however, are very distant and appear bright only because of their enormous intrinsic luminosity. In contrast, the luminosities of many stars in the sun's neighborhood are low, and these nearest stars therefore appear inconspicuous.

In 1717, Edmund Halley had gone over the positions of stars given in Ptolemy's Almagest. The observations cited there had been made by Timocharis around 300 B.C., and by Hipparchus and perhaps Ptolemy himself a century and a half later. These positions Halley compared to positions determined in his own times. When he corrected the difference for a precession of Earth's polar axis in the sky, three of the stars he examined, Sirius, Arcturus, and Aldebaran, showed residual displacements amounting to half a degree. This appeared to be their proper motion across the sky in 2,000 years.* As a check, Halley was able to compare the shift of Sirius in the century since Tycho Brahe had recorded its place in the firmament. It had shifted by several minutes of arc—more than any error Tycho might have made. The conclusion was clear: The stars moved relative to each other.[38]

Nearly sixty-six years after Halley's discovery, William Herschel, in 1773, examined the combined proper motions of some thirteen stars and deduced that the sun's motion, assuming these stars to be fixed, lies roughly along the direction toward the star Vega. This estimate of the sun's motion through its immediate neighborhood of the Galaxy proved to be surprisingly accurate.[39]

In 1792, Piazzi had noted an unusually large proper motion for the star 61 Cygni, amounting to 5.2 arc seconds per year. Bessel again called attention to this star twenty years later, in 1812, realizing that high proper motion (figure 2.7), rather than great brightness, could be a sign of proximity and that a parallax measurement on this star carried much promise. Bessel did succeed in measuring this parallax, but his observations, initiated in 1837, indeed were difficult.

The choice of stars with high proper motion was crucial. But the success of Bessel and the partial success of Struve in finally detecting the parallax of a star can also be attributed to two excellent achromatic refractors produced respectively for the Königsberg and Dorpat observatories by the Munich Optical Institute under Joseph Fraunhofer's skilled supervision. The Dorpat lens measured 9-½ inches in diameter, was by far the largest achromat constructed up to that time, and brought Fraunhofer a title of nobility and sole management of his institute.

Thomas Henderson's work succeeded despite considerably poorer equipment. But he was fortunate in having picked Alpha Centauri which has a parallax of 0.75 arc seconds, is more than twice as close as 61 Cygni, and 6 times closer than Vega. Henderson, however, did not publish until his work had been independently verified and then only two months after Bessel.

Bessel's achievement did not rest on excellent equipment alone. Earlier in his career he had carefully worked out a theory of observing errors and his observations took into account some of the most refined nuances through a program of careful checks and cross-checks. His finally derived parallax differs by no more than $\frac{1}{20}$ of an arc second from the currently cited value. None of his contemporaries came close to

---

* For Sirius and Arcturus, Halley's findings were approximately correct. Aldebaran, however, would only have moved $\frac{1}{10}$ of a degree across the sky in the available time.

Figure 2.7 *Proper Motion as an Indicator of Proximity*

The two photographs reproduced here show the motion of Barnard's star, the star whose proper motion, 10 seconds of arc per century, is the largest known for any star. While Barnard's star is not our nearest neighbor, it lies only 6 light-years away, not much more distant than the Proxima Centauri and α Centauri systems, our closest neighbors in the Galaxy. The two exposures, made in 1894 and 1916, show that Barnard's star is quite faint, but as indicated by the arrows, it moved an appreciable distance relative to other stars during the twenty-two years between exposures.

Photograph courtesy of Yerkes Observatory

matching that performance. Struve's parallax for Vega was too high by more than a factor of 2, by $\frac{1}{7}$ of an arc second, while Henderson's value was too high by about 0.4 seconds of arc.[40] Henderson, however, was well aware of the probable size of his errors, which, at least, were smaller than the parallax he had found; but Struve's error was larger than the star's actual parallax. His was not much of a measurement.[41]

After Bessel's initial success, three-quarters of a century had to pass before a sufficient number of stellar distances could be determined for the next big step. In the early years of our century, just before World War I, Henry Norris Russell of Princeton and Ejnar Hertzsprung of Potsdam independently started entering the brightness and the coloration or temperature of stars in a diagram now known as the Hertzsprung-Russell diagram.[42]

In order to derive a relationship between the color and intrinsic brightness of a star, both men needed data on the colors of stars, stellar distances, and apparent brightness. The coloration of stars was available to them mainly through the work of E. C. Pickering and his colleagues at Harvard University who had catalogued spectra from some 10,000 stars in the decade before 1890. Both Hertzsprung and Russell made use of Pickering's spectra, which had been made possible by photographs of exceptionally large fields of view, no less than a 100-square degrees in the sky. These were taken with an objective prism spectrograph, a technique in which a thin prism was placed in front of the objective lens of a telescope. The objective prism, initially used half a century earlier by Fraunhofer, had been coupled to a wide field photographic astrograph which permitted Pickering to capture the spectra of hundreds of stars in a single photographic exposure.[43]

For relative distances of stars, three indicators were available: For isolated nearby stars, either proper motions or parallaxes could serve to indicate distance. For stars that were members of well-defined clusters, identical distance could be assumed for all the cluster members. Armed with these data, Hertzsprung and Russell were able to tabulate a relationship between the brightness and coloration of stars, and this eventually led to a diagrammatic presentation, a Hertzsprung-Russell diagram, much as in figure 2.8.

Hertzsprung and Russell independently noted that by far the largest number of stars fall along a diagonal strip, the main sequence on this diagram. The sun is a main sequence star. Like all stars on the main sequence, it shines through the aeons because it is slowly converting hydrogen into helium deep in its central regions. This conversion liberates the nuclear energy the star needs in order to shine. Russell and Hertzsprung, however, were unaware of these nuclear processes when they constructed their diagrams. A full account of how the hydrogen-to-helium conversion takes place was not to emerge until three decades later, when Hans Bethe of Cornell University published a detailed explanation in 1939.[45]

Stars belonging to the main sequence are so diverse that they might well be considered to represent at least two, and more probably three, distinct phenomena. At one extreme are young massive stars formed

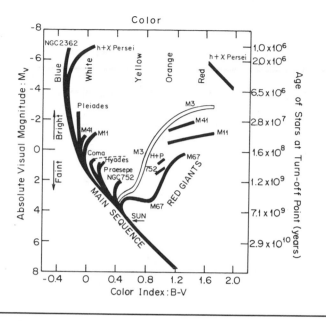

Figure 2.8 *The Color and Brightness of Stars*

Stars found in the same cluster exhibit a brightness related to color. If the star's absolute visual magnitude, M_v, is plotted against color index, given by the ratio of light seen through a blue filter, B, and a yellow filter, V, most stars fall along a sloping line called the *main sequence*. A high color index denotes a red star; a low index, a blue star. The colors of stars are indicated at the top of the diagram. The drawing shows the principal brightness and color regimes in which stars from different clusters are found. Most of the clusters shown here are galactic clusters (dark bands). The light band represents stars from the globular cluster M3. The region denoted Red Giants contains stars whose colors are the same as those of stars directly below on the main sequence. But the giants are far brighter than the corresponding main sequence stars. The point at which the red giant branch turns off the main sequence is called the *turn-off point* and is an age indicator for a cluster. Young galactic clusters like those denoted by catalogue numbers NGC 2362 and h + χ Persei are just a few million years old. The Pleiades cluster shown in figure 2.12 is a hundred million years old, and the globular cluster M3 is a few billion years old. The color index is a measure of the star's surface temperature. A high color index corresponds to a low temperature, and a low index to a high temperature. Diagrams of the type shown, and similar-looking diagrams that plot the luminosity of stars as a function of temperature, often are referred to as Hertzsprung-Russell diagrams, after the two astronomers who first constructed these figures. (Based on a drawing by Dr. Allan Sandage[44].)

within the last few million years. Their masses may be more than twenty times as great as the sun's, their luminosities a million times greater, their life spans some ten thousand times shorter, and their speeds of rotation ten times as great. At the other extreme are stars one hundred times fainter, some three times smaller and, by all indications, much longer lived than the sun. If the universe were to become twice as old as its present apparent age of fifteen billion years, these small faint stars would still be shining steadily, oblivious to all the changes the cosmos had seen. The sun, on the other hand, would appear unrecognizably aged.

8. Subgiants and Red Giants

The main sequence is not the only richly populated region in the Hertzsprung-Russell diagram. If all the stars that belong to a single globular cluster—a large spherical aggregate comprising some hundred thousand stars—are displayed on such a diagram, the upper main sequence is entirely missing. There is only a short stub of a main sequence running upward to the left. Where it breaks off, there is a fainter trail continuing upward but veering back to the right. A star exhibiting a given temperature may therefore correspond either to a faint main sequence star or else to a far larger giant thousands of times brighter. If the colors of these two stars are identical, we know their surface temperatures must be the same: Red stars are cooler than orange, which in turn are cooler than yellow stars like our sun. The color temperature also tells us the star's surface brightness, the quantity of light emitted by any given area at the star's surface.

If two stars are at the same distance from us, have the same color temperature but differ in brightness, then one of them must actually be larger than the other. The giant stars, in fact, do live up to their names. If the sun's radius were to expand two hundredfold so as to swallow the orbits of Mercury, Venus, and Earth, it would still fit handily into a red giant the size of Betelgeuse in the constellation Orion. Such stars are large and cool and therefore appear both bright and red. The low surface temperature permits atoms to form molecules like titanium oxide, carbon monoxide, or zirconium oxide in the star's atmosphere. These gases are readily identified by spectral bands apparent even at quite low wavelength resolution.

The light from giant stars can vary somewhat erratically. The brightness of Betelgeuse increases and decreases irregularly over a period of years. Its maximum brightness is more than twice the minimum. Betelgeuse, like other members of its class, is also losing mass by blowing away its outer layers. We observe this spectroscopically. The spectral lines in the cooler gases more distant from the star's surface are Doppler shifted, and the gas is seen to approach us. The loss of mass is no small effect. If it persists for many millions of years, the star must waste away. As a probable result of the mass loss, circumstellar clouds of dust are also seen around these stars. The dust grains evidently condense in the outflowing stream as the gas cools on receding from the star's hot surface.

9. Pulsating Variable Stars

On October 19, 1784, John Goodricke of York—both deaf and dumb and fated to die before his twenty-second birthday—started "A Series of Observations on, and a Discovery of, the Period of the Variation of the Light of the Star Marked δ by Bayer, Near the Head of Cepheus," on which he reported in the *Philosophical Transactions of the Royal Society of London*.[46]

Goodricke, who had initiated a systematic study of variable stars during the preceding two years, did not then realize that this set of measurements dealt with a phenomenon different from other variables he had observed. These others, he correctly suspected, changed brightness as they were eclipsed by companions. Delta Cephei is different. It varies intrinsically by successively expanding and contracting, completing a full cycle with a period of nearly five and a half days. At its brightest, Delta Cephei is about 3 times brighter than at its faintest appearance, and such a brightness variation is readily detected by eye if stars that shine steadily can be used for comparison.

Goodricke, incidentally, was not the first to observe a pulsating star, though his observations probably were more systematic than any that had been previously undertaken. Nearly two centuries earlier, on August 13, 1596, David Fabricius, a Lutheran pastor and amateur astronomer from East Friesland, had detected a star which disappeared by the following October. Later, it appeared again; its periodicities were noted and Johannes Hevelius of Danzig named it Mira—the Wonderful.[47] Its somewhat irregular period spans nearly a year.

In 1912, more than a century after Goodricke's careful observations, Henrietta Leavitt found a second outstanding feature of pulsating stars of the Delta Cephei type.[48] When she observed the brightness variation and pulsation periods of twenty-five of these Cepheids, all within the Small Magellanic Cloud and at nearly the same distance from us, she found a simple relationship linking brightness and period. The brighter the star, the longer the period. This discovery had great impact. The Cepheid variables are intrinsically bright and can be seen at large distances. The new relationship meant that the distance of any group of stars containing even one Cepheid could be compared to the distance of any other grouping containing this type of variable. The Cepheid variables could be used as beacons of standard brightness for mapping our galaxy.

Interestingly, the Cepheid variables were long considered to be a form of eclipsing binary star, and it was not until 1914, more than three centuries after their original discovery, that Harlow Shapley offered a roughly correct model for these objects.[49] Several properties common to many variables, such as a Doppler shift toward the violet associated with increasing brightness, or Henrietta Leavitt's period-luminosity relation, had strongly suggested to Shapley that the variability in each case was associated with one star, not with two: Enormous pulsations, successive expansions and contractions on a scale rivaling that of the entire solar system, seemed to fit all the major observed characteristics. Four years later, in 1918, the pulsation theory was put on a sound theoretical footing by the English astrophysicist Arthur Stanley Eddington, and has been accepted ever since.[50] By now, we recognize many different types of pulsating stars, exhibiting differences in brightness or brightness variation, color or color changes, and regularity or irregularity of period.

Dwarf Cepheids like the star SX Phoenicis can have a period as short as eighty minutes. At the other extreme are long-period variables with periods of several hundred days. These slowly pulsating stars are red, large, and luminous; and there is no sharp boundary separating them from the red giants.

The bright blue Beta Canis Majoris type stars pulsate with extreme regularity. They can exhibit periods that vary less than one part in a thousand million in successive cycles. In contrast, the red semiregular or irregular giants show essentially no constancy at all.

The faintest of the pulsating variables are the dwarf Cepheids, typically some twenty times brighter than the sun. They are dwarfed by the classical Cepheids which can be more than 1,000 times brighter. The range of brightness variations can similarly be quite extreme. The star χ Cygni has a visual brightness that is more than 10 thousand times greater at maximum than at minimum, while typical Beta Canis Majoris stars brighten and dim by but a few percents.

Very crudely, the many different known classes of pulsating variables divide into three extreme groups:

1. The highly regular, short-period, faint stars
2. The regular, bright, short- and intermediate-period stars
3. The slowly pulsating, more or less irregular, bright yellow or red stars

This last group can also radiate strongly in the infrared.

10. Multiple Stars

In 1672, Gemiano Montanari of Padua published a report on changes in brightness of the star Algol—the Ghoul. Perhaps the most surprising aspect of this observation is that no one had reported these variations centuries earlier. The Arab astronomer Al Sufi had observed this star as early as the tenth century A.D. and listed its brightness as second magnitude, without noting any variability even though its brightness can drop by more than a magnitude during the course of a single night. That Montanari's discovery should have come as late as it did is therefore puzzling, particularly because Algol is a bright star, readily apparent to the naked eye. Its variations also are pronounced. At its brightest, the star appears some 3 times brighter than at minimum. How could such a change have escaped Al Sufi's notice?

Be that as it may, we now know that the brightness variations in Algol are due to eclipses of a brighter star by a fainter one. This hypothesis was first offered in 1783 by John Goodricke, who had noted the regular period and the short interval of lowered brightness. He suggested the interposition of a large dark satellite as the most plausible explanation.

If we are to see an eclipse of one star by its companion, the orbital plane of the stars must be seen nearly edge on. If we do not view the system along a line of sight that lies precisely in the orbital plane, we will see an eclipse only if the two stars are very close to each other, at most a few stellar diameters apart.

If we cannot see an eclipse, there are still other ways of establishing that a pair of stars orbit each other. We can select stars of approximately the same brightness that lie near each other in the sky and observe them to see whether they appear to move about each other. It was through this method that William Herschel first gave incontrovertible proof to the existence of double stars.

Herschel's discovery actually resulted from yet another unsuccessful attempt to measure the distance to the nearest stars. Noting the failure of others who had attempted to measure the annual parallax, Herschel resolved to study the relative proper motions of large numbers of star pairs that appeared close in the sky. He assumed that the pairings were chance superpositions of stars, some of which might be widely separated along the line of sight. In the course of decades, he suspected that some of the nearer stars might then appear shifted relative to more distant companions. Such stars might then also be examined for annual parallaxes.

Herschel had been periodically observing these double stars for a quarter of a century when he published his findings on their orbits in 1803. These observations left no doubt that double stars were gravitationally bound systems, not just chance superpositions of stars in the sky. The observations also established that gravitational forces are active beyond our own solar system and are of dominant cosmic importance.

Algol is an example of an eclipsing binary, while Herschel's double stars, which were further apart, belong to the class of visual binaries (figure 2.9). There are several ways of observing star pairs, and each

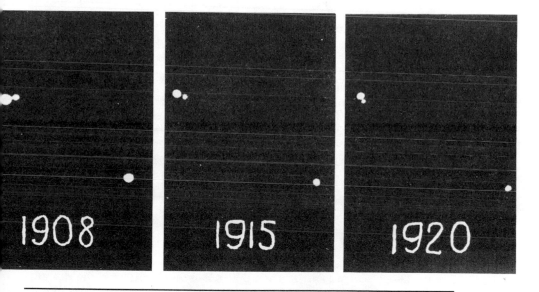

Figure 2.9   *The Visual Binary Krüger 60*

The star Krüger 60 is a visual binary, only 11 light-years from Earth. Its two components complete an orbit about each other every forty-five years. The photograph shows the pair executing approximately a quarter revolution over a period of twelve years between 1908 and 1920.

Photograph courtesy of Yerkes Observatory

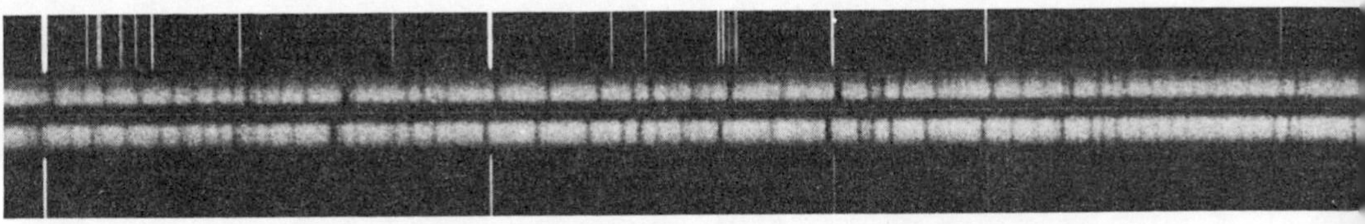

Figure 2.10   *The Single-Line Spectroscopic Binary α' Geminorum*

Two stars that orbit each other often have their orbital motions periodically directed toward and away from Earth. When one of the two stars is much brighter than its companion, we are likely to detect the spectrum of just the brighter star, whose spectral lines are successively Doppler shifted toward shorter and longer wavelengths. Such a configuration of stars is called a single-line spectroscopic binary. The binary shown here is catalogued as α' Geminorum. The spectral shift between the dark absorption lines in the bright star's spectrum can be seen by the displacement between the two central traces on the spectrogram which were obtained by observing the binary at different orbital epochs. Bright vertical lines at top and bottom are laboratory lines superposed on the spectrogram as fiducial marks.

Photograph courtesy of Lick Observatory

method lends its name to the type of binary observed. Spectroscopic binaries (figure 2.10) are stars that are too close to be resolved by telescope, yet do not eclipse. If these stars in their orbital motion alternately move toward and away from us, their spectra will be Doppler shifted and we will see periodic changes in the spectrum of the binaries.

Astrometric binaries are pairs in which one companion is far fainter than the other. The only method for establishing the existence of a dark companion is through the orbital motion induced in the brighter star by the gravitational attraction of its faint companion. Sirius is part of such an astrometric binary. Its companion when first discovered caused a great stir in astronomy.

Binaries are only a special case of small groupings of gravitationally bound stars. Three, four, and higher membered systems are quite common. Algol, for example, is not an isolated binary; the two stars are part of a triple system.

At times, the stars in a binary can approach so close to each other that matter is drawn from the surface layers of one of the stars onto the other. Such stars can then also give rise to radio or X-ray emission.

11. White Dwarfs

In 1834 Bessel was struck by apparent irregularities in the motion of Sirius, the brightest star in the sky. He gathered accurate positional observations that had previously been made on Sirius, and when he arranged these data sequentially he noted that the star seemed to be moving across the sky in a wavelike motion. This result was so startling that Bessel decided a verification was needed. After ten years of regularly conducted precision measurements, he was sure that Sirius must be part of a binary star system that takes some fifty years to complete a full revolution. The companion to Sirius had to be very faint indeed, since Bessel could detect no light at all from it. Procyon similarly appeared to have an invisible companion that also completed an orbit each half century.[51]

There the matter had to rest for twenty years. In the meantime, Alvan Clark, a portrait painter of Cambridgeport, Massachusetts, was gaining renown for the fine quality lenses he ground as a hobby. Around 1860 the University of Mississippi placed an order with him for the construction of an 18-inch objective lens, 3 inches larger than any constructed up to that time. On the night of January 31, 1862, Clark's son, Alvan G. Clark, was testing the definition of the refractor. He looked through it at Sirius and was surprised to find the companion.[52] At the time such a measurement was only possible with a well-corrected lens. The speculum reflectors then in use scattered too much light, and silvered glass mirrors were not to come into use until the 1870s.[53] The companion was exactly where predicted, and its period turned out to be fifty years. Procyon's companion was not found until 1896; it has a period of forty years.

Clark had the good fortune of looking at Sirius when the two companions were near greatest separation. Nevertheless, the observation was an extraordinary feat requiring the detection of an extremely faint point of light only 10 seconds of arc away from a far brighter source (figure 2.11).

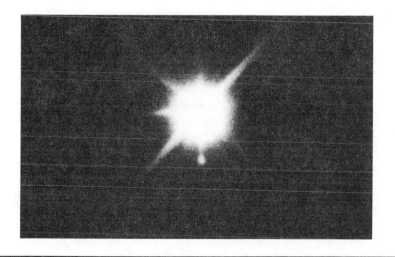

Figure 2.11 *Sirius and Its Faint White Dwarf Companion*

The star Sirius actually is a binary star consisting of a visually bright component, Sirius A, and a far fainter, tiny, but white hot companion, Sirius B, called a white dwarf. In this picture the faint companion looks like a droplet hanging down from Sirius A. Other rays pointing outward from the brighter star are artifacts produced by diffraction within the telescope. Interestingly, when Sirius A and B are photographed at X-ray wavelength, the pattern seen in the picture shown here is reversed: Sirius B is bright, and Sirius A—the bright visual component—is relatively faint in its X-ray emission.

Photograph courtesy of Lick Observatory

Sirius B is 10 thousand times fainter than its bright partner, Sirius A, and 500 times fainter than the sun. Since it is so faint, it was expected to be red, because red stars are cool and therefore can be relatively faint; but in 1862 its coloration could not yet be determined.

In 1915 when the companion again appeared at its greatest orbital separation from Sirius, Walter Adams of the Mount Wilson Observatory in California undertook observations to determine its color.[54] To everyone's surprise, he found that the star was white, rather than red, and that both its surface temperature and its surface brightness must be greater than the sun's. To appear so faint, as well as white, the star had to be far, far smaller than the sun; yet its mass, judged from the orbital motion of Sirius A, was not far different from the sun's. This was a real puzzle.

We now know of many white dwarf stars similar in size and mass to Sirius B and Procyon B. Typically, they appear as faint stars whose proximity to us is convincingly demonstrated by their unusually high proper motions across the sky. Their nearness is proof of intrinsic low luminosity. At greater distances, these same stars would not even be detected since it is usually their large proper motion that calls them to attention.

White dwarf radii are 100 times smaller than the sun's, and their masses range from $\frac{1}{10}$ of a solar mass to about 1.2 solar masses. Their densities therefore range around a million times the density of water— roughly a ton for a thimbleful.

12. Galactic Clusters

Several diffuse patches of light can be discovered in the sky with the unaided eye. Some of these had been known for a few centuries when, in the latter half of the eighteenth century, Charles Messier, one of history's most ardent discoverers of comets, began to compile a catalogue of nebulae, diffuse sources that could be mistaken for faint comets and therefore interfered with his prime searches. In contrast to the comets that came and went, these nebulae did not move, and Messier recorded positions for over a hundred of them. The list contains virtually every type of diffuse object that can be visually recognized, but Messier, with the instruments then available, could not tell them apart. Even William Herschel, whose superior telescopes permitted the compilation of a list of 2,500 nebulosities over the years, was unable to classify many of these unambiguously. Some clearly could be resolved into stars, but others could not; and it was there that classification was most difficult.

Among Messier's nebulous sources there are two kinds that comprise aggregates of stars and lie within our galaxy. The first are the galactic clusters, sometimes called open clusters. These are irregularly shaped groupings of hundreds of stars that always lie close to the Milky Way plane and often contain an unusually high fraction of blue stars. The

Figure 2.12 *The Pleiades Cluster of Stars*

The Pleiades comprise the most easily recognized galactic cluster in the sky. They contain bright blue, luminous stars embedded in tenuous reflecting dust clouds. This reflection nebulosity remained unnoticed until 1885, when it was discovered by two French brothers, Paul and Prosper Henry of the Paris Observatory. The brothers Henry were then pioneering astronomical photography and first noted the reflection nebulae on their photographic plates. The discovery of these clouds signalled an impressive victory for the new photographic technique, and within a few years—certainly by the start of the twentieth century—most optical work in astronomy was being done photographically rather than by eye.

On this photograph, as in figure 2.11, the fine rays emanating from the bright stars are artifacts caused by diffraction within the telescope. Diffraction effects similar to these dictate ultimate limitations on the angular resolution we can obtain with any telescope. These limitations lie in the nature of the light waves that bring us information from the stars and can never be entirely overcome with instruments or arrays of instruments of finite size.

loosely assembled stars typically are spread across distances of 10- to 30-light-years. Together they shine as brightly as a hundred thousand suns. The best-known galactic cluster stars are the Pleiades (figure 2.12), which were known to the Babylonians in antiquity and are readily seen in a glance at the sky.

Two decades before Messier had begun to compile his list, the French abbé Nicolas Louis de Lacaille had traveled to the Cape of Good Hope in South Africa to compile a catalogue of nearly 10,000 Southern Hemisphere stars. During this expedition he also came across some 42 nebulosities which he subdivided into three classes. Not all of these nebulosities were later confirmed, but the 9 members of his second class of nebulae, a group he called "nebulous clusters," all have turned out to be galactic clusters.[55] These apparently were sufficiently resolved with his ½-inch aperture telescope. Other galactic clusters found their way into Lacaille's Classes I and III, called "nebulae" and "stars accompanied by nebulosity," where they appear mixed in with globular clusters, spiral galaxies, and diffuse, gaseous nebulae. At any rate, Lacaille's distinction in 1754 of these Class II nebulae may be considered the first definite indication of the existence of galactic clusters as a separate class of astronomical sources.

Stars that constitute a cluster are thought to have had a common origin. They move through the Galaxy with the same velocity. But some of the more loosely bound clusters may slowly be disintegrating and billions of years from now might no longer be recognizable aggregates.

13. Globular Clusters

Globular clusters constitute the second kind of large-stellar aggregate we see in the Galaxy. They are far more massive than galactic clusters and can contain as many as a million stars in an easily recognized spherical distribution. In contrast to galactic clusters, they contain short-period pulsating variable stars, as well as predominantly red stars and Cepheids. Cepheids are bright stars that can be detected all the way across our galaxy; the other, shorter-period variables also are quite bright. The short-period cluster variables were first discovered to be widely prevalent in 1899 when S. I. Bailey of Harvard made observations at Arequipa in Peru, using photographic plates to find eighty-five of these stars in the globular cluster Messier 5 (figure 2.13).[56]

Later, using the Cepheids as distance indicators, Harlow Shapley at the Mount Wilson Observatory, during the years of the First World War, was able to determine the size of our galaxy.[57] As a doctoral student at Princeton, Shapley had developed a method for determining the distance and brightness of eclipsing binary stars. From the Doppler-shifted spectra of these stars he was able to determine their orbital velocities, and from the duration of an eclipse he could determine the sizes of the individual stars. This was a straightforward application of Newton's laws of motion. The spectrum and coloration of a star, however, also gave Shapley the star's surface temperature, and this permitted calcula-

Figure 2.13 *The Globular Cluster Messier 5*

Globular clusters had been known to a number of astronomers as early as the middle of the eighteenth century; and the French astronomer Charles Messier, as well as William Herschel in England, described the appearance of several of these clusters that seemed to be perfectly round. In 1899 S. I. Bailey of Harvard discovered an unusually high number of variable stars in these clusters. Bailey had been sent to set up an observing station at Arequipa in Peru and there made observations of Messier's cluster M5. On successive exposures he noted as many as eighty-five variable stars in the cluster. Twenty-five years earlier such a discovery would have been impossible: Sufficiently reliable photographic techniques did not yet exist. A globular cluster contains somewhere between a hundred thousand and a million stars. By eye, the brightness of each of these stars would have had to be measured separately on successive nights, and those few stars showing variability would have had to be culled. With the photographic plate, on the other hand, all the stars could be examined simultaneously, and the variable stars could be quickly discerned by comparing the brightness of stars on successive plates.

Photograph courtesy of Kitt Peak National Observatory

tion of the actual amount of light emitted by each unit of area on the star's surface. When he further multiplied this surface brightness by the star's dimensions obtained from eclipse durations, Shapley obtained the actual luminosity of the star. By comparing this absolute magnitude with the apparent magnitude seen from Earth, he was then able to compute the binary's true distance.

Having completed his thesis in 1913, Shapley went West to work at the Mount Wilson Observatory. There, he soon calibrated the true distances of the globular cluster Cepheid variables in terms of binary star distances and then went on to check the distribution of globular clusters in the sky.

He found, unexpectedly, that the several hundred known globular clusters occupy a roughly spherical volume and that the sun is quite far removed from the center. Our best current estimates confirm these findings. The Galaxy's center is 30 thousand-light-years away, and the Galaxy stretches 100 thousand light-years across.

The motions of the globular clusters take them far out to the periphery of the Galaxy. But gravitation tugs them back until they return to the center. Each complete orbit to the edge of the Milky Way and back may take half a billion years. The globular clusters are thought to have formed when the Galaxy was young, 15 billion years ago, and since that time each cluster has completed some 30 orbits about the Galaxy's center.

14. Planetary Nebulae

On November 13, 1790, William Herschel recorded in his observing log:

> A most singular phenomenon! A star of about 8th magnitude, with a faint luminous atmosphere, of a circular form. . . . The star is perfectly in the center, and the atmosphere is so diluted, faint, and equal throughout that there can be no surmise of its consisting of stars.[58]

In presenting to the Royal Society his observations on this and other planetary nebulae, as he called them, Herschel remarked that it appeared to be "a star which is involved in a shining fluid of a nature totally unknown to us." He was right. The envelopes of planetary nebulae consist of ionized gas, a form of matter that was not known at the time.

That planetary nebulae have a highly unusual spectrum was first discovered in 1864 by William Huggins, an English gentleman of independent means, in collaboration with W. Allen Miller, professor of chemistry at Kings's College, London. These two researchers had undertaken some of the earliest spectroscopic studies in astronomy, following Gustav Robert Kirchhoff and Robert Wilhelm Bunsen's discovery of a connection between Fraunhofer's dark spectral lines in the solar spectrum and bright lines emitted by flames.

Spectroscopy applied to astronomy was not new. Fraunhofer had initially discovered spectral lines in the sun and in bright stars as early as 1815; but the application of the spectroscope as a tool for chemical analysis in astronomy was not realized until the work of the physicist Kirchhoff and the chemist Bunsen in 1859 clearly identified the chemical recognition marks provided by spectral lines. The two German scientists had been able to identify some of the chemical elements that make up the sun and further pointed out that

the method of spectrum-analysis . . . opens out the investigation of an entirely untrodden field, stretching far beyond the limits of the earth, or even
of our solar system. For, in order to examine the composition of luminous
gas, we require, according to this method, only to see it; and it is evident
that [this] mode of analysis must be applicable to the atmospheres of the
sun and of the brighter fixed stars.[59]

Kirchhoff and Bunsen had shown that minute traces of different elements introduced into a flame could be identified by the unique set of
spectral lines that fingerprint each element. So powerful was this technique, that the scientists had to devise special means for introducing
only the tiniest controlled amounts of the chosen material into their
flame. To estimate the amount of strontium needed to produce clearly
identified lines they describe the following procedure:

For the purpose of examining the intensity of the reaction, we quickly heated
an aqueous solution of chloride of strontium, of a known degree of concentration, in a platinum dish over a large flame until the water was evaporated
and the basin became red-hot. The salt then began to decrepitate, and was
thrown in microscopic particles out of the dish in the form of a white cloud
carried up into the air. On weighing the residual quantity of salt, it was
found that in this way 0.077 grm. of chloride of strontium had been mixed
in the form of a fine dust with the air of the room, weighing 77000 grms.
As soon as the air in the room was perfectly mixed, by rapidly moving an
open umbrella, the characteristic lines of the strontium-spectrum were beautifully seen.

With all this in mind, Huggins had first used his spectroscope to
examine the spectra of stars. But then, in pointing the instrument at
the planetary nebula in Draco (figure 2.14), he found himself baffled.
Instead of seeing the normal broad continuum spectrum of a star, he
saw a single bright, sharp spectral line. Huggins also found two fainter
lines separated by some distance from this brightest component, and
he identified the strong line with nitrogen and one of the weaker two
with hydrogen. These identifications were made on the basis of laboratory spectra that Huggins had previously analyzed.[60]
    The hydrogen line identification was correct, but the other two lines
were not to be correctly identified for another six decades, when the
American astronomer Ira Bowen first showed them to be forbidden lines
of oxygen atoms stripped of two electrons. The lines are forbidden in
the sense that the radiation is seldom emitted and has never been seen
in the laboratory where the oxygen ions interact too rapidly with walls
of their containers before having time to emit the forbidden lines. Huggins therefore was unable to identify these particular lines. But the important point of his work was that he did make the attempt to find a
correspondence between cosmic spectral features and terrestrial chemical elements and that he was at least partially successful.
    Not all nebular sources that Huggins examined showed line struc-
ture. A bright globular cluster of stars and the Great Nebula in Andromeda, now known to be a spiral galaxy like our own, showed no

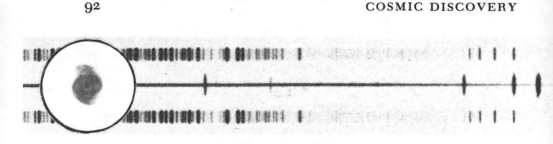

Figure 2.14   *The Planetary Nebula NGC 6543 in Draco and Its Spectrum*

In the early 1860s William Huggins and H. Allen Miller devised a spectroscope that could be mounted on the telescope in Huggins's private observatory at Tulse Hill near London. At first Huggins looked at the spectra of bright stars and found some of the same elements that the German researchers Gustav Kirchhoff and Robert Bunsen had already identified on the sun.[61] At that stage an analysis of nebular sources suggested itself, and Huggins selected a bright planetary nebula in Draco. Of his first look at the nebula Huggins wrote:

> On August 29, 1864, I directed the telescope armed with the spectrum apparatus to this nebula. At first I suspected some derangement of the instrument had taken place; for no spectrum was seen, but only a short line of light perpendicular to the direction of dispersion. I then found that the light of this nebula, unlike any other ex-terrestrial light which had yet been subjected by me to prismatic analysis, was not composed of light of different refrangibilities, and therefore could not form a spectrum. A great part of the light from this nebula is monochromatic, . . .[62]

The bright central portions of this planetary nebula are shown here superposed on part of its optical spectrum. The entire picture is a negative, the darkest portions corresponding to the brightest features of the planetary nebula. The top and bottom spectral traces are laboratory spectra used as fiducial marks. The central trace shows four strong lines and a weaker one. From left to right the strong ones are the atomic hydrogen lines Hγ and Hβ, and two lines due to doubly ionized oxygen. Huggins only saw the last three of these and identified the Hβ line correctly. The other two remained unidentified for more than half a century. The spectrum shown here was obtained by Dr. C. Roger Lynds. The superposed picture of the nebula is a Kitt Peak National Observatory photograph.

indication of bright lines. The globular cluster spectrum looked much like that from a red star.

Planetary nebulae are not always beautifully symmetrical. Many are quite irregular in appearance, and neither their visual nor their radio appearance then differs remarkably from that of ionized hydrogen regions like the Orion Nebula. However, when the distance to the nebulae is ascertained, a rather sharp difference becomes apparent. The planetary nebulae are far smaller. The total mass content of the envelope is a fraction of a solar mass in contrast to the thousand solar masses of the larger ionized hydrogen regions. Correspondingly, the size of the planetary nebulae is also smaller, although the density of gases is about the same in the two types of nebulae.

In contrast to the giant ionized nebulae, we find that planetaries appear quite isolated from large aggregates of gas and dust. This leads to the supposition that the planetaries represent an evolved form of star that has already gone through its main sequence and giant phases and is now shedding its outer layers. The central star that remains behind probably contracts drastically and terminates its life as a white dwarf.

Radio observations of planetary nebulae were not possible until the early 1960s when instruments first became sufficiently sensitive to detect the faint radio fluxes. The earliest reliable observations by C. Roger Lynds were carried out in 1961 at the National Radio Astronomy Observatory in Greenbank, West Virginia.[63] Today, both the radio continuum emission and the subsequently discovered radio recombination lines permit identification of planetary nebulae through radio observations about as readily as by visual means.

15. Ionized Hydrogen Regions

The central star in the Sword of Orion (figure 2.15), when viewed through a large telescope, resolves into bright nebulosity partly masked by dark patches (figure 2.16). A hint of this was first discovered by Christian Huygens in 1656, and successive improvements in technique have yielded increasing amounts of information on this nebula. The patches are dense clouds of gas and dust. The bright regions are ionized gases in the form of huge clouds whose atoms have been torn into electrons and positively charged ions by the ultraviolet radiation from a number of bright hot stars embedded in the nebula. The gaseous nature of this nebula first was established spectroscopically by Huggins and Miller in 1865, though the concept of ionization was unknown at the time and had to await the construction of an atomic theory of matter four decades later.[64] The first ionized regions to be discovered at radio wavelengths were found in 1954. They involved clear-cut sightings of the Orion Nebula and the Omega Nebula (which lies in the Milky Way plane at southern declination). These two sources were observed by the U.S. Naval Research Laboratory scientists F. T. Haddock, C. H. Mayer, and R. M. Sloanaker with a 50-foot paraboloid they had constructed to work at wavelengths as short as 9.4 centimeters.[65]

We recognize the ionized gases by a thermal continuum spectrum and by spectral line radiation emitted when the rapidly moving electrons and ions collide or recombine to form atoms. Some of the spectral lines lie at visible wavelengths. Others, which had been predicted in a remarkable paper written by the Soviet astrophysicist N. S. Kardashev in 1959, are apparent only at radio wavelengths.[66]

The Orion Nebula is several light-years in diameter, contains approximately a thousand electrons and ions in each cubic centimeter (several hundred solar masses in all), and has a temperature comparable to the surface temperature of the sun. Dust embedded in the region absorbs most of the light emitted by the stars and then reradiates this energy, in the infrared, at wavelengths 100 times longer than those of visible light. At these long wavelengths radiation readily passes through dust clouds without further hindrance. So strong, however, is the dust absorption of visible light that some of the stars cannot be seen directly, but are evident only through re-emitted infrared radiation.

The Orion region appears to be a birth place of stars now forming within our galaxy. Many of the stars we see are radiating so much power that they must soon consume their total energy budget and evolve into a different form. At best, they can be no older than a few million years.

Figure 2.15   *Photograph of the Orion Region Obtained in the Light of the Hydrogen Hβ Line*

In this photograph, obtained by Dr. Syuzo Isobe of the Tokyo Astronomical Observatory, north is at the top and east is left. The circular arc is known as Barnard's loop, a shell of gas that may have been produced some millions of years ago in an explosion. The row of three stars slanting upward and to the right, just above the center of the photograph form the belt of Orion. The central star and the star on its right, respectively, are ε Orionis and δ Orionis. When examined at high spectral resolution in 1903, the star δ Orionis first revealed to the German astronomer Johannes Hartmann that there exist tenuous clouds of cold gas dispersed between the stars in our galaxy.

The Sword of Orion, a set of three stars projecting vertically downward just below the center of the photograph, is brightest in its central star. At higher angular resolution (figure 2.16) this central source is actually seen to consist of ionized gases energized by several bright hot stars. This is the Orion Nebula.

Reprinted with the permission of Dr. Syuzo Isobe and the *Publications of the Astronomical Society of Japan*[67]

Figure 2.16   *The Orion Nebula*

This region, consisting of turbulent ionized gases energized by bright hot stars, lies at the edge of a large cloud of cold gas and dust. Smaller remnants of dust obscure part of the radiation from the ionized cloud and appear as dark patches in the photograph. Within the large cold cloud new stars are believed to be forming. Some of these stars will ultimately shine as brightly as those found at the center of the luminous nebula and will similarly ionize the gas surrounding them.

Photograph courtesy of Hale Observatories

16. Cold Gas Clouds

During the winter months of the years 1900 through 1903, the German astronomer Johannes Hartmann of the Potsdam Astrophysical Observatory was carrying out high resolution spectral observations on the binary star Delta Orionis, in Orion's Belt. Early in 1900, the French astronomer Henri Alexandre Deslandres had reported rapid changes in the positions of spectral lines of this star, and Hartmann had set out to repeat these observations. But his results disagreed with those of Deslandres. Not only did Hartmann find far slower periodic changes in the Doppler shift of these lines, he also found that a calcium spectral line in the extreme violet—the line that Fraunhofer had discovered in the solar spectrum and labeled K—did not "share in the periodic displacements of the lines caused by the orbital motion of the star" (figure 2.17).[68] From this and from a similar set of observations on a nova that had erupted in the constellation Perseus in 1901, he was able to conclude that the calcium line was produced by atoms belonging to an interstellar cloud along the line of sight between Delta Orionis and the sun. This was a surprising discovery since there seemed no reason to expect any atoms, let alone calcium, between stars.

That Hartmann's discovery should have been made at the Potsdam Observatory is not surprising: Although Huggins had discovered Doppler shifts of spectral lines as early as 1868, his earliest measurements really just told of stars that approached the sun and others that receded. And even those measurements were not altogether reliable. Stellar velocity determinations remained uncertain by tens of kilometers per second until 1887, when H. C. Vogel and Julius Scheiner of the Potsdam Observatory constructed a spectrometer that was properly stabilized against thermal expansion and contraction and against flexure when the telescope pointed in different directions. With this instrument, reliably repeatable observations of Doppler shifts could be made, and for the first time the earth's orbital motion about the sun at a speed of 30 kilometers a second was clearly measurable as a seasonal change in the shift of stellar spectra.[69] The installation of an 80-centimeter-aperture photographic refractor at Potsdam in 1900 had further made it possible to obtain accurate spectra for hundreds of stars fainter than second magnitude, including Delta Orionis.[70]

Hartmann's finding constituted the start of a long line of visual observations and eventually led to the radio astronomical studies of interstellar gases that were to follow half a century later and still persist today. The radio advances came in two main steps. On March 25, 1951, H. I. Ewen and E. M. Purcell of Harvard first detected the 21-centimeter-wavelength radio emission from interstellar hydrogen atoms.[71] Purcell had been actively involved in wartime work at the Massachusetts Institute of Technology (MIT) Radiation Laboratory, and the instrumentation used was heavily dependent on techniques that had been developed during World War II, as well as on surplus military equipment.[72]

The existence of the 21-centimeter spectral line had been predicted seven years earlier by H. C. van de Hulst, then a young student in Holland. In a prophetic paper written toward the end of the war, he had

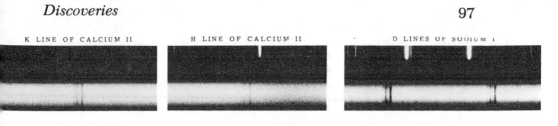

K LINE OF CALCIUM II   H LINE OF CALCIUM II   D LINES OF SODIUM I

Figure 2.17 *Interstellar Absorption Lines Seen Along the Line of Sight to the Star ε Orionis*

In 1903, Johannes Hartmann of the Potsdam Astrophysical Observatory observed the spectrum of the binary star δ Orionis in the belt of Orion, (figure 2.15). While the stellar spectral lines shifted to and fro as the stars orbited each other, Hartmann noted that an absorption line due to calcium remained stationary and evidently did not participate in the orbital motions of the stars. Hartmann correctly concluded that he was seeing evidence for an interstellar cloud of atoms along the line of sight to δ Orionis.

The star ε Orionis lies only a degree southeast of δ Orionis. Its spectrum shown here also provides evidence for the existence of interstellar gases. There are five distinct dark spectral lines superposed on the bright stellar continuum, in a part of the spectrum in which the sun has a dark spectral line that Fraunhofer had labeled K. The calcium H line and the sodium D lines also are Doppler split into five components. The Doppler shifts for the K and H lines and for the D line pairs correspond to five absorbing clouds whose Doppler shifts respectively signify recession velocities of 3.9, 11.3, 17.6, 24.8, and 27.6 kilometers per second with respect to the sun.

Photograph courtesy of Hale Observatories

noted that a neutral hydrogen atom contains an electron and a proton, each having a definite direction of spin. When the spins are parallel, the atom has an energy that is higher than when the spins are opposed.[73] The hydrogen atom can therefore emit radiation as the spin alignment changes from parallel to antiparallel, and the emission wavelength characterizing this transition lies in the 21-centimeter radio band. Van de Hulst predicted that this radio emission line should be detectable through astronomical observations.

On May 11, 1951, less than seven weeks after Ewen and Purcell's discovery, C. A. Muller and Jan H. Oort in Leyden were able to confirm the 21-centimeter line's existence.[74] Oort, the leading astronomer in The Netherlands, had recognized the immense contributions that radio astronomy might be expected to make and had started a vigorous Dutch effort immediately after the war. He was the only established astronomer anywhere to have successfully initiated a leading program in radio astronomy.

Let us return to the gas clouds between the stars. By the late 1960s, a number of diatomic interstellar molecules had been known for some decades. Andrew McKellar at the Dominion Astrophysical Observatory in Victoria, British Columbia, in Canada had written about CH radicals and their association with a visual absorption line at 3877.6 Angstrom units as early as 1940.[75] Here was a puzzle. No one could explain how these molecules were formed, and at any rate, molecules consisting of more than two atoms were not anticipated.

Nevertheless, a radio astronomical advance which started around 1968 culminated in the discovery of increasingly complex interstellar molecules of water, ammonia, carbon monoxide, formaldehyde, iso-

cyanic acid, acetaldehyde, and so on.[76] The radio spectra of some of
these substances had been previously measured in the laboratory, and
their potential interest to astronomy had been pointed out as much as
a decade earlier by Charles H. Townes, one of the leaders in the search
for interstellar molecules.[77] Spectral lines due to such molecules are
largely found at millimeter and centimeter radio wavelengths. They
tend to be sharp and well defined since the molecules largely exist in
cold interstellar clouds in which thermal and turbulent velocities are
low and spectral lines show little Doppler broadening. The discovery
of these lines introduced a whole new area of theoretical studies—the
analysis of chemical reactions in highly tenuous gases, where atoms
are buffeted by shocks, irradiated by ultraviolet and visible starlight,
and occasionally collide with grains of dust that float in the spaces between stars.

17. Interstellar Dust and Reflections Nebulae

In January of 1919, Edward Emerson Barnard of the Yerkes Observatory in Wisconsin published an article "On the Dark Markings of the
Sky—With a Catalogue of 182 Such Objects."[78] Barnard felt that the dark
patches represented absorption by intervening clouds which extinguished the radiation from more distant stars (figure 2.18).

Despite Barnard's beautiful photographic evidence, final verification of his views did not come for another decade when Robert Trumpler
at the Lick Observatory on Mount Hamilton, in California established
that the light from more distant stars in the Milky Way appears fainter
than we should expect on the basis of their great distance alone.[79] Trumpler compared the observed brightness and diameters of galactic clusters
of stars in the plane of the Milky Way. He assumed that, at least statistically, the clusters all might be considered roughly equal in size. The
angles they subtended in the sky should provide a measure of their
relative distances which could be compared to the distances derived
from the apparent brightness of individual cluster stars. For the hundred
clusters he examined, Trumpler found a systematic drop in the intrinsic
brightness he computed for the more distant cluster stars. This suggested
that intervening matter must obscure part of their light. He further
noticed that when he identified individual stars by characteristic lines
in their spectra, they appeared no different from normal stars in the
sun's neighborhood; but when the overall coloration of the stars was
examined, the more distant stars seemed reddened in comparison to
nearer objects.

Trumpler concluded that the obscuring clouds acted selectively to
extinguish blue light more than red. The extinction had to be highly
concentrated in the Milky Way plane since globular clusters and external
galaxies did not show a systematically redder coloration. He figured
that fine dust grains sparingly distributed in interstellar space could
be responsible for such effects.

Another twenty years were to pass before W. A. Hiltner, working
at the McDonald Observatory in 1948, and John S. Hall of the U.S. Naval

Figure 2.18   *The Dark Globule Catalogued as Barnard 335*

In the first two decades of the present century, E. E. Barnard of the Yerkes Observatory catalogued nearly two hundred dark patches in the sky. The more he looked at these regions, the more he became convinced that some form of obscuring matter must be hiding stars at distances beyond these patches. He did not believe, as Herschel had more than a century earlier, that these simply were gaps in the Milky Way through which we were looking out into an empty universe beyond. We know now that Barnard's view was correct and that the obscuration is due to dense clouds of interstellar dust. Barnard 335 lies at a distance of 1,000 light-years and measures 2 light-years across. The photograph shown here was obtained by Dr. Bart J. Bok at the Steward Observatory of the University of Arizona. (I thank Dr. Bok for permission to use this picture.)

Observatory, accidentally discovered that light from stars in the plane of the Milky Way is slightly polarized. Professor S. Chandrasekhar of the University of Chicago had predicted a polarization that could arise in the atmospheres of very young stars, and Hiltner and Hall had been searching for this effect when they discovered polarization of much greater prevalence.[80] For some stars the polarization exceeded 10 percent. Hiltner noted that the light systematically appeared polarized parallel to the galactic plane, and work by Hall showed that the increases in polarization were correlated with increased reddening of the stars so that the same mechanism responsible for the obscuration of starlight also seemed active in polarizing the radiation.

Within a few years of this discovery, a variety of mechanisms were proposed that might align slightly elongated interstellar dust grains with their long axes preferentially oriented perpendicular to the galactic plane. A starlight component, polarized perpendicular to the plane, would then be perferentially absorbed on penetrating through an interstellar dust cloud. The unabsorbed radiation that passed through the clouds and reached our telescopes, in turn, would show the observed excess polarization parallel to the Milky Way.

While this explanation is widely accepted today, there still are doubts about the actual mechanism that determines the alignment of dust. Perhaps several different processes combine to produce the complex pattern of alignment we observe, but magnetic fields in interstellar space are widely held responsible for the bulk of alignment and for the resulting polarization of starlight.

The extinction and polarization of starlight is not the only evidence for cosmic dust clouds: In 1914, V. M. Slipher had examined the spectrum of the nebula NGC 7023 and found it to be similar to that of the star embedded in the nebula.[81] He had already noted that a nebulosity, which had been photographically discovered in France by the brothers Paul and Prosper Henry in 1885 and is associated with the star Merope in the Pleiades cluster (figure 2.12), also seemed to have the same spectrum as Merope. By all indications such nebulae shone by reflected light.

We now know that Slipher's conjecture was correct. In reflection nebulae, starlight is scattered in all directions by fine dust grains, and in this process the spectrum of the light remains essentially unchanged.

18. Supernovae

A little before dinner on the evening of November 11, 1572, Tycho Brahe was amazed to see a dazzling new star in the sky, close to the constellation Cassiopeia.[82] It was so bright that it could be seen with the naked eye during daytime when other stars are lost against a bright blue sky. While this star surprised all Europe, in China, such guest stars, which came, stayed, and left, had been known for at least thirteen centuries. David H. Clark and F. Richard Stephenson in their fascinating book, *The Historical Supernovae,* have analyzed the ancient Chinese records on the first star, which by its description fits the pattern of supernovae. Its appearance corresponds to a date of December 7, 185 A.D. A translation of the Chinese historical summary of this event states:

2nd year of the Chung-p'ing reign period (of Emperor Hsiao-ling), 10th month, day *kuei-hai,* a guest star appeared within . . . *Nan-mên.* It was as large as half a mat; it was multicoloured (lit. "It showed the five colours") and it scintillated. It gradually became smaller and disappeared in the 6th month of the year after next. . . . According to the standard prognostication this means insurrection. When we come to the 6th year, the governor of the metropolitan region Yüan-shou punished and eliminated the middle officials. Wu-kuang attacked and killed Ho-miao, the general of chariots and cavalry, and several thousand people were killed.[83]

Nan-mên is one of the constellations constructed by the ancient Chinese to locate positions in the sky. Clark and Stephenson associate this constellation with the stars Alpha and Beta Centauri. These stars lie so far south as to be no more than a few degrees above the horizon at Lo-yang, the ancient Chinese capital where the guest star was observed. The reference to coloration and scintillation tells us that the source of light was stellar (because extended sources, such as comets, do not scintillate) and confirm a location near the horizon where atmospheric dispersion can particularly enhance color effects. The dates given tell of a duration of twenty months during which time the star could be seen. Only a supernova remains bright and stationary for that long a period. The constellation indicated and the scintillation, taken together with the location of the horizon seen from the observing site in Lo-yang, provide us with a reasonably accurate position for the star in the sky. The star's apparent brightness may have been a magnitude brighter than Venus at its brightest as judged from its date of appearance, which places it in the dawn glow and close to the sun. The position indicated and the inferred brightness of the star suggest an association with a current supernova remnant known as RCW 86 to visual astronomers and as G 315.4-2.3 to radio observers. This remnant, like others described below, is a source of radio emission and shows arc-shaped filaments on photographs. Its distance is somewhat uncertain but believed to be several tens of thousands of light-years, and such a distance could be in agreement with the star's reported brightness.

The discovery of supernovae by the ancient Chinese court astrologers was not a matter of luck. Their observations were systematic in the extreme: Five observers would spend each night on a special observing tower. One would look toward the zenith and each of the others into one of the four directions, north, south, east, and west, to see that nothing in the sky escaped notice. They noted winds, rain, the air, unusual phenomena such as eclipses, the conjunction or opposition of planets, meteors, fires, and other useful information. A strict account was kept of all this and registered the following morning in the Office of the Surveyor of Mathematics. This procedure went on throughout the rise and fall of dynasties until the modern republic was established in 1912.[84]

Supernovae visible to the naked eye appear in our galaxy no more frequently than once every century or two. The last one recorded is Kepler's supernova of 1604. Ordinary novae, bright enough to be similarly detected, also are quite rare. The recognition and naming of these guest stars as a definite class of known celestial objects, therefore, came about only because the ancient Chinese persisted with their continuous and extensive observations over so many centuries.

This is an important point to note. There are likely to be other astronomical phenomena that occur only with extreme rarity but are nevertheless important; and these will be discovered only if we can persevere with systematic observations stretching over periods lasting many centuries or even millennia. Phenomena of this type may well be the most difficult to discover.

In Europe, after Tycho's discovery, stars that suddenly appeared came to be known as novae—new stars. At first no one realized that there are several different types of new stars. Ordinary novae are very bright at maximum—certainly bright enough to be readily seen in nearby galaxies with modern telescopes. In 1934, Walter Baade and Fritz Zwicky of the Mount Wilson Observatory noted that most of the novae found in any given galaxy are roughly comparable in brightness.[85] In our nearest large neighbor, the Andromeda Nebula, the twenty or thirty novae that erupt each year have an apparent brightness that does not greatly deviate from seventeenth magnitude, corresponding to a brightness of some hundred thousand suns. In 1885, however, one nova in the Andromeda galaxy had lit up to a brightness another 10 magnitudes, or 10 thousand times greater. Baade and Zwicky called such enormous explosive events *super-novae*—and by now the hyphen has been dropped.

Supernova explosions are believed to occur at the end of a star's existence, when it has used up all its available fuels and faces ultimate collapse (figure 2.19). In this collapse the nuclear contents fuse into one giant mass of neutrons, and at the same time an enormous burst of liberated energy hurls the outer layers of the star into space in an explosion that rivals the brightness of ten thousand million suns.

The 1885 eruption in Andromeda, first noticed by the German astronomer Ernst Hartwig at the Dorpat Observatory in Estonia, caused a great deal of excitement. Its brightness led many to believe they were witnessing a transformation of the entire nebula.[86] The true magnitude of this eruption, however, was slow to be recognized. In 1917, G. W. Ritchey at Mount Wilson first noticed two fainter explosive events in the Andromeda Nebula, which strongly resembled ordinary galactic novae. At that point the truly enormous distance of the Andromeda spiral became apparent; and when Edwin Hubble in 1924 determined the brightness and periods of thirty-six Cepheid variables in Andromeda, this distance was accepted.[87] Only then was the superlative brightness of the explosion of 1885 clearly grasped.

Curiously, Harlow Shapley had been quick to understand the implications of Ritchey's work on the distance of spiral galaxies within weeks of Ritchey's 1917 discovery. He estimated a rough distance to the Andromeda Nebula, based on the apparent brightness of what he judged to be globular clusters, on the lack of resolved stars, and on the brightness of the novae. These methods gave consistent results which led Shapley to the correct estimate that the explosion of 1885 had exceeded a luminosity of a hundred million suns.[88]

However, Shapley soon dropped this view and accepted the erroneous notion that spirals are local. He was misled by claims of the astronomer Adriaan van Maanen who thought he had demonstrated the motions of stars in spirals. Large proper motions, if real, would have placed the spirals much closer and made them smaller.[89] Shapley believed van Maanen's results, though they were faulty, and gave them greater credence than all the other evidence he himself had compiled. Later he blamed this error of judgment on his friendship with van Maanen which blinded him to reality.[90] The error in van Maanen's work was quite

SN

Figure 2.19   *The Supernova Explosion of 1974 in the Galaxy NGC 4414*

When a supernova explodes, it can rival the light output of an entire galaxy containing a billion stars. In this picture the supernova lies just below the Galaxy and is labeled SN. A supernova that explodes within a few thousand light-years from the sun can be seen with the unaided eye during the day.

Supernovae appear to brighten to peak luminosity over a period of a few days. Their subsequent decline takes place over a few months. Next to nothing is known about the precursor in which the explosion originates.

Photograph courtesy of Drs. Bruce Patchett and Roger Wood and the Royal Greenwich Observatory subtle and rested on differences in photographic technique employed by Ritchey, who had photographed the spirals earlier, and van Maanen, who had prepared his comparison plates years later.[91]

Baade and Zwicky's 1934 argument, though based on the same data as Shapley's original calculation, now found general acceptance because by 1934 van Maanen's error had largely been recognized.

19. Eruptive Variables

While supernovae dwarf the more frequently seen ordinary novae in all respects, nova outbursts still are immensely impressive. Before exploding, the star has a blue continuum spectrum. As the explosion proceeds, the emitted light can increase by a factor of one to ten thousand in a single day (figure 2.20). Doppler-shifted and broadened emission lines in visual spectra show the exploding material to be streaming out of the star at speeds of 1,000 kilometers a second. The subsequent decline takes weeks, and during this time the star's infrared luminosity may increase. Perhaps dust grains are forming from cooling ejecta; this dust can absorb the star's visible light and re-emit the energy in the infrared.

Ancient novae certainly were recorded by Far Eastern astronomers in pretelescopic times. Clark and Stephenson cite guest stars recorded in A.D. 837 and A.D. 1437 as likely novae;[92] other guest stars, however, may just have been bright comets.

Not until the late 1860s were novae studied spectroscopically and a search into their nature begun. Novae often are recurrent: They explode, relapse, and explode again decades later. They emit X rays and are components of binary stars. Visual spectra suggest that the eruptions result from a symbiotic relation between the two stars. In that respect, the novae resemble other, lesser-eruptive variables.

Such symbiotic pairs generally consist of one hot blue star and a red giant or supergiant or else a cool dwarf. There are many eruptive variants named after such archetypes as the stars Z Andromeda, U Geminorum, and Z Camelopardalis, which brighten only ten to a hundredfold. The rise time may be even faster than for novae, and the active cycle typically repeats after a period of several hundred days; but the flare-up of a nova is millions of times brighter than that of a U Geminorum star.[93]

Figure 2.20   *Nova Herculis 1934*

During a nova explosion a star's brightness can increase by as much as a factor of 10 thousand in a single day. Most novae are known to be recurrent. They explode in a bright outburst of light, eventually relapse to their prenova luminosity, and then erupt again decades later. Novae generally appear to be components of a stellar binary system and are emitters of X-radiation. The photograph shows the nova which exploded in 1934 in the constellation Hercules as it appeared some months later in March of 1935, and again as it appeared eight weeks later in May of that year. Its brightness had considerably diminished.

Photograph courtesy of Lick Observatory

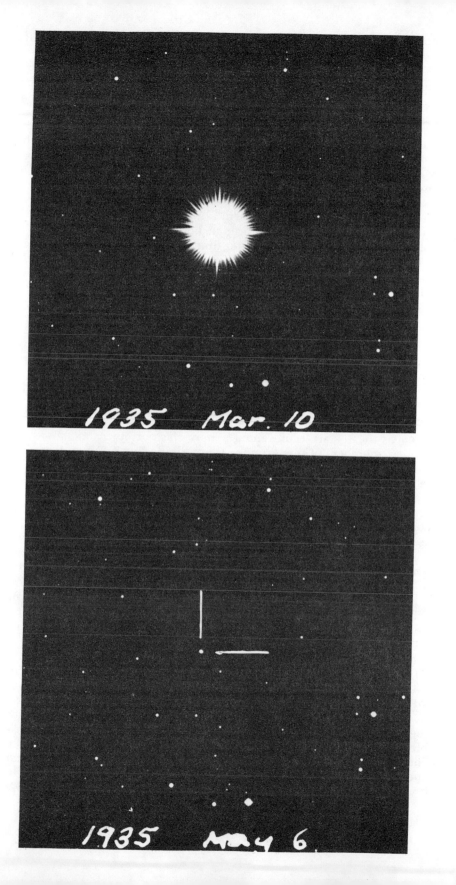

1935   Mar. 10

1935   May 6

R Coronae Borealis is the archetype of a class of stars that undergo a complementary change. Normally these stars are bright—several thousand times brighter than the sun. But bright periods that may last as long as ten years may be interrupted by brief periods when the star becomes a thousand times fainter. The faint phase which lasts several weeks is followed by a slow recovery. Stars of this type also exhibit strong infrared emission, and this suggests the stars' light is absorbed by dense dust during periods of decline. The dust again re-emits its absorbed radiation at infrared wavelengths.

Based on their appearance, the novae, symbiotic stars, and R Coronae Borealis type stars seem to represent a number of quite fundamentally differing phenomena: The R Coronae Borealis stars are intrinsically more than a thousand times brighter than the symbiotic stars; and while the R Coronae Borealis stars are bright during 90 percent of their life, the still more luminous novae remain bright less than 1 percent of their active cycle.

20. Variable Stars Associated with Nebulosity

A few Variable Stars associated with nebulae were first noticed in the latter half of the nineteenth century. Not very many of these stars were known, and they were not intensively studied as a group until 1945 when the Mount Wilson astronomer Alfred Joy recognized that eleven of these stars represented a new class of object.[94] One of the brightest members of this group is the variable star T Tauri. It suffers brightness changes over a range of approximately 40:1. The changes are irregularly spaced but may take only a few hours and then persist erratically for a few days or weeks when they do occur. T Tauri is the archetype of a group that shows brightness variations by a factor of about ten, strong emission lines, and spectral evidence for ejection of gases at speeds of a few hundred kilometers per second (figure 2.21).

Dust clouds surrounding such a star may be loosely joined or have openings in them, so that the star's light may not reach us directly, but may instead be seen in reflection off a dusty cloud. That reflection will then mimic the light variations of the star. Objects suggestive of such reflections were originally observed by Guillermo Haro and by George Herbig in the 1950s,[95] but the connection between observations and explanation is still quite tenuous. We know all too little about these classes of stars. They tantalize because they seem to represent stars just barely formed from cosmic gas and dust, still quite unsettled in their behavior, and not quite ready to start the long steady period of hydrogen-to-helium conversion that characterizes main sequence stars.

Since the stars listed by Joy in 1945 ranged from ninth to twelfth magnitude at maximum brightness, we might expect that nebular variables could have been recognized as a separate group as much as a century earlier. All the necessary instrumentation was available at that time. In fact, that is just what did take place. As early as 1852, J. R. Hind discovered a small nebulosity associated with the star T Tauri.[96]

Figure 2.21 *The Variable Star T Tauri with Hind's Variable Nebula NGC 1555*

On October 11, 1852, J. R. Hind, the director of a private observatory founded by the Englishman George Bishop in Regent's Park, London, discovered a small nebula in the constellation Taurus. Later that same year, an adjoining star, now known as T Tauri, was found to be variable as well. Over the next decade Hind's Nebula vanished completely, as far as any visual observations could tell. Then, by the end of 1861 it had reappeared. Since then it has vanished and returned erratically changing its shape, its brightness, and its location relative to T Tauri. Nebulae such as this appear to shine by reflected light from the star. The star itself varies in brightness as gas and dust clouds ejected periodically block its radiation from our line of sight. These clouds can also block the radiation from more distant dust clouds that could reflect the star's radiation toward us. Those distant clouds therefore behave like reflection nebulae whose source of reflected light is periodically turned on or off.

Photograph courtesy of *Astrophysical Journal*

But the number of these nebulae and of stars associated with them did not quickly increase with this discovery. In 1921, C. O. Lampland of the Lowell Observatory still was only able to enumerate three. It was this lack of sufficient sources to be studied that delayed the recognition of nebular variables as a distinct class of stars interacting with their surroundings.[97] Observational astronomy on several occasions has harbored little-known, pathological-appearing objects—oddities that later were recognized as important new phenomena. This appears to have been true of the nebular variables as well.

21. Infrared Stars—Circumstellar Dust Clouds

A number of stars that are barely detectable with the best optical techniques are extremely bright in the infrared. Other infrared objects seem to have no apparent optical counterpart at all; and any of these objects could be called infrared stars because all except a negligible fraction of their energy is concentrated in the infrared band. One star of this kind is known only by its number in a catalogue of sources bright at a wavelength of 2 microns in the infrared. Its designation is IRC+10216, where IRC stands for infrared catalogue.

The infrared catalogue *Two-Micron Sky Survey* was the result of a major effort launched by Robert Leighton and Gerry Neugebauer, both physicists at the California Institute of Technology.[98] Leighton had had the idea of constructing an inexpensive 1½-meter telescope out of a spun-up dish of epoxy that would harden into a parabolic shell while spinning. This curved lightweight shell could then be aluminized to make a crude telescope with a blur circle of about 3 minutes of arc and a 1-minute arc precision in determining positions of bright stars in the sky. The atmosphere transmits well and emits little radiation of its own at a wavelength of 2 microns; this determined the choice of wavelength for these infrared observations.

Before the start of their systematic search, there was no way of knowing what kind of new source, if any, Leighton and Neugebauer were likely to find. One evening, shortly after the telescope had been installed on Mount Wilson and was systematically surveying the sky, the two scientists were talking near their instrument when they noticed that the chart recorder had registered an enormous pulse of infrared radiation and very little accompanying visible light. There seemed no doubt that the telescope just had passed across a very bright infrared source in the sky. This caused great excitement, not only because the source was so unexpectedly bright but also because it had been found so soon after operations had started. The portion of the sky surveyed so far had been tiny, and this suggested that there might be a huge number of other equally luminous infrared stars. A heady prospect indeed!

As the days went by, however, Leighton and Neugebauer found that there were not all that many other bright infrared stars in the sky. The early discovery of their first intense source had been somewhat of a statistical fluke. Nevertheless they did later find several stars that were brighter than their first; and by the time the two-micron survey had been completed, it had led to a tabulation of 5,000 relatively bright infrared stars. Many of the stars listed had very little emission at visible wavelengths and clearly constituted a new class of stars embedded in heavy blankets of dust.[99]

At 2 microns the giant Betelgeuse is the brightest star in the night sky. At a wavelength of 5 microns IRC+10216 is the brightest object we see outside the solar system. IRC+10216 varies in brightness over periods of roughly two years and is enveloped by so much light-absorbing dust that less than 1 percent of its visual emission passes through without conversion into infrared radiation.

Many such sources are now known. Dust shells evidently can accompany a variety of stars, usually bright stars that seem to be ejecting matter from their surfaces into interstellar space—sometimes quietly, sometimes, as in the case of novae, explosively. One such object is Eta Carinae, the most luminous single source we recognize in the Milky Way. It has a luminosity that exceeds one million times that of the sun. More than 99 percent of its emission is radiated in the infrared, most of it beyond 10 microns. In visible light the source appears as a brilliant red condensation about 2.5 arc seconds in diameter, surrounded by an elliptical nebula about 4 times larger.

For some 200 years prior to 1830, Eta Carinae seems to have been known as a fairly bright star. But around 1840 its brightness rose to well above the level of ordinary novae. Instead of dying out after a few months, as ordinary novae do, it remained bright for about a decade before diminishing by a factor of several hundred. The infrared radiation observed from Eta Carinae today is not far lower than the total visual luminosity at peak brightness during the last century. This suggests that the radiation from a bright central source simply is being absorbed by a dust shell that has condensed around it and that the absorbed energy is being reradiated at infrared wavelengths.

We still know too little about circumstellar dust clouds to tell whether they represent a phenomenon that arises in only one way, for example, through the ejection of material by unstable stars. We suspect that very young stars just forming in interstellar space are similarly surrounded by dust clouds. In fact, we observe infrared stars deeply imbedded in huge interstellar dust complexes, and T Tauri stars, also, are well-recognized emitters of infrared radiation. However, we do not yet have a clear way of distinguishing by appearance alone very young and very old stars embedded in a cocoon of dust, and we have no certain way of distinguishing two such different classes of infrared stars.

22. Flare Stars

Stars that are particularly close to the solar system, and especially white dwarfs which are so faint that they cannot be seen when far away, are best recognized by their rapid (seconds of arc per century) motion across the sky. In 1926 the Bruce Proper Motion Survey was planned with the aim of recording the proper motions—displacements across the sky—that can be discerned for these nearby stars in the course of several decades. At that time 1,009 pairs of plates were planned to cover the entire Southern Hemisphere, but since early epoch plates taken between 1896 and 1911 already were available for 950 of these regions, only the remaining 59 required immediate attention. Willem J. Luyten set out in 1929 to take plates still needed for twenty-five areas in the sky south of declination −10 degrees. These same areas were then photographed again a quarter of a century later, and the positions of stars compared.

A comparison of this kind is made by a blinking mechanism that first shows the viewer the picture taken earlier and then immediately switches to the plate taken later. The stars which do not move across the sky appear fixed on blinking back and forth between the early epoch and the later plates, but stars that have moved between exposures appear to jump back and forth as the machine blinks between the two photographic plates.

In the course of such an examination carried out on these plates late in 1947, a southern declination star was found with a proper motion larger than that of any other star discovered in the survey. On closer examination it was accidentally found to be a binary system in which the fainter of the two stars could suddenly flare up in brightness (figure 2.22).[100]

Luyten's star is now designated UV Ceti, and other stars subsequently found to flare in a similar way are often called UV Ceti variables—or simply flare stars.

Flare stars generally are faint; and faint stars normally are not examined through a rapid sequence of short exposures. This may explain why flare stars were not discovered several decades earlier. That they could have been is candidly explained in a letter written by Professor Luyten, in 1979.

> I believe that the first time such a thing as a flare from a star was discovered, was at Harvard. This was in the middle twenties, I was there at the time, and the person who found it was Ejnar Hertzsprung. What he did see was that on one plate, taken with a double exposure, there appeared two identical, bright images of a star which did not show on any other similar plate taken with the same telescope. Because of the two images there was no question about the reality of the object, but at the time no one knew what it meant. I do not even know whether this was ever published.
>
> The second time was when Van Maanen, at Mt. Wilson, noticed that on one of his parallax plates a star which was normally faint, was much brighter. I believe this was one of the nearby stars with large proper motion. . . . No one knew what this meant, and it was even suggested that it must mean that that particular portion of the emulsion must have been unusually sensitive: you know how people will invent all kinds of fancy solutions in order to avoid having to accept a new phenomenon.
>
> But undoubtedly, the first time that an actual flare was observed, and seen for exactly what it was: a temporary burst of sudden brightening of a star, of very short—minutes only—duration, was on the plate which Carpenter took for me, with FIVE exposures, for parallax measurement.

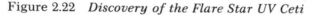

Figure 2.22 *Discovery of the Flare Star UV Ceti*

In 1947 W. J. Luyten was completing a photographic survey aimed at finding stars with high proper motion and found one star that had moved 3 seconds of arc, during a twenty-five-year interval between successive exposures of the same part of the sky. He asked Edwin Carpenter of the Steward Observatory in Tucson, Arizona, to take a closer look at the star, in preparation for observing the star's parallax. Carpenter photographed the star several times in quick succession on a single plate, slightly displaced on successive exposures, in order to define the positions of nearby background stars against which the high velocity star's parallax could be measured over the next several months. The picture on the right shows a single exposure of the star field. The high velocity star is indicated by an arrow and other stars are labeled with letters. On the left, the multiple exposure plate is shown. Each star has been exposed five times, and the plate has been moved in an L-shaped pattern between successive exposures. UV Ceti, the star just right of center, can be seen to have suddenly brightened at the point where the L turns the corner. The other stars on the same plate do not show this sudden change in brightness. The interval between the first and the last of the five exposures was about twenty minutes, so the star had suddenly flared and subsided to its original brightness in just a few minutes.

Photograph courtesy of Professor Willem Luyten and the *Astrophysical Journal*

. . . Also, after the discovery of this, the first flare star, Shapley, at Harvard, looked up their multitude of plates on Proxima Centauri, and found that it occasionally flared. I do not know how old the oldest plate was on which he observed this, but probably it, too, went way back into the early 1900's. . . .[101]

Flare stars are found not only in the sun's neighborhood. They also occur wherever a group of T Tauri stars is seen—in dusty young stellar aggregates whose estimated ages are no greater than a few million years. During an outburst which tends to occur every few hours the flare star becomes several times brighter than normal. Some flare stars can normally be as bright as the sun while others are as much as a few thousand times fainter.

Flare stars in the sun's neighborhood are fainter than those in young clusters. Generally they are thousands of times less luminous than the sun; but their total visual brightness can suddenly increase by as much as 100 times within a minute or less. The decline is only slightly slower, and the normal state is resumed after several minutes. The flares appear to occur at random, and their amplitudes vary considerably.

In the late 1950s, A.C.B. Lovell, who had brought about the construction of the first giant radio telescope—the 250-foot steerable telescope at Jodrell Bank near Manchester—decided to search for possible radio flaring in UV Ceti variables. Joined by Fred Whipple, the director of the Smithsonian Astrophysical Observatory in Cambridge, Massachusetts, he found flaring to occur at similar times at visual and radio wavelengths.[102] Later, the simultaneity of radio and optical outbursts was established more firmly, although the bulk of the flare's energy was found to reside in the visible part of the spectrum. These early observations were the first indications that radio emission from individual stars beyond the sun could be observed.

In many ways the flares we observe in UV Ceti stars are similar to those frequently seen on the sun; but what makes them so spectacular in these particular stars is the enormous relative brightness of the flare. While the sun hardly brightens perceptibly when a flare erupts, in flare stars the eruption for a brief interval completely dominates the star's energy output.

The first radio star of any kind to be discovered was the sun. It was independently detected by J. S. Hey, Grote Reber, and G. C. Southworth during the years of World War II.[103] In 1946, in a letter to the British journal *Nature,* Hey reported:

> It is now possible to disclose that, on one occasion during the War, Army equipments observed solar radiation of the order of [a hundred thousand] times the power expected from the Sun. . . . This abnormally high intensity of solar radiation occurred on February 27 and 28, 1942. . . . The main evidence that the disturbance was caused by electromagnetic radiations of solar origin was obtained by the bearings and elevations measured independently by the [radar] receiving sets, sited in widely separated parts of Great Britain (for example Hull, Bristol, Southampton, Yarmouth). . . .[104]

At the time of these observations strong solar flares were observed at the French observatory at Meudon, and it soon became clear that solar radio emission is enhanced when the sun is active. Because its radio emission is faint, the sun would not as yet have been detected at radio wavelengths if it were at the distance even of our nearest neighboring stars. Only a few extraordinary stars, such as the flare stars, and the masers, pulsars, and X-ray stars to be described below, are sufficiently powerful emitters of radio waves, to be detected as radio sources at any appreciable distance across our galaxy.

23. Magnetic Stars

In 1946 Horace Babcock of the Mount Wilson Observatory first detected a strong magnetic field on a distant star, 78 Virginis. The great solar observer George Ellery Hale had already discovered strong magnetism in sunspots as early as 1908, but while the general solar field is only 1 or 2 gauss—2 to 3 times stronger than Earth's—78 Virginis has a magnetic field of 500 gauss, and we now know stars that have fields in the 30- to 40-thousand-gauss range.

Stellar magnetism is detected by observing a splitting of spectral lines emitted in the star's atmosphere. In the laboratory this type of splitting had first been discovered in 1896 by the Dutch physicist Pieter Zeeman.

The stellar observation consists of separating the light received into its left- and right-circularly polarized components. If a strong magnetic field directed along the line of sight is present, spectral lines observed in one polarization are slightly displaced in wavelength relative to the lines viewed in the oppositely polarized sense. Spectra obtained in the two polarized light beams can therefore be compared, and as Zeeman had shown, the measured displacement is directly proportional to the magnetic field strength while the direction of displacement depends on the direction of the field.

Writing some years after his discovery of magnetic fields in sunspots, Hale explained:

> If a sun spot is an electric vortex, and the observer is supposed to look along the axis of the whirling vapor, which would correspond with the direction of the lines of force, he should find the spectrum lines double, and be able to cut off either component with the polarizing attachment of his spectroscope.
>
> I applied this test to sun spots on Mount Wilson in June, 1908, with the 60-foot tower telescope, and at once found all the characteristic features of the Zeeman effect. Most of the lines of the sun spot spectrum are merely widened by the magnetic field, but others are split into separate components, which can be cut off at will by the observer. Moreover, the opportune formation of two large spots, which appeared on the spectroheliograph plates to be rotating in opposite directions, permitted a still more exacting experiment to be tried. In the laboratory, where the polarizing apparatus is so adjusted as to transmit one component of a line doubled by a magnetic field, this disappears and is replaced by the other component when the direction of the current is reversed. In other words, one component is visible alone when the observer looks toward the north pole of the magnet, while the other appears alone when he looks toward the south pole. If electrons of the same kind are rotating in opposite directions in two sun-spots vortexes, the observer should be looking toward a north pole in one spot, and toward a south pole in the other. Hence the opposite components of a magnetic double line should appear in two such spots. The result of the test was in harmony with my anticipation. . . .[105]

The field strengths Hale observed in sunspots had ranged up to 4,500 gauss; but the sunspots cover only a small portion of the solar surface, and the fields of groupings of spots tend to cancel. The overall magnetic field of the sun, therefore, is far smaller. Babcock's discovery of a strong field in the star 78 Virginis, therefore, came as a surprise. He had used a method which, though well established in solar observations, required considerable improvement for application to stellar work. The main difficulty in stellar observations was an insufficiency of light for the necessary high resolution spectrograms. Observations of this kind were made possible through the installation, in the late 1930s, of the new spectrograph at Mount Wilson which was allowing Alfred Joy to discover new T Tauri stars, and for similar reasons—high resolving power and high light gathering power—enabled Babcock to detect magnetic stars.[106]

On his plates Babcock noted a differential displacement of spectral lines circularly polarized in opposing senses. It amounted to no more than a small fraction of a millimeter—in fact, no more than $\frac{2}{1,000}$ to $\frac{3}{1,000}$ of a millimeter, but this was sufficient to tell him of the star's large total magnetic field.[107]

All high stellar magnetic fields are found to be variable with a period in the range of one to twenty-five days. Some of the observed fields vary regularly. As the star rotates, we see the magnetic field projected differently along our line of sight: The observed Zeeman splitting reverses, and we see changes both in the appearance of the spectrum and in the strength of the field component directed along our line of sight. Small brightness variations also are perceived as the rotating star alternately faces a magnetic pole or an equatorial region toward the observer.

24. Cosmic Masers

The *maser*, a device that can amplify very precisely defined microwave frequencies, was invented in 1954 by Charles Townes and his coworkers at Columbia University. Maser, their choice of a name for their device, is an acronym for Microwave Amplification through Stimulated Emission of Radiation. Microwaves are radiowaves whose wavelength is short compared to those used in broadcasting. The microwave band covers the wavelength range roughly from 1 millimeter to 1 meter.

Within ten years of its invention, the principles used in constructing the maser had been applied in many different ways, most strikingly at visible wavelengths where an optical analogue, the *laser*, came into widespread use.

In 1964, only a decade after the invention of the maser, a Berkeley, California, group, headed by Harold Weaver, was searching for emission of microwave radiation at frequencies of 1665 and 1667 MHz and discovered Galactic regions in which a bewildering set of spectral lines was strongly radiated.[108] The Berkeley group was particularly perplexed by the complexity of the spectra they detected, and they referred to emission by a substance they called "mysterium." Soon this emission was identified with a transition of the diatomic hydroxyl radical OH.

Microwave emission of this kind shows extreme spectral purity— very narrow spectral lines. Often the radiation is strongly polarized, variable in strength, and the sources of radiation are highly compact, frequently having an angular extent of only one-ten thousandth of a second of arc. These traits are reminiscent of laboratory masers, and the observed sources have subsequently, in fact, become accepted as cosmic masers. OH, silicon monoxide (SiO), and water vapor molecules (H_2O) are the best identified cosmic maser gases, but observed lines identified with other substances may also represent maser action.

Two types of associations are known. Masers have been found in cool clouds near interstellar regions of hot ionized gas; they have also been detected coming from some types of red giant and variable stars that are strong emitters of infrared radiation. It is not understood what makes some of these stars OH masers and others not, but the infrared radiation is believed to produce conditions favorable for maser emission.

The discovery of cosmic masers is not a tidy story; no key individual or group can be singled out and credited. Groups at MIT in the United States and at Jodrell Bank in England were involved in the initial discovery of OH radicals and subsequently in the discovery of strong polarization of the OH lines from compact sources. The realization that maser action was playing a role in the emission from these sources gradually forced itself upon astronomers over a period of two or three years, as high polarization, high spectral purity, and highly compact source size successively became apparent through the use of radio techniques capable of high resolving power and of sorting out different modes of polarization.[109]

There is a possibility that OH masers could have been discovered ten years prior to their actual discovery, if laboratory data showing the precise frequency of OH radical transitions had existed in the mid-1950s. At that time a search for cosmic OH lines was undertaken at the U.S. Naval Research Laboratory by A. H. Barrett and A. E. Lilley, but the attempt ended in failure because too wide a spectral band had to be surveyed.[110] In this particular instance the requisite instrumentation appears to have been available, and the main factor delaying discovery was a lack of basic physical data.[111]

25. Pulsars

In October of 1967, Jocelyn Bell, a graduate student at Cambridge University, noticed "a bit of scruff" amounting to half an inch of recording on a 400-foot chart of radio noise representing one complete coverage of the sky by her antenna array. The array Bell was using was designed to detect the twinkling, or scintillation, of radio sources as streams of gas emitted by the sun passed between the source and her receivers. The twinkling was recorded on long paper rolls. There was a great deal of other radio noise recorded on these charts, some of it man-made interference from radio and television transmitters. The man-made noise had to be weeded out, and Bell could easily have rejected, along with all the other noise, this scruff she had just noticed. Still, somehow it looked familiar, and she looked back through other records.

The scruff had first appeared on a record on August 6, and she noticed that it always came from the same part of the sky, proving its astronomical origin (figure 2.23). But no astronomical source known at the time showed just this type of behavior. Jocelyn Bell had discovered the first radio signals from a pulsar. There was some discussion about whether these signals could be man-made, or might perhaps originate from a civilization across the Galaxy.

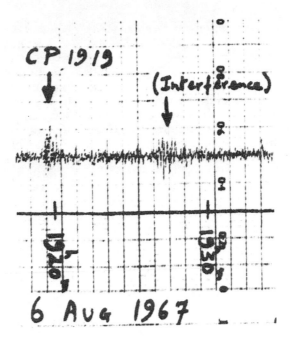

Figure 2.23 *A Bit of Scruff*

The first record of a pulsar's signal was found in October of 1967, by Jocelyn Bell, a graduate student in radio astronomy at Cambridge University, who noticed what she later described as "a bit of scruff." The signal had first appeared on a chart two months earlier, on August 6, and she soon realized that it always came from the same part of the sky and therefore shifted its time of appearance by some four minutes a day. Man-made interference was unlikely to do that.

The pulsar detection in this photograph of the August 6, 1967, record, is indicated by its subsequently assigned catalogue number CP 1919. The radio interference right next to it does not look strikingly different. (I thank Dr. S. Jocelyn Bell Burnell for permission to reproduce this chart record.)

Over the next few months Bell, her thesis adviser, Anthony Hewish, and three of their Cambridge colleagues made detailed observations. The pulses were observed to be regularly spaced about 1⅓ seconds apart, keeping time with an accuracy that could be faulted by no clock (figure 2.24). Just before Christmas, Bell found a second source with a slightly different period, about 1¼ seconds.[112]

We now know that pulsars radiate strongly at wavelengths between 1 and 100 meters. Some pulse only once every 3 seconds. The fastest known pulsar emits 30 pulses a second (figure 2.25). The pulses can be highly polarized, and their arrival time at Earth is different at differing radio wavelengths. The long wavelengths travel more slowly through the ionized gases in interstellar space and can arrive seconds later than the short wavelengths. This is not much of a delay considering that the total travel time may be ten thousand years, but it does imply a dispersion of the pulses unless they are only observed over a rather

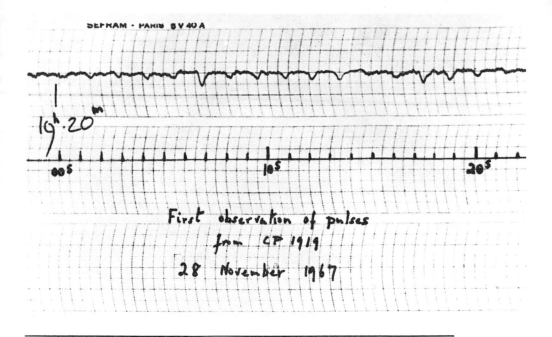

Figure 2.24 *First Observation of Pulses from the Pulsar CP 1919*

After it had become clear to Jocelyn Bell and her thesis supervisor, Anthony Hewish, that a celestial source was giving rise to the ragged-looking signals which occasionally appeared on their recording charts, a higher speed chart recorder was obtained to record the shape of the signal. On November 28, 1967, when the source again was to pass through the antenna array's field of view, a recording was obtained and showed a series of deflections (top trace) that were evenly spaced 1⅓ seconds apart. Reference time marks appear on the bottom trace. The radio source was seen to pulse regularly, each 1⅓ seconds. Nothing like it had ever appeared in radio observations before.

Here was a new phenomenon!

(I thank Dr. S. Jocelyn Bell Burnell for permission to reproduce this record.)

narrow wavelength band. For the more distant pulsars this effect increasingly interferes with detection.

Pulsars appear to be rapidly rotating neutron stars—stars in which matter is so strongly condensed that the entire star acts like a giant atomic nucleus consisting wholly of neutrons. A neutron star is so small that it could not even shade a city the size of London or New York. Yet, it is more massive than the sun and a million times more massive than Earth. A thimbleful of matter from this star would weigh a billion tons. Neutron stars are a billion times denser even than white dwarfs.

Embedded in the neutron star is a magnetic field 10^{12} times stronger than Earth's. And as the star rotates it pulls along the magnetic field which in turn accelerates electrons and protons that are trapped in the outermost atmospheric envelope of the star. It is these accelerating charged particles that radiate toward us once each revolution of the rapidly rotating neutron star. As the neutron star ages, its rotation gradually slows. Over a year the rate may slow down by one part in a few thousand or even more slowly, and the observed pulse rate also slows down.

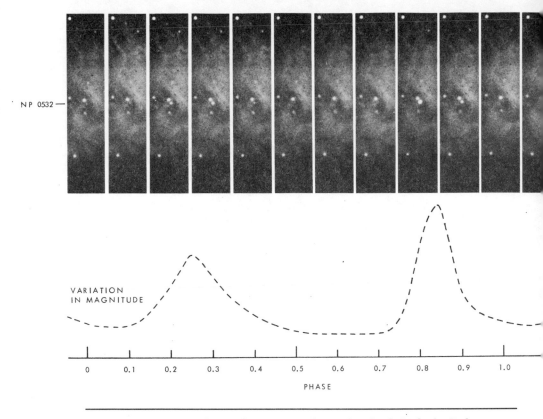

NP 0532 —

VARIATION
IN MAGNITUDE

| 0 | 0.1 | 0.2 | 0.3 | 0.4 | 0.5 | 0.6 | 0.7 | 0.8 | 0.9 | 1.0 |

PHASE

Figure 2.25   *Optical Emission from the Crab Nebula Pulsar*

The two most rapidly pulsating pulsars are known to emit not only radio but also gamma-ray and faint optical pulses. The first lies at the center of the Crab Nebula supernova remnant (figure 2.26); the second, in a supernova remnant in Vela, (figure 2.36). The Crab Nebula pulsar pulses each $\frac{1}{30}$ of a second. In this series of twelve frames, each representing one *phase* of the pulsation cycle, that period is further subdivided into ten evenly spaced intervals and the pulsar brightness is recorded separately for each interval. The last two frames are identical with the first two and represent a continuation into the next repeating cycle.

The pulsar designation NP 0532 shows the position of the pulsar on the strips. The star appears brightest in the fourth strip from the right. The frames shown here actually represent carefully timed exposures over a large number of successive pulses. A single pulse would not have provided enough light to produce a picture. The dashed curve traces the brightness of the pulsar through the twelve frames and shows the initiation of a new cycle in the last two frames.

Photograph courtesy of Kitt Peak National Observatory

26. X-ray Stars

Astronomical investigations are seldom undertaken on purely rational grounds. The observer may be willing to invest his time and energies in the pursuit of an idea that might lead nowhere. Writing in 1974 about the discovery of X-ray stars, Bruno Rossi was to recall:

It was about fourteen years ago that some of us became intrigued with the idea of searching the sky for X-ray and gamma-ray sources other than the

Sun, the only celestial emitter of high-energy protons known at that time. It was, of course, clear that an effort in this direction would not have been successful unless there occurred, somewhere in space, processes capable of producing high-energy photons much more efficiently than the processes responsible for the radiative emission of the Sun or of ordinary stars. The possible existence of such processes became the subject of much study and discussion. . . .

The theoretical predictions did not provide much encouragement. While several "unusual" celestial objects were pin-pointed as possible, or even likely, sources of X rays, it did not look as if any of them would be strong enough to be observable with instrumentation not too far beyond the state of the art. Fortunately, we did not allow ourselves to be dissuaded. As far as I am personally concerned, I must admit that my main motivation for pressing forward was a deep-seated faith in the boundless resourcefulness of nature, which so often leaves the most daring imagination of man far behind.[113]

Rossi was a professor of physics at MIT and also chairman of the board of the American Science and Engineering Corporation, (AS&E) of Cambridge, Massachusetts, that had been founded by Rossi's former student, Martin Annis. As it happened, AS&E had just hired a young Italian cosmic-ray physicist, Riccardo Giacconi, to start a space science program within the recently formed company, and Rossi suggested to him that a program in X-ray astronomy might be well worth pursuing. Giacconi therefore set to work with the MIT physicist George Clark, a consultant to AS&E, to investigate both the technical questions of instrumentation for X-ray astronomy and, together with Bruno Rossi, the questions of the promise of X-ray observations of the sun, the moon, normal stars, magnetic stars, and the remnants of supernova explosions. Giacconi's first proposal for support in this venture was sent to the National Aeronautics and Space Administration (NASA) but was turned down. However, a group at the Air Force Cambridge Research Laboratories, which was engaged in lunar work, expressed interest, and Giacconi wrote a proposal suggesting a search for secondary X-ray emission from the moon. AS&E's president, Martin Annis, had been instrumental in solving a problem for the U.S. Air Force that involved X rays emitted in nuclear bomb bursts. With his help it was arranged that the rocket flight to search for lunar X-ray fluorescense would take place under Air Force sponsorship.[114]

In June 1962, after two earlier unsuccessful flights with a newly developed highly sensitive detector, the first important achievements were recorded.[115] Later Riccardo Giacconi was to recall the two most important results of that flight.

The intensity of the radiation was estimated to be approximately 5 photons $cm^{-2}s^{-1}$ in the direction of the peak. The existence of this source and of a diffuse isotropic background completely obscured the possible contribution from fluorescent X-rays from the Moon. . . .

The truly surprising aspect of the discovery . . . was not that cosmic X-ray sources existed, but that they emitted radiation at a level several orders of magnitude greater than until then estimated. It is not surprising that a few weeks were spent in satisfying ourselves that we were indeed observing

X rays from celestial sources and not low energy electrons or X rays from the upper atmosphere, and in speculating on the possible source for this radiation. We were puzzled in particular by the fortuitous coincidence of the X-ray peak with the local azimuth of the magnetic field.[116]

The last sentence is instructive. Scientists always are worried about extraneous or spurious effects. A major discovery leaves lingering doubts that can only be removed through further investigations.

Thus far nothing has been said about a team of researchers led by Herbert Friedman at the U.S. Naval Research Laboratory (NRL). This group had been successful in the mid-1950s in rocket observations of solar X-ray emission. Writing in *Scientific American* in June 1959, Friedman was able to state, "Emboldened by the progress in solar-rocket astronomy, our group at the Naval Research Laboratory set out in 1956 to investigate the ultraviolet and X-ray emission of other stars." Later in the article he concedes, "Rocket-astronomy has not yet undertaken the observation of celestial objects in the X-ray spectrum. Such cosmic-ray sources as the Crab Nebula, however, have high priority in experiments now being designed and instrumented."[117] This turned out to be a shrewd prediction. In 1964 the NRL group was able to show clearly that the Crab Nebula is an extended X-ray source, thus providing the first unambiguous identification of a cosmic X-ray source beyond the solar system.

One of the investigators on the original AS&E team that discovered cosmic X-ray sources was Francis R. Paolini; and as Bruno Rossi recalled:

> Most of the detectors used for the study of solar X-rays had been Geiger counters provided with small (about 1 cm²) very thin windows to admit X-rays into the sensitive volume. Frank Paolini at AS&E developed Geiger counters with thin mica windows of much larger area (about 20 cm², a difficult achievement at that time).[118]

A second important innovation was the use of anticoincidence devices that the group's leader, Riccardo Giacconi, had already known from his experience in cosmic-ray observations. These two items underscore the almost imperative need for significant technological innovation. Few of the truly important astronomical discoveries appear possible without such advances. The requirements for theoretical preparatory work, on the other hand, are not nearly as clear. Rossi's interpretation again is interesting.

> The reason why no one had expected any significant result in extrasolar X-ray astronomy from experiments carried out with the rather rudimentary means available in 1962—and the reason why our findings were received at first with great skepticism by much of the astronomical community— was not a failure of the theory (although sometimes theories are proved incorrect). It was rather the reluctance to consider the possible existence of celestial objects entirely different from those already known. The X-ray emission from the Sun, above 0.5 kev, is never more than one part in one million of the visible emission, even during large solar flares; it is about one part in ten billion in the absence of activity; whereas, as it turned out,

Sco X-1 radiates about one thousand times more energy in the form of X-rays than in the form of light.

It took about one and half years, and several additional rocket flights, by the AS&E group and other groups, and the observation of a few other X-ray sources, before the early misgivings about the credibility of our discovery were dispelled. Particularly important, in this respect, was a rocket flight carried out in April 1963, by the NRL group headed by Herbert Friedman, which confirmed the existence of the source discovered by the AS&E group and pinpointed its position in the sky more precisely, placing it in the constellation of Scorpio; hence the name of Sco X-1 by which this source became known.[119]

I have quoted Rossi's recollections so extensively because the association of Giacconi and Rossi in the discovery of X-ray stars represents one of the most fruitful collaborations to be found in science. We see a combination of Rossi's experience, brought to the task after many decades of pioneering work in cosmic physics, combined with the energy and inventiveness of the much younger Giacconi who, in the following two decades, was to play a leading role in the further development of X-ray astronomy.

One more point might be mentioned before going on. Herbert Friedman recalls a 1957 rocket flight in which he and his colleague James Kupperian, Jr. thought they had detected a celestial X-ray source.[120] But when they tried to repeat the observations on a second flight, the signal was not there. Friedman suspects they might have seen one of the occasionally erupting transient X-ray sources at the time or perhaps the source they had originally seen was below the horizon the second time. The results of the first flight therefore were never published. Nevertheless the AS&E group was aware of them. Their 1960 planning document states:

> No extra-solar X or γ-radiation had been reported to date (January 1960) in the literature, although Kupperian and co-workers recently observed X-rays from a limited region of the sky during a moonless night flight.[121]

Bright X-ray stars of the kind represented by Sco X-1 are quite rare. In our galaxy only a couple of hundred are known. Most of these stars have been optically identified as components of close binaries. Many of the binaries eclipse, and the optical eclipses are synchronized with a decline of the X-ray brightness. Sometimes the star that emits the X-rays radiates pulses toward us at well-defined intervals. Perhaps the star rotates and only exposes a bright X-ray emitting spot once per revolution. At any rate the otherwise regular periodicity slightly changes as the X-ray star approaches and recedes in its orbital motion about its companion. The pulse frequency is Doppler shifted in the same way as the spectral lines are in the binary star spectrum.

The emission of X rays is produced by matter falling toward the surface of the smaller companion from an extended atmosphere about the larger of the two stars. If the compact companion is a white dwarf or a neutron star, the in-falling matter acquires such high speeds that X rays are emitted when the particles are suddenly stopped by collisions at the star's surface or just above.

A few times each year a new X-ray source flares brightly in the sky and disappears, sometimes prevailing for less than an hour, sometimes lasting for several weeks. By now such sources have been observed over a period of years and a few have been found to flare repeatedly. Perhaps two stars are orbiting each other in highly elongated orbits, and X rays are emitted only when the stars are near closest approach.

Both radio and infrared radiation are emitted by some X-ray stars.[122] Quite generally different X-ray stars appear to exhibit a sufficient range and variety of properties that we may be dealing with several distinct phenomena not with just one single class of X-ray emitters. For example, series of observations made from spacecraft bearing X-ray telescopes have shown the existence of sources that emit bursts of radiation. These bursters typically have a rise time less than a second and a decay time lasting several seconds. Sometimes the emission spikes are separated by only tens of seconds; at other times by minutes. One source exhibits a separation between spikes that is roughly proportional to the intensity of the preceding spike, as though a constant charging rate were returning the source to a certain breakdown level. Some of the bursters appear to lie in globular clusters of stars. Do they represent a new type of source peculiar to the dense interior regions of these clusters?

27. Supernova Remants

What happens to matter that bursts out when a supernova explodes? In 1942, Nicholas U. Mayall at the Lick Observatory and Jan Oort in Leyden provided a partial answer. Together with the expert on Chinese historical records, J.J.L. Duyvendak, they had summarized data from Chinese astronomical records dating back nine hundred years, analyzed these jointly with contemporary records concerning the observed expansion of the Crab Nebula, and shown that the Crab must be the remnant of the powerful supernova observed in China in 1054 A.D.[123]

The Crab Nebula is a filamentary interstellar cloud (figure 2.26). In 1939 John C. Duncan of the Mount Wilson Observatory compared photographic plates of the Crab taken in 1909, 1921, and 1938, and found an expansion amounting to about $\frac{1}{5}$ second of arc each year.[124] He concluded that the nebula must be some eight hundred years old. Duncan's first efforts in 1921 had been prompted by the work of C. O. Lampland of the Lowell Observatory, who had found variations in the filamentary structure of the Crab that indicated brightness variations, motions of the filaments, or both.[125] In fact, both of these changes do occur, and Duncan, who had found evidence for the expansion on photographs taken in 1909 and 1921, returned to this study in 1939 to obtain a longer-term confirmation. Two years earlier Mayall had determined a radial expansion velocity from the Doppler splitting of spectral lines near the center of the nebula (figure 2.27). Mayall concluded that one of each pair of lines corresponded to the approaching part of an expanding shell, while the other represented the receding part.[126] The observed expansion velocity was enormous—over 1,000 kilometers a second. While the dis-

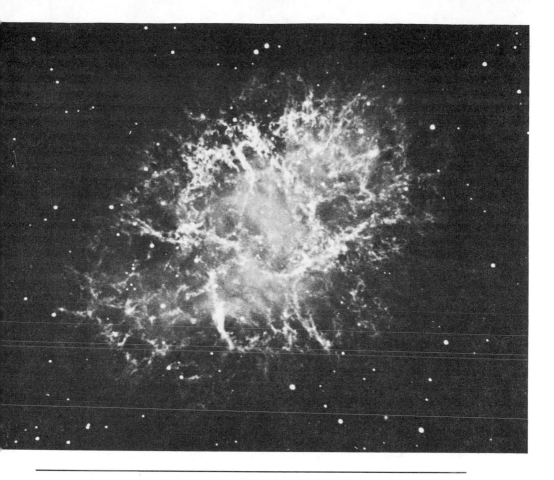

Figure 2.26   *The Crab Nebula Seen in Red Light*

The Crab Nebula is a cloud that looks filamentary when seen in the spectral lines of hydrogen and nitrogen, as in this photograph. It also emits a continuum spectrum; and in continuum radiation its structure is amorphous, as shown in figure 2.28.

In 1939 J. C. Duncan of the Mount Wilson Observatory compared photographic plates of the Crab Nebula taken in 1909, 1921, and 1938 and found the nebula to expand at a rate of ⅕ of an arc second each year.

In 1937 N. U. Mayall of the Lick Observatory had measured the spectrum of the nebula and found that the nebula was expanding at a velocity of over 1,000 kilometers per second (figure 2.27). If the distance to the nebula was 5,000 light-years, the expansion velocity and angular size of the nebula indicated an explosive origin nine hundred years earlier, a date that agreed with the suspected origin of the Crab in the supernova outburst of 1054 A.D. recorded by Chinese chroniclers. The stellar remnant of that outburst is believed to be the pulsar seen in figure 2.25.

Photograph courtesy of Lick Observatory

tance to the Crab Nebula was not accurately determined until two decades later, Duyvendak, Mayall, and Oort suggested that a distance of 5,000 light-years would make the observed expansion velocity and angular size agree with an origin nine hundred years earlier. In an article on the history of the Crab, Mayall points out that the split spectral lines of the nebula had already been noted by Slipher as early as 1915; that Duncan had seen the radial motion in 1921; and that Hubble had guessed

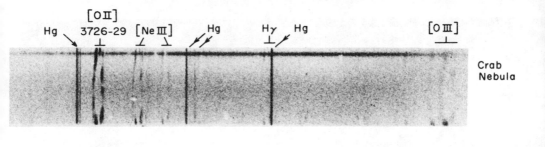

Figure 2.27 *Spectrum of the Expanding Crab Nebula*

The portions of an expanding nebula that face the observer contain gas that streams directly toward him. Gases on the far side of the nebula stream away from the observer along the line of sight, and so a look at the central portions of such a nebula reveals both a strongly blue-shifted and a strongly red-shifted component. At the edges of the nebular disk, gas also streams out of the center, but along a direction perpendicular to the observer's line of sight; it is, therefore, hardly spread in a Doppler shift at all.

In a spectrum of the Crab Nebula obtained by N. U. Mayall at the Lick Observatory in 1937, the lines emitted by the nebula appeared strongly bowed. In the center of the nebula, the split between the red- and the blue-shifted lines was greatest. At the periphery of the cloud, the shift was minimal. The lines, therefore, looked like highly elongated elipses. The maximum separation between the two Doppler-shifted components amounted to a velocity of expansion in excess of 1,000 kilometers per second—an expansion velocity larger than any that had been observed for a source up to that time. The straight lines marked Hg in this spectrum do not emanate from the Crab Nebula, but rather are due to reflected light from mercury vapor streetlamps scattered by the earth's atmosphere. The other markings denote a number of lines due to hydrogen (H) and ionized oxygen (O) and neon (Ne), emitted by the Crab Nebula.

in 1928 that the Crab was nine hundred years old.[127] As far as technical capabilities were concerned, the nebula's explosive origin might therefore have been documented a decade or two earlier than Duyvendak's, Mayall's and Oort's pioneering papers.

Several years later, in 1949, John G. Bolton and Gordon Stanley in Australia found that the Crab Nebula is a strong radio source, and in 1954 Viktor A. Dombrovsky and M. A. Vashakidze in the Soviet Union found the nebula to be strongly polarized at visible wavelengths. Two years later these results were followed up by Jan Oort and Theodore Walraven in Leyden and by Walter Baade, who worked with the Mount Palomar 200-inch telescope (figure 2.28).[128]

The polarized emission in the nebula comes from an amorphous glow outside the filamentary structure that exhibits spectral lines of hot ionized gases. From this it appeared that both the radio waves and the visible light were generated by highly relativistic particles—electrons or nuclei moving at almost the speed of light. When such charged particles move through a magnetic field, their paths become curved and the particles emit synchrotron radiation.

In addition to radio waves, the Crab Nebula in 1964 was shown to also emit X rays. In a classical rocket observation Herbert Friedman and his colleagues at the U.S. Naval Research Laboratory were able to show that X rays from the Crab Nebula gradually died out during an occultation, as the moon slowly moved across the source.[129]

This proved that at least an appreciable fraction of the X-radiation came from the nebulous source. A purely stellar source would have been suddenly extinguished as the moon's limb passed across. Later observations, however, showed that a strong stellar X-ray component also is present in the Crab. It is the Crab pulsar at the nebula's center. The Crab Nebula's relativistic particles are now believed to be accelerated largely by the pulsar. As the pulsar rotates, it pulls with it magnetic fields in which charged particles are initially enmeshed. These particles are whirled to extremely high energies before release into space.

Several dozen supernova remnants are known today. They are expanding clouds of gas with filamentary structure and generally emit radio waves and X rays.

28. Interstellar Magnetic Fields

In 1949 the physicist Enrico Fermi suggested that cosmic-ray particles reaching us from beyond the solar system might be accelerated through a large number of reflections off magnetic fields embedded in moving interstellar clouds of gas.[130] The energetic particles were to be accelerated in a chain of reflections much as a Ping-Pong ball is made to go faster and faster when bounced back and forth between two paddles that rapidly approach each other. Fermi also speculated that the observed velocities of cosmic gas clouds were related to the presence of magnetic fields. With this assumption he calculated a field strength of about 5 microgauss (5×10^{-6} gauss), roughly a hundred thousand times weaker than the magnetic field at the earth's surface.

This speculation opened a floodgate of new ideas and new explanations for observed cosmic phenomena: The polarization properties of starlight were explained in terms of the selective absorption of light by magnetically aligned interstellar dust grains;[131] and polarized radio waves were explained by postulating synchrotron emission—a form of radiation known to be emitted by electrons and charged nuclei moving at extremely high relativistic velocities across a magnetic field.[132]

Perhaps the first definite proof of the presence of magnetic fields in interstellar spaces followed the 1953 prediction by the Soviet astrophysicist I. S. Shklovsky that both visible and radio continuum radiation emitted by the Crab Nebula supernova remnant could be produced by the synchrotron mechanism. The following year I. M. Gordon and V. L. Ginzburg independently pointed out that this radiation should be polarized, and V. A. Dombrovsky actually detected polarization at visual wavelengths. All of this work was done in the Soviet Union and went almost unnoticed in the West.[133]

In 1957 a team of physicists and engineers at the Naval Research Laboratory set up equipment to study the polarization of radio waves whose wavelengths were as short as 3 centimeters. The direction of polarization of such short waves was expected to change but slightly on passage through potential interstellar fields; indeed, the direction of polarization of the Crab Nebula at these radio wavelengths proved to be almost identical to the direction of visual polarization.[134] This story of success is remarkable for being one of the very few cosmic phenomena discovered as a consequence of correct theoretical predictions.

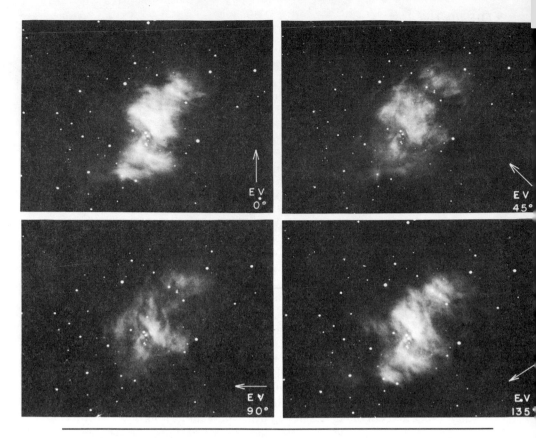

Figure 2.28 *Synchrotron Radiation from the Crab Nebula*

In 1953 the Soviet astrophysicist I. S. Shklovsky suggested that the radio emission from the Crab Nebula might be produced by cosmic-ray particles trapped in magnetic fields that might be permeating the nebula. The magnetic fields would cause the cosmic-ray particles to gyrate, just as highly energetic particles describe circular paths in a synchrotron. Like particles in such an accelerator, the cosmic rays would then also emit synchrotron radiation, since any electrically charged particle forced to travel a circular trajectory has to radiate photons.

Shklovsky also felt sure that the optical continuum radiation in the Crab was produced by this same mechanism of synchrotron emission. The Soviet astrophysicists I. M. Gordon and V. L. Ginzburg independently pointed out that Shklovsky's ideas would require the emitted radiation to be plane polarized, with the same orientation for the plane of polarization at optical and radio frequencies. Subsequent observations fully bore out these predictions. Through those observations, the presence of appreciable magnetic fields in the spaces between stars first became firmly established.

In the four panels of the photograph, the amorphous continuum of the nebula is photographed through a polarization analyzer placed so that the recorded radiation's electric field is oriented along the arrow. It is quite clear that different patches become successively brighter and darker as the analyzer is rotated, showing that the light is polarized.

Photograph courtesy of Hale Observatories

Even after interstellar magnetic fields had become definitely recognized, it still took a decade before unambiguous measurements of the magnetic field strength could be made. There are two such tests. One involves the splitting of spectral lines into differently polarized components—the Zeeman splitting that George Ellery Hale had used to detect magnetic fields in sunspots. For interstellar fields this effect was first observed in 1969 by Gerrit Verschuur, who found two cool interstellar clouds in which the field strengths were remarkably close to, although higher than, those that Fermi had postulated. Verschuur established this by noting that the 21-centimeter spectral line radiation of hydrogen in these clouds could be observed actually to have slightly different wavelengths in its two circularly polarized radiation components.[135] This wavelength difference amounted to no more than half a millionth of a centimeter, less than one part in ten million of the total wavelength.

Three years later a different technique developed by Richard Manchester gave somewhat lower results for the general interstellar medium. He observed the direction of polarization for different radio wave frequencies reaching us from pulsars in our galaxy.[136] On the assumption that these waves all are emitted with the same direction of polarization but that the direction of polarization is Faraday rotated by the galactic magnetic field, Manchester was able to calculate the field strength. In Faraday rotation the direction of polarization of a wave travelling along a magnetic field is rotated in proportion to the square of the wavelength. The polarization direction for long radio wavelengths rotates strongly; for short waves the rotation is weak. With this method Manchester found a field strength of about 3 microgauss in surprisingly good agreement with Fermi's twenty-year-old conjecture.

Interestingly, Fermi's originally suggested mechanism for the acceleration of cosmic-ray particles by interstellar magnetic fields no longer appears to be significant. Sometimes theoretical ideas lead to useful observations even when the ideas later proved to be wrong.

29. Galaxies Containing Gas

In 1845, William Parsons, third earl of Rosse, had finished the construction of a reflecting telescope at Birr Castle, Parsonstown, Ireland. The 72-inch-diameter ground speculum mirror provided it with the largest aperture of any instrument constructed up to that time. Indeed, it was to remain the largest telescope to be erected in the nineteenth century. Its nominal magnification was 6,000—roughly 300 times greater than that of Galileo's original spyglass, and the amount of light it gathered was more than 1,000 times greater.[137]

In April, 1845, within two months after the erection of the Parsonstown telescope, a nebula in the constellation Canes Venatici was seen to have a spiral structure. Soon other nebulae were found that also looked like giant whirlpools—quite different in appearance from the more ring- or disk-shaped planetary nebulae, or the amorphous clouds of the type seen in Orion. The distance to the spiral nebulae, however, was to remain uncertain for three-quarters of a century. Ordinary novae were found in the Andromeda spiral (Messier 31) by Ritchey in 1917 (figure 2.29). Ritchey had discovered a nova in the spiral nebula NGC 6946 that year and had then searched older photographic plates at his disposal for evidence of two earlier novae that had erupted in M31. By 1922 James C. Duncan also had found three variable stars in another nebula, Messier

Figure 2.29   *The Andromeda Nebula (M31)*

Seen from far away in the universe, our galaxy would appear to be part of a binary system in which the Andromeda Nebula is our close companion. These two galaxies have much in common. They both have spiral structure; they are comparable in size; and they both have smaller satellite galaxies, like the two small companions seen respectively just above and to the lower left of the Andromeda Nebula.

Photograph courtesy of Lick Observatory

33. He noted the variability of these stars on photographic plates dating back twenty-five years.[138] However, the periods of the stars eluded him. Only three years later Edwin Hubble had data on forty-seven variables in M33 and another thirty-six in M31. Many showed classical features of Cepheid stars, and Hubble concluded that the Andromeda Nebula was at a distance of 930,000 light-years—clearly outside the Milky Way.[139] Though subsequent work by Walter Baade and others showed the actual distance of M31 to be quite a bit larger, Hubble's main conclusion that the spirals lie outside the Galaxy has been fully vindicated.[140]

One question of importance was whether the Milky Way was similar to these spirals, or whether it had a unique structure of its own. The answer to that question was not to be provided until 1927 when the young Dutch astronomer Jan Oort provided convincing dynamical proof.[141] Following an idea that had been developed by the Swede Bertil Lindblad in 1924–25, Oort argued that if the sun and its neighboring stars indeed were far from a massive galactic center, then a rotation about that center should be apparent. The stars orbiting this massive nucleus should behave just like the planets orbiting the sun: The inner planets which feel the gravitational pull more strongly orbit more rapidly. Mercury, the innermost planet, completes a solar orbit in 88 days and moves with a velocity 10 times greater than the outermost planet, Pluto, which requires 248 years to complete a single revolution about the sun. In a similar way, Oort expected stars nearer the galactic center to orbit more rapidly than more distant stars. This is called differential rotation.

By analyzing the Doppler shift in the spectral lines of Milky Way stars lying in different directions from the sun, Oort was able to derive their rotation rates about Shapley's galactic center position. In this way, he proved that the Galaxy is in differential rotation, a crucial point because it showed that our galaxy must be very similar to the spiral nebulae, whose curved arms strongly implied this same rotation.

Oort's discovery showed once again how unpriviledged a position Man occupies: Earth is not at the center of the solar system; Copernicus had shown that. Neither is the sun at the center of the Milky Way as Shapley's study of globular clusters had shown. Finally, not even the Milky Way can boast any special significance. There are myriad galaxies just like it!

External galaxies roughly divide into two kinds, those that contain gas and those that don't. They represent two rather different phenomena. In the gas-containing galaxies we see both cool gas with dust and hot ionized gas clouds. Bright blue stars mark the outlines of frequently well-developed spiral arms. Cosmic rays abound. This active life is reflected in a variety of radio and visual emissions quite unmatched by the galaxies devoid of gas.

This division of galaxies into two groups was not recognized until 1939, when it was documented in some depth by Nicholas Mayall then working at the Lick Observatory. Mayall had constructed a special, ultraviolet sensitive spectrometer with which he was able to obtain spectral observations at shorter wavelengths than had previously been possible. With this instrument he noticed that a number of galaxies exhibit a strong ultraviolet spectral line produced by ionized oxygen. When he systematically checked a large number of galaxies for this feature, he found that spiral and irregular galaxies generally emit a strong line and elliptical galaxies usually show no emission. Mayall correctly concluded that the ionized oxygen was to be found in the interstellar spaces in these galaxies and the spiral galaxies evidently contained larger amounts of interstellar material.[142]

We now know that regular spirals, barred spirals, and irregular galaxies contain significant amounts of interstellar gas. So do the giant elliptical galaxies that frequently sit at the center of large aggregates. These massive galaxies appear to rob their smaller companions of their gas whenever the smaller galaxies come too close. In the tidal contest that ensues, the gravitational attraction of the bigger galaxy prevails, and gas is pulled from the smaller companion to fatten the larger.

The gas-containing galaxies are now as readily studied by radio techniques as by visual observations. Throughout the 1950s the identification of radio sources as galaxies depended on close collaboration between radio and optical astronomers. At the same time, 21-centimeter studies of the atomic hydrogen gas in galaxies became possible on a sufficiently fine angular scale to permit direct observation at radio wavelengths of the rotation of these galaxies. Along a given direction across the source, the Doppler shift of the hydrogen line would increase systematically showing that one edge of the galaxy was moving away from us more rapidly than the other. Observations of this kind were first carried out on the two Magellanic Clouds by Frank J. Kerr, J. V. Hindman, and B. J. Robinson in Australia. The Magellanic clouds are so close to the Galaxy that their diameters subtend angles of many degrees, and even the relatively crude radio telescopes available in 1953 were capable of resolving such large objects and establishing their rotations.[143] Because the Magellanic Clouds are small, rotation velocities range around fifty kilometers per second, and for large galaxies containing gas, the rotational velocities typically reach a few hundred kilometers per second.

30. Galaxies Devoid of Gas

Normal elliptical and globular galaxies, as distinct from the giant ellipticals which appear to be ten times more massive, contain little if any, gas or dust in the space between the stars. While elliptical galaxies are about as massive as spirals, containing somewhere between a hundred million and a hundred thousand million stars, they show none of the signs of active star formation.

The stars seen in elliptical galaxies are largely faint red stars found also in globular clusters. Entirely absent are the bright blue stars found in the arms of spiral galaxies. There are no spiral features; there are no dark dust lanes; there is virtually no radio emission.

It is not clear whether elliptical galaxies were formed as they now appear or whether they have at times contained extensive amounts of gas, typical spiral arms, and active regions of star formation. If an elliptical galaxy was active once, it may now simply be advanced in age. Alternately, it might be resting, while its stellar population evolves and returns gas to the interstellar space. That may prepare the way for a further epoch of star formation from accumulated gases and to a reappearance of a normal gas-containing spiral.

We know so little about the birth and evolution of galaxies that we can only speculate—or observe further in order to learn.

31. Clusters of Galaxies

In 1934 Edwin Hubble published a set of observations on "The Distribution of Extragalactic Nebulae."[144] He had counted 44,000 galaxies on photographic plates taken with the Mount Wilson 60-inch and 100-inch telescopes pointed in more than a thousand different directions more or less uniformly distributed over the northern three quarters of the sky.

Figure 2.30   *The Cluster of Galaxies in Coma Berenices*

In 1785 William Herschel was able to write about "that remarkable collection of many hundreds of nebulae which are to be seen in what I have called the nebulous stratum of Coma Berenices. . . ."[147]

Although Herschel was able to recognize this cluster of galaxies, he had no real idea of the nature of galaxies; he had no means of distinguishing between a globular cluster and the giant elliptical galaxy at the center of the photograph. Nevertheless, he could tell that here was an aggregate quite different from the normal random distribution of nebulae in the sky. This discovery, though premature, was not so different from discoveries being made today: Though the discovery of quasars dates back to 1963, we still are not sure of the nature of quasars or of the mechanisms responsible for their high luminosities. But we recognize that quasars exist, that they somehow are quite different from other astronomical sources, and that they deserve a name of their own.

Photograph courtesy of Kitt Peak National Observatory

Except for obscuration from dust clouds in the galactic plane, he found a rather uniform distribution of galaxies in all directions, although there was a small degree of clumping of galaxies into individual clusters. Hubble found this tendency to hold both for small groupings—double and triple galaxies—and for the large compact clusters whose membership occasionally was as great as several hundred galaxies.

These findings were by no means new. Sir John Herschel already had been aware of a strong clustering of nebulae in certain regions about a century earlier. Herschel for many years had been compiling a catalogue of all nebular sources known in his time, and his general catalogue of nebulae and star clusters which appeared in 1864 contained more than 5,000 sources, most of which had been observed either by his father, William Herschel, or by himself.[145] The compilation of this catalogue brought out the striking absence of nebulae near the plane of the Milky Way and the occasional clusters of nebulae that could be seen nearer the Galactic poles—the directions most distant from the Milky Way plane (figure 2.30).[146] Hubble's refinements of these views were important primarily because his sample of galaxies was ten times larger than that known to Herschel, and because by 1934 a clear distinction had been drawn between the different types of nebulae. Thus Hubble was able to confine himself to cataloguing galaxies, while Herschel's catalogue contained clusters of stars, planetary nebulae, and ionized hydrogen regions in addition to galaxies. The statistical conclusions that Hubble was able to draw concerning the clustering of galaxies therefore were more persuasive. But there is no doubt that both Herschels, father and son, recognized the clustering of nebulae without, however, having a way of knowing that those nebulae for certain were Milky Ways just like our own.

In the forty years since Hubble's work, clustering on a larger scale has been sought, and the findings indicate that there could be a hierarchy of clustering on an ever-increasing scale. However, aggregation on the largest scales is not as easily noted. The largest clear-cut clusters we see are a hundred million light-years across. Such a dimension roughly corresponds to $\frac{1}{100}$ of the distance to a cosmic horizon that separates us from regions so distant that no information can ever reach us, no matter how long we wait.

32. Radio Galaxies

We have already seen how Karl Jansky came to discover the radio waves emitted by our galaxy. If his first paper noting the unexpected radio emission was still hesitant, his second article, published the following year, left no doubt about his findings.[148]

This second paper unwittingly shows the magnitude of Jansky's discovery. A good portion of the manuscript is a capsule course in elementary astronomy aimed at radio engineers. Evidently radio specialists had had little cause to concern themselves with astronomy, and astronomers probably thought of radio techniques as unrelated to their work.

Figure 2.31 *The Radio Galaxy Cygnus A*

The first extragalactic radio source to be discovered was Cygnus A, found in 1946 by J. S. Hey and his co-workers who had run a trouble-shooting operation for the British wartime radar network. At the end of hostilities, they made use of radar receivers to map the radio noise first noticed by Jansky fifteen years earlier.

In scanning the radio emission from the galaxy, they came across a secondary peak in radio brightness. But the signal received from this source fluctuated appreciably on a time scale of several seconds, while the signal seen elsewhere seemed much steadier. Later, it became clear that this was a twinkling similar to the visible scintillation of stars and that this fluctuation in brightness was just a sign that a compact radio source had been discovered. Extended sources do not twinkle; and that is why we tend to see stars twinkling, even when the light from planets, which have more extended angular diameters, does not. For Cygnus A, the radio fluctuations were being produced in the earth's ionosphere, but that was not to be realized until 1950, four years after the discovery of the source.[150]

Cygnus A, shown at the center of this photograph, is roughly as bright at radio as at optical wavelengths. Its X-ray luminosity, however, is even greater than its radio emission.

Photograph courtesy of Hale Observatories

At any rate, there was no rush into radio astronomy for another fifteen years.

The total radio emission from our galaxy is of the order of one millionth of the radiation emitted in starlight. This is quite insignificant compared to the emission of some of the powerful radio sources discovered in the years following World War II.

The first extragalactic radio source, Cygnus A (figure 2.31), a bright source in the Swan constellation, was discovered by James Stanley Hey and his co-workers in 1946.[149]

By 1954 a more precise radio position had been obtained. This allowed the optical astronomers Walter Baade and Rudolph Minkowsky at Mount Palomar to bring the 200-inch telescope into play, and to identify Cygnus A with a source on their photographic plates that appeared to consist of two colliding spirals in a cluster of galaxies.[151]

Cyngus A is the strongest extragalactic source of radio waves, roughly equally bright in its radio and optical luminosity. Its X-ray brightness is far greater still.

The radio emission seen from a typical spiral galaxy is quite different from that of a strong radio source. A substantial part of the emission from a spiral galaxy may be in the form of radiation from hot ionized gases, atoms, and molecules. In contrast the most powerful sources like Cygnus A, or the bright radio galaxy Virgo A, emit radiation through synchrotron emission produced by highly energetic cosmic-ray particles forced to spiral along magnetic fields in the galaxy. These cosmic-ray particles probably are emitted in a series of powerful explosions characterizing the present stage in the galaxy's development. How these explosions originate is uncertain.

We recognize two largely distinct groups of radio galaxies—extended and compact. The compact sources sometimes remain unresolved even with instruments capable of resolving objects whose angular diameters are no larger than $\frac{1}{1000}$ of a second of arc. Sources in this class are dense and therefore opaque over at least part of the radio spectrum. They tend to be brightest at short wavelengths, and many of them show intensity variations on a time scale of weeks.[152]

In contrast, extended sources show no time variations and exhibit a spectrum that peaks at the long wavelengths where many of the earliest post-World War II radio observations were made. Extended radio galaxies have dimensions of the order of a hundred thousand light-years, while some compact sources may be smaller than a light-week across— less than 10^{12} kilometers.

The compact radio galaxies may well represent two quite distinct classes, a fairly stable group and a set of sources that seems to consist of components that rapidly expand and move at velocities that, incredibly enough, appear to exceed the speed of light. We shall return to these superluminal sources below.

33. Unidentified Radio Sources

Many radio sources correspond to no visually detected objects. In that sense they appear to differ markedly from other radio sources that appear bright on photographic plates. In 1974 Allan Sandage and his collaborators Jerome Kristian and Basil Katem searched for forty-seven previously unidentified or poorly identified sources in the Third Cambridge University catalogue of radio sources.[153] Using the most powerful optical instruments available, a combination of the 48-inch Schmidt and the 200-inch Hale reflector on Mount Palomar, they still failed to find optical counterparts for twenty-one of the radio sources.

Little is known about these objects, except that their radio properties are not markedly different from identified radio galaxies. B. J. Harris finds that about 60 percent scintillate—twinkle—at radio wavelengths.[154] Normally this is characteristic of sources that appear compact. The Mount Palomar observations show that many of these unidentified objects lie off the galactic plane and therefore probably have no connection at all with the Milky Way. That suggests powerful radio sources quite far away across the universe.

34. Expansion of the Universe

In 1912 V. M. Slipher of the Lowell Observatory obtained the first well enough defined photograph of the spectrum of a galaxy to determine a velocity from Doppler-shifted spectral lines. Slipher had been observing our nearest large spiral galaxy, the Andromeda Nebula (figure 2.29) and found that it approaches at a speed of 300 kilometers a second.

By 1925 Slipher had compiled the radial velocities of forty-one galaxies. They ranged from an approach velocity of 300 to a recession of 1,800 kilometers a second, and most of the velocities were directed away from us.[155] These velocities were of a much higher order than any observed within our own galaxy. At first it seemed the velocities of these galaxies were random, and a number of astronomers attempted to see whether there was some mean velocity to this apparently random mix of speeds. If such a mean velocity could be found, it would represent our own Milky Way's motion relative to the average speed of all other galaxies in the universe.[156]

In 1929, when radial velocities were known for forty-six galaxies, Edwin Hubble of the Mount Wilson Observatory noticed a relationship between the velocities and the distances of galaxies as defined by the apparent brightness of the brightest individually recognized stars. The relation is very simple: Galaxies recede from us at a rate proportional to their distance; the further away the galaxy, the faster it moves.[157] The same relation had already been vaguely noted by other astronomers around 1924–25, like C. Wirtz; and the Dutch astronomer W. de Sitter, noting that Einstein's general relativistic equations could also describe an expanding universe, had suggested such an expansion as early as 1917, citing Slipher's earliest results in support of this contention.[158] Hubble, nevertheless, is usually credited with this discovery. Credit for a discovery often goes to those whose arguments are clearest, and Hubble's presentation was convincing and simple.

Using the 100-inch telescope at Mount Wilson, Milton L. Humason was able to increase the list of galaxies to nearly 200 by 1935. He extended the range of observations out to distances characterized by velocities as high as 42,000 kilometers per second—$\frac{1}{7}$ the speed of light.[159] Figure 2.32 shows a shift of 61,000 kilometers per second in the spectrum of the Hydra cluster of galaxies.

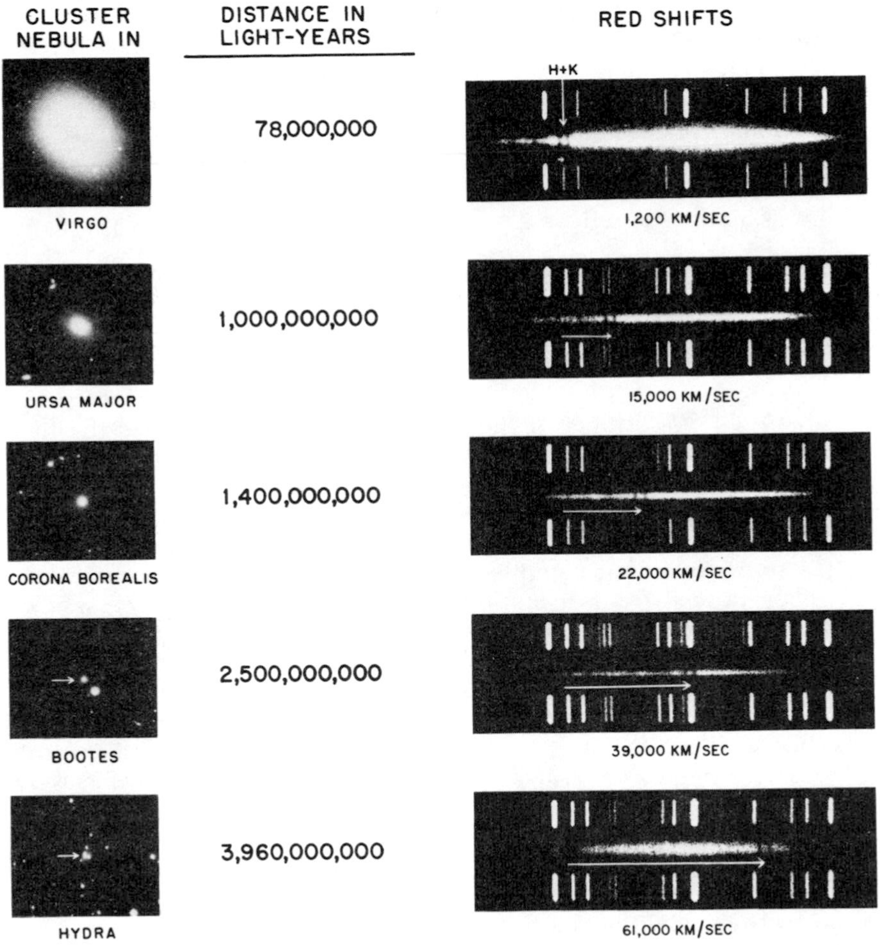

CLUSTER NEBULA IN	DISTANCE IN LIGHT-YEARS	RED SHIFTS
VIRGO	78,000,000	1,200 KM/SEC
URSA MAJOR	1,000,000,000	15,000 KM/SEC
CORONA BOREALIS	1,400,000,000	22,000 KM/SEC
BOOTES	2,500,000,000	39,000 KM/SEC
HYDRA	3,960,000,000	61,000 KM/SEC

Figure 2.32   *The Relation Between Red Shift and Distance for Extragalactic Nebulae*

In this photograph the pictures of bright galaxies in increasingly distant clusters are shown together with their spectra. The spectra are the cigar-shaped, horizontal traces sandwiched between sharply defined laboratory spectra, shown for reference above and below each nebular spectrum. A pair of dark absorption lines—the H and K lines of calcium—can be discerned in each spectrum. The more distant the galaxy, the smaller it appears and the more red shifted is its spectrum. Galaxies at increasing distances are expanding away from us and from each other at ever increasing speeds.

Photograph courtesy of Hale Observatory

During the 1960s Doppler velocities for galaxies became available also through radio astronomical observations. The 21-centimeter spectral line of atomic hydrogen was found to be shifted in the same proportion as the calcium absorption lines appearing in the visual spectra of galaxies. Radio and visual observations of the recession velocities were in good agreement.

In another few years we expect to have a velocity-distance relationship based entirely on radio data. Perhaps a detailed comparison of visual and radio information will provide new insights heretofore overlooked.

35. Quasars

In the late 1950s, the British radio astronomer Cyril Hazard set out to measure angular diameters of small radio galaxies. The original aim of these measurements was the mapping of compact radio sources, to obtain accurate positions and detailed structural features that might reveal the nature of these objects. One way of doing this was to pick galaxies that occasionally are occulted by the moon. A small source dims rapidly in such an occultation, while an extended source disappears quite slowly. The time required for complete occultation then is a measure of the source angular diameter. Precise timing of just when the occultation takes place can also provide an accurate position of the source in the sky.

Among the sources that Hazard wanted to study was the 273rd entry in the Third Cambridge University catalogue: 3C 273. He and two coworkers in Sydney, Australia, followed the decrease in brightness of the source during occultations by the moon on April 15, August 5, and October 26, 1962. They determined that the source actually consists of two components, some 20 seconds of arc apart.[160] The positional accuracy obtained for these sources was so high—better than 1 second of arc—that an optical identification was assured. The first tentative identification is described by Edge and Mulkay, who have portrayed in some detail the parts played in the discovery of quasars by a number of different optical and radio astronomers.[161] In particular they quote "one of our respondents, who was at Sydney at the time and was involved in this work."

> . . . 3C 273 just happened to be occulted (by the moon); nothing was known about it, although it was "identified" at that time with a galaxy about 1' of arc or so preceding the [quasar]. Because of this (previous incorrect "identification") a (Mount Palomar) 200-inch picture was available at Sydney: Minkowski [Rudolph Minkowski, optical astronomer at Mount Wilson and Palomar observatories], who was visiting Sydney at the time, had one of the region with him. With a preliminary position we looked at the region and saw the "star". . . and more particularly the "jet," although we didn't know what it was; but Minkowski remarked it would be very strange if it was an edge-on spiral (galaxy), as no such object had ever been identified with a strong [radio] source.[162]

Later, when Maarten Schmidt at the California Institute of Technology obtained a photographic plate of the region with the Mount Palomar 200-inch telescope, he confirmed the starlike object of thirteenth magnitude accompanied by a faint wisp or jet (figure 2.33). The starlike source's spectrum was quite unusual, having a red shift corresponding to a recession velocity of 47,400 kilometers per second.[163] If this velocity indicated the source's distance across the universe, the total brightness would exceed that of the brightest radio galaxies then known by a hundredfold!

The Australian group and Schmidt published their results in the March 16, 1963, issue of the British journal *Nature,* and Schmidt's colleagues, Jesse Greenstein and Thomas A. Matthews, added a paper on

Figure 2.33   *The Quasar 3C 273*

In 1962 the British radio astronomer Cyril Hazard and two colleagues began observations on the radio source 3C 273 in Australia. They observed occultations of the source by the moon and were able to derive its position and shape quite accurately. They found the source to consist of two slightly separated components which at optical wavelengths respectively have a stellar and a jetlike appearance, as seen in this photograph. Once the radio position had been accurately determined, the optical counterpart for this source was quickly found, and a spectrum obtained by Maarten Schmidt at Mount Palomar showed a Doppler shift corresponding to a recession speed of 47,400 kilometers a second.

the compact radio source 3C 48, whose spectrum had puzzled them until Schmidt had arrived at a high red-shift interpretation of the spectrum of 3C 273. Greenstein and Matthews found that 3C 48 had a Doppler velocity of 110,200 kilometers per second—more than one-third the speed of light.[164] These quasi-stellar appearing radio sources soon came to be known simply as "quasars."

Mention of the source 3C 48 calls attention to a rather more complex history: As early as December 29, 1960, Allan Sandage had presented an unscheduled paper at the 107th meeting of the American Astronomical Society held in New York City. He reported that the source appeared starlike. It was accompanied by a faint, luminous wisp, approximately 5 by 12 seconds of arc in dimension, and showed a spectrum of strong emission and absorption lines unlike those of any other known star. The authors of the paper reported that they failed to find a red shift that would fit the spectrum of 3C 48. The radio astronomers Matthews and J. G. Bolton, and the optical astronomers Greenstein, G. Münch, and Sandage had collaborated on this study, and the radio position used had been provided by Matthews working at the Owens Valley Radio Observatory.[165]

Significantly, perhaps, the work on 3C 48 was never published at the time. A paragraph summarizing Sandage's talk appeared in the popular magazine *Sky and Telescope* in March 1961. But astronomers rarely rely on such reports since the information is secondhand and involves none of the responsibility that a scientist normally assumes for his own papers. At any rate, Cyril Hazard in Australia, does not recall ever even having heard of this work, and according to him it had no effect on the subsequent identification of 3C 273 and the discovery of its large red shift.[166]

To assign proper credit, however, it is worth noting that the small radio diameter of the source 3C 48 had previously been determined interferometrically as early as 1960 by a group of observers headed by H. P. Palmer at Jodrell Bank in England and was known to be less than 4 seconds of arc in diameter.[167] While this early effort showed the existence of extremely compact radio sources with a starlike visual appearance, the actual discovery of quasars is usually associated with the events three years later that led to the correct identification of the large spectral red shift of 3C 273.

The key elements in the discovery of the quasar as a distinct new phenomenon seem to have been the measurements at high angular resolution at radio wavelenghts. These observations, which first became possible around 1960, established the surprisingly high surface brightness of the quasars at these wavelengths and provided accurate positions which also brought optical astronomers into the search.

Since these early observations, quasars have been studied in great detail. Their light emission varies over months and years; the radiation may be partly polarized; powerful X-ray emission is observed; and many quasars show families of optical absorption spectra, each family corresponding to a different velocity shift.

Although first discovered at radio wavelengths, quasars need not necessarily be bright radio sources. In 1965 Allan Sandage began finding quasars that were not notable radio emitters and the term *radio quiet quasar* was coined. Like all quasars, these objects are highly luminous.[168]

The most fundamental questions on quasars are still not properly understood: Are these sources really as distant as their red shifts indicate? Or are they more local red-shifted objects? If they are truly distant, they are fantastically bright. If they are local, their high red shift would have to be explained. Quasars remain a puzzle more than fifteen years after their discovery, although evidence is gathering that each quasar is associated with a galaxy or cluster of galaxies that shows the same red shift. Previously these galaxies could not be detected since the quasar was so much brighter and outshone nearby galaxies. More recent results, however, suggest that the quasar may represent a bright eruption involving many stars in a galaxy and, if verified, will clearly confirm a cosmic distance for quasars.

A class of sources that have many of the properties of quasars is represented by the archetype BL Lacertae which used to be considered a variable star. The BL Lacertae objects are compact sources with high surface brightness. Their visual, infrared, and radio emission varies with time and shows some linear polarization. Their visual spectra are much smoother and their spectral lines more difficult to discover than those seen in quasars. Nevertheless, a number of spectral features in BL Lacertae objects have by now been analyzed and indicate red shifts comparable to those of the quasars.

Some types of more normal galaxies also can show high surface brightness, compact nuclei, and time variability reminiscent of quasars. Chief among these are a set of galaxies first recognized by Carl K. Seyfert of the Mount Wilson Observatory in 1943. Seyfert noted that the spectrum of the nuclear regions of these galaxies showed emission lines of hot ionized gases moving at velocities of the order of thousands of kilometers a second. X-ray emission from these galaxies appears similar to the X-ray emission from quasars. Because of these overlapping characteristics some similarities in the physical properties of quasars, BL Lacertae objects, and compact galaxies may be expected.

36. Superluminal Radio Sources

Radio telescopes placed thousands of miles apart are capable of angular resolving powers measured in milliseconds of arc. Two sources of radio waves placed one light-year apart can be resolved even when they are at distances exceeding a hundred million light-years. With this technique it became apparent, around 1970, that some radio sources in the universe are seemingly flying apart at speeds greater than the speed of light. This is reflected in their name. They are called superluminal sources.

In 1966, Martin Rees, a young theorist at Cambridge University, had published a paper suggesting that light variations in quasars might be produced through explosions. In such outbursts the fragments flying apart at high velocities could appear to be receding from each other at speeds greater than the speed of light, c, even though the actual velocity of recession remained lower than c. The apparent superluminal velocity is an artifact of the way in which the observer computes the velocity he measures.[169]

If we were to watch fireworks going off at some distance, D, and saw colored flares shooting apart with an angular velocity $\dot{a}$, we would normally conclude that the individual flares were receding from each other at a speed $D\dot{a}$, or that their expansion velocity from the center of the burst was half that velocity, $D\dot{a}/2$. At speeds of light, this simple calculation no longer holds. We have to worry about the directions in which the different sources are moving. When we view two rapidly expanding sources, light from the nearer one will have been emitted considerably later than light emitted from the more distant source, even though the observer views both these sources simultaneously. While the light has been on its way toward us, the more distant source may have moved in some unknown fashion, relative to the nearer source, and we are therefore not viewing actual motion of these two sources, but rather something like a distortion of the true change of separation. If both sources are approaching us at high speeds a further distortion takes place: The rate at which photons are emitted at each of the sources is much slower than it appears to the observer; the emitted photons reach him closely bunched together. For this reason, an approaching source also appears considerably brighter than a receding component. Such relativistic effects have to be unscrambled if a correct interpretation of observations is to be obtained.

Although Rees's paper appeared in 1967, it was only one among many theoretical studies and did not directly influence the subsequent course of observations.

The discovery of superluminal sources is generally attributed to a team of researchers led by Irvin I. Shapiro of the Massachusetts Institute of Technology. His team of colleagues from MIT joined forces in late 1970 with observers from NASA's Goddard Space Flight Center, from the Jet Propulsion Laboratory in Pasadena, California, and from the University of Maryland. Their aim was to use widely separated radio telescopes to form a very long baseline interferometer (VLBI) with which to observe a completely different effect—the gravitational bending of radiation passing near the limb of the sun. For this purpose the group planned to observe two quasars, 3C 273 and 3C 279, which lie only a few degrees apart, in a portion of the sky through which the sun passes each October. As 3C 279 approaches the sun's limb, its apparent position relative to 3C 273 changes because its radio waves are gravitationally bent as they pass by the sun. The extent of this bending is predicted by Einstein's general theory of relativity; and the team of observers was hoping to measure this effect to test the validity of Einstein's predictions.

The observations were carried out in October 1970 and made use of the Goldstone radio antenna in Southern California and the Haystack antenna in Massachusetts to obtain a transcontinental baseline. Since the separation between these radio telescopes was 3,900 kilometers, and the wavelength 3.8 centimeters, the angular resolving power was better than one part in 10^8—the ratio of separation to wavelength. Angles that could be resolved were approximately $\frac{1}{1,000}$ of an arc second. At this resolving power it became possible to measure accurately the bending of radiation by the sun. The observations also showed the 3C 279 exhibited a double structure with a separation between components approximately $\frac{1}{1,000}$ of a second of arc. The quasar 3C 273 also exhibited fine structure on this scale.

While the reports on the gravitational bending were being written and sent to the journal *Science,* a further set of observations on the quasar structure was undertaken. The results were startling: In a note added to the *Science* article in proof, the authors wrote:

> Recent observations have revealed significant time variations in the structure of both quasars; in terms of the two-point source model, the separation of the components of 3C 279 appears to have increased by about 10 percent in 4 months.[170]

If the quasar was at the large distance indicated by its high red shift, expansion at this rate indicated an expansion velocity 10 times in excess of the speed of light.

When he had first become aware of the fast expansion of 3C 279, Shapiro's reaction, he tells, was to think that these observations would help to set an upper limit to the distance of quasars. If the quasars were thought to be very far away, the observed angular expansion would translate into an apparent explosion at speeds greater than c. At first glance such a finding appeared inconsistent with the laws of relativity. It was only then that Shapiro became aware of Rees's article and realized that the situation was more complex.[171]

Subsequent observations both by the MIT group and by a group led by Marshall H. Cohen at the California Institute of Technology have confirmed that 3C 279 and a number of other quasars are expanding superluminally.[172]

The story of this discovery would be simple were it not for a nearly identical set of observations carried out by a group led by D. S. Robertson in Australia and by Alan T. Moffet at the Owens Valley Radio Observatory in California. The observations they made were carried out at a wavelength of 13 centimeters, where the angular resolution is only one-fourth as clear as at the shorter wavelength employed by Shapiro's colleagues. Nevertheless, in 1967 the group published an article in the British journal *Nature.* The paper reports on both 3C 279 and 3C 273 and notes a superluminal expansion of 3C 273 with an apparent expansion velocity more than 3 times the speed of light.[173] The best current estimate, a dozen years later, is that the expansion velocity appears to be roughly 5 times the speed of light, in reasonable agreement with the earlier results. However, in a publication that appeared in 1972 Moffet, Robertson, and their colleagues retract the earlier work on 3C 273, but report a superluminal velocity for 3C 279.[174] Moffett has explained that the retraction was justified despite the appearance that the data originally presented seem to stand rather well on their own merits and appear to be in reasonable agreement with subsequent findings of other groups.[175] He also points out that the 3C 279 result presented at an International Astronomical Union Symposium held in Uppsala, Sweden, in August 1970 gave the first correct report of a superluminal expansion. Shapiro's group, however, has generally been credited with the discovery, perhaps because their independently obtained results appeared in print first and because their results were presented in far greater detail. Robertson and Moffet were clearly aware of Rees's paper and cite it in their 1967 article.

The interaction of theoretical prediction and observations is therefore not clear in this case, particularly since the retraction by Robertson and Moffet leaves the MIT group in the apparent position of prime discoverers of the effect. This group's preoccupation with a relativisitic test, in which they were hoping to use 3C 279 and 3C 273 as point sources to measure the gravitational bending of radiation, apparently made them unaware of the discussions that had been taking place on the structure of these quasars.[176]

37. X-ray Galaxies and Clusters

Early in 1966 E. T. Byram, T. A. Chubb, and Herbert Friedman of the Naval Research Laboratory in Washington announced the discovery of two extragalactic X-ray sources detected on an Aerobee rocket flight launched the previous April.[177]

Observations carried out over the next few years were to verify the NRL group's results. The radio galaxies Messier 87 (figure 2.34) and

Figure 2.34   *The X-ray Galaxy Messier 87*

Early in 1966 Herbert Friedman and his Naval Research Laboratory colleagues, E. T. Byram and T. A. Chubb, announced a set of startling results obtained with an Aerobee rocket bearing an X-ray telescope. While aloft, the payload had scanned the positions of the radio source Cygnus A and the giant elliptical galaxy Messier 87. Both of these sources yielded observable amounts of X-ray emission. So bright are these sources that most of their energy is being radiated in the X-ray portion of the spectrum. These are not just visible galaxies that also emit some X rays. They can more correctly be considered X-ray galaxies that also emit part of their energy at visible and radio wavelengths. Messier 87 exhibits a jet whose emission appears to be produced by cosmic-ray particles generated within the galaxy. The sources of energy responsible for the jet, the radio emission, and the X-ray emission still are unknown.

Photograph courtesy of Lick Observatory

Cygnus A, which they had observed, are X-ray sources that emit more than 10 times as brightly in the X-ray domain as in the radio wavelength region. Cygnus A, the brightest known radio source, is also approximately 10 times more luminous at X-ray frequencies than in the optical domain.

A number of galaxies and perhaps all quasars emit powerfully at X-ray wavelengths. Radio galaxies and galaxies that are characterized by a bright nucleus in which ionized gases are observed moving at speeds of the order of 1,000 kilometers a second—the Seyfert galaxies—are the brightest X-ray galaxies. But quasars, if they are at the distances indicated by their red-shifted spectra, emit 1,000 times more brightly. And clusters of galaxies sometimes exhibit X-ray emission that appears to come from regions not directly linked to any particular galaxy. Emission from clusters tends to be comparable to the total X-ray luminosity of a single bright X-ray galaxy.

The mechanism by which the X-ray photons are produced is not yet properly understood, but highly energetic cosmic-ray particles in these galaxies are almost certain to play an important role in the process.

38. Infrared Galaxies

In a paper published in 1966 Harold L. Johnson of the University of Arizona wrote,

> From the beginning of our infrared work, we have hoped to measure the infrared radiation from galaxies. We recently obtained from Infrared Industries an extremely sensitive PbS cell. . . . This detector has made possible the measurement of galaxies. . . .[178]

Johnson goes on to describe the results of observations on the nuclei of ten galaxies at wavelengths going out to 3.4 microns.

In 1968 Eric Becklin and Gerry Neugebauer of the California Institute of Technology reported on a search for near-infrared emission from the center of our Milky Way galaxy. Working at wavelengths 3 to 7 times longer than visible light, they found that they were able to overcome the obscuration by dust. At visible wavelengths the dust clouds between us and the galactic center provide a formidable barrier which passes only one quantum of light out of every ten billion emitted at the center. But at 1.65, 2.2, and 3.4 microns in the infrared progressively more radiation is transmitted, and this enabled the two CalTech observers to identify the Galactic center.[179] They also were able to detect the center of the Andromeda Nebula, the nearest large galaxy, at a wavelength of 2.2 microns.

A year later William F. Hoffmann and Carlton Frederick, working at NASA's Goddard Institute for Space Studies in New York, sent a tiny—1-inch-aperture—balloon-borne telescope into the stratosphere and observed a bright glow emanating from the center of the Galaxy at wavelengths of 100 microns—$\frac{1}{10}$ of a millimeter.[180] These wavelengths, which are 30 times longer than the longest waves sensed by Becklin and Neugebauer, normally are completely absorbed by the water vapor in our atmosphere, and therefore observations had to be carried out at high altitudes.

The observed flux from the Galactic center region was about 3 percent of the total radiation believed to be emitted by the Milky Way. Subsequent observations confirmed that the Milky Way may be emitting about one-tenth of all its radiation at these far-infrared wavelengths. There are, however, galaxies that are far more powerful emitters of infrared waves. Starting around 1967 Frank J. Low, first working with

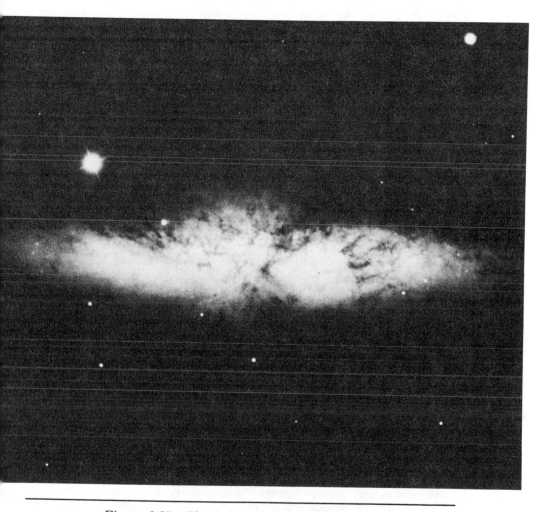

Figure 2.35 *The Infrared Galaxy Messier 82*

Starting in the late 1960s F. J. Low, first with D. Kleinmann and later with G. H. Rieke of the University of Arizona, unexpectedly found that a number of galaxies emit far more radiation in the infrared than at all other wavelengths combined. These galaxies generally are characterized by a high content of dust, occasionally by X-ray emission, and generally also by fairly strong radio emission. Messier 82, the brightest far-infrared extragalactic source known, at one time was thought to be an exploding galaxy because hydrogen is seen streaming out of its central portions at velocities of 1,000 kilometers a second. Energetic processes that we do not yet understand appear to be active in this galaxy.

Photograph courtesy of Hale Observatories

Douglas Kleinmann and later with George H. Rieke of the University of Arizona, conducted studies of galaxies in discrete wavelength bands ranging from 2 to 25 microns. Surprisingly, the group found that a number of galaxies emit far more radiation in the infrared than at all other wavelengths combined.[181] Messier 82 (figure 2.35) is the brightest of these. These infrared galaxies are characterized by a great deal of dust, occasionally by X-ray emission, and generally also by fairly strong radio emission.

Infrared galaxies may be objects in which a spate of young stars has just appeared during the past few million years. The more massive of these have already completed their main sequence sojourn and are now exploding as supernovae whose remnants are active emitters of X-rays and radio waves. Other young stars are still largely surrounded by the dust clouds from which they formed. Their radiation is completely absorbed by the dust and reradiated by these grains as infrared—thermal—radiation. This explanation still is speculative and will have to await observational confirmation—or disproof.

39. Gamma-ray Bursts

Remarkably little is known about gamma-ray bursts. Several times a year gamma rays reach Earth in strong outbursts lasting a few seconds. Then there is quiet for weeks or months before another burst arrives from a direction we cannot yet predict.

In 1973 Ray W. Klebesadel, Ian B. Strong, and R. A. Olson of the Los Alamos Scientific Laboratory published the first paper announcing the discovery of the bursts.[182] With Klebesadel's permission I reproduce here portions of a letter in which he describes the Los Alamos group's recollections of the discovery.

October 24, 1978

A history of the events leading to the discovery of the gamma-ray burst phenomenon must begin with a consideration of the Vela Nuclear Test Detection satellite (NTDS) system which was responsible for the observations leading to that discovery. The Vela NTDS system was sponsored by the Advanced Research Projects Agency in response to a mandate from the U.S. Congress to develop a capability to monitor for clandestine nuclear weapons tests in outer space, conducted in violation of the Geneva Limited Nuclear Test Ban treaty that was then being considered. Six pairs of Vela satellites were successfully launched in the course of the program over the period 1963 to 1970. Those launched in the years 1969 and 1970 are still in service in mid-1978.

All Vela payloads included instruments designed to respond to gamma rays emitted promptly in nuclear explosions as well as to those emitted upon decay of the fission products. From early in the Vela program, data from the gamma-ray detectors were searched for enhancements in the response at times of reported supernovae. This was done at the instigation of Drs. Stirling Colgate and Edward Teller, both of whom predicted the generation of gamma radiation in the early phases of the expansion of supernovae. These searches were not fruitful in that no response in the gamma-ray detectors was found associated with reported supernovae.

The instrumentation for the detection of the fission-product gamma radiation was improved on successive satellites to provide more sensitive monitoring for nuclear tests. . . . It was in 1969 that the search of the data from the launch 4 satellites revealed several events that occurred in near-coincidence. Of these, only one was sufficiently intense to produce records exhibiting significant and distinctive temporal structure. The records of this event from the two satellites were virtually identical. Although this single event aroused considerable interest, it was obviously premature to attribute it to a previously unobserved astrophysical process.

. . . Preparation of the computer data base and automated scanning routines, and preparation and postlaunch operations for Vela 6, delayed success in the continued search for simultaneous events until 1971, when a dozen such events were resolved in quick succession. Among these dozen events, however, was one event which seemed clearly of solar origin. An immediate concern was whether these events all originated at the sun or within the solar system.

In an attempt to identify the sources of the bursts, or at least to eliminate some of the candidates, the transit time of the signal wavefront between the various spacecraft was used to determine the directions to the sources of the bursts. In those cases where only two spacecraft observed a given event, the directions could be resolved only to a circle of position, and, where data from three or more spacecraft could be employed in the determination, directions could be resolved to two regions of several tens of square degrees in extent. This degree of accuracy was found to be adequate to eliminate the major members of the solar system as source objects for the bursts. Finally, in 1972, we felt that the amassed data provided sufficient evidence for the existence of a previously unreported astrophysical phenomenon. A report of the observations was submitted for publication in the Astrophysical Journal Letters, and the discovery was announced at the 140th Meeting of the American Astronomical Society held at Columbus, Ohio, June 1973.

Although there has been a proliferation of theories proposing various mechanisms for the generation of these bursts, none of those yet presented is capable of reproducing all of the details of the observations. Neither the indicated directions nor coincidence in times of occurrence have yet established an association between these bursts and any other reported astrophysical phenomena. Even today, 1978, with 71 bursts cataloged and with improved directional resolution available, the sources of these bursts remain unidentified, without even a strong suggestion of the class or classes of objects responsible.[183]

40. Microwave Background Radiation

In 1940 Andrew McKellar of the Dominion Astrophysical Observatory in Victoria suggested an identification for a puzzling interstellar absorption line which Theodore Dunham and Walter Adams had discovered in their observations at Mount Wilson.[184] Using unprecedentedly high spectral resolution, Dunham and Adams had found photographic evidence of a very narrow, extreme violet line—much in the way that Hartmann, several decades earlier, had made his discovery of cold interstellar gas.

McKellar suggested that the new line might be due to absorption by the cyanogen molecule, CN, which consists of a carbon and a nitrogen atom chemically bonded together. If this identification were correct, McKellar predicted, a second spectral line might be found, corresponding to CN molecules that are rotating rather than at rest. If this new line were about one-third as strong as the line produced by nonrotating molecules, this would imply that the molecules were endowed with an energy equivalent to that obtained by heating the gas to a temperature of 2.7 degrees Kelvin, a temperature only 2.7 degrees Celsius above absolute zero.*

Shortly after this prediction Adams found the specified line.[185] Its intensity was roughly a third of the originally discovered line's intensity but the idea that the molecules were being kept at a temperature near 3 degrees Kelvin seemed too simple to be taken seriously. There were too many other conceivable ways in which such a ratio of line intensities might be produced. Given what was known at the time, it is not surprising that Gerhard Herzberg, who later was to win a Nobel prize for his contributions to chemical spectroscopy, concluded: "From the intensity ratio of the lines . . . a rotational temperature of 2.3°K follows, which has of course a very restricted meaning."[186] At the time, little else could have been said or done. The instrumental means were largely lacking.

A quarter of a century after Adams's measurements, Arno Penzias and Robert W. Wilson of Bell Telephone Laboratories in Holmdel, New Jersey, were testing a new horn-shaped radio antenna originally designed for signal reception from communication satellites. Their receiver was sensitive to wavelengths of 7½ centimeters—radio waves 200 times shorter than those used by Karl Jansky, who had worked in the same laboratories three decades earlier. Jansky had found excessive radio noise which he had been able to attribute to emission from the Milky Way. In almost the same way Penzias and Wilson found unaccountable excess noise that was unpolarized and completely isotropic.[187] Effectively the noise corresponded to radiation at a temperature of about 3 degrees impinging on Earth from all directions.

Such a cosmic radiation bath had already been suggested in the late 1940s by George Gamow and two young associates, Ralph Alpher and Robert Herman of the Johns Hopkins Applied Physics Laboratory, Silver Springs, Maryland. They argued that if the universe had had an explosive origin, this vestigial radiation should still persist.[188] In 1965 Robert H. Dicke, P. J. E. Peebles, R. G. Roll, and D. T. Wilkinson at Princeton, not knowing of the earlier work of Gamow, Alpher, and Herman, independently interpreted the radiation discovered by Penzias and Wilson as this remnant of the birth of our cosmos.[189]

The Princeton group had hypothesized the existence of the background radiation and had set about to search for it just before becoming aware of the Penzias-Wilson measurements. The two Bell Laboratories researchers, on the other hand, seemed not as sure of the cosmic interpretation. They cautiously titled their paper "A Measurement of Excess Antenna Temperature at 4080 Mc/s" and would state only that

> this excess temperature is within the limits of our observations, isotropic, unpolarized, and free from seasonal variations (July 1964–April 1965). A pos-

---

* See Glossary Figure G.2 for a comparison of temperatures.

sible explanation for the observed excess noise temperature is the one given by Dicke, Peebles, Roll, and Wilkinson . . . in a companion letter in this issue.[190]

Later, Penzias recalled the discovery in this way:

When, in the late Spring of 1964, we finished the installation of the 7.3 cm system in the horn, we found that the antenna was considerably "hotter" than expected. This "excess antenna temperature" proved to be a very stable phenomenon. It had no measurable sidereal or solar variation, was not polarization or antenna position dependent, and was apparently completely noiselike in character. Since it interfered in no way with our work, which was entirely incremental in character, we merely noted the existence of the phenomenon and periodically measured its size. We were unable to dismantle the throat of the antenna, which was a suspected source of attenuation because we did not wish to invalidate an elaborate calibration of the effective collecting area of the antenna. This was required for a flux measurement which was not completed until the beginning of 1965.

Upon the completion of the flux measurement we dismantled the throat section of the horn, replaced suspected components, and cleaned its interior surfaces. . . . These efforts had only a small effect upon the measured antenna temperature, reducing it by a few tenths of a degree Kelvin to just under three and one-half degrees. Having explored and discarded a host of terrestrial explanations, and "knowing," at the time, that no astronomical explanation was possible, we frankly did not know what to do with our result. . . .

I mentioned our problem to Bernard Burke during a casual telephone conversation on another matter. He replied that a preprint from Princeton had come across his desk shortly before, predicting a ten-degree background at 3 cm. I called Professor Dicke, who sent me a copy of the paper in question. It was written by P. J. E. Peebles and predicted, using certain assumptions, a thermal background of radiation as a residuum of the hot, highly condensed early state of the evolution of the universe.

After a visit to our antenna by Dicke and his co-workers, it was agreed that we should publish our results in a letter accompanying one by his group outlining their theoretical work modified to take account of our result. Little did any of us suspect that George Gamow had gone quite a long way down the road we were now traveling. Some time after the papers appeared, I found that Gamow's work two decades earlier had indeed covered the same ground. I sent him a draft of a paper enlarging on our original work.[191]

In 1978 Penzias and Wilson were awarded a share of the Nobel prize in physics for the discovery of the microwave background radiation. We might still ask how early the microwave background might have been discovered if Gamow's prediction had been energetically pursued.[192]

We have already seen how the work of McKellar and Adams was not enough to suggest the existence of the radiation in any persuasive way in 1940–41. A postwar measurement by Robert Dicke and co-workers, then active at MIT, showed that at wavelengths of 1.00, 1.25, and 1.50 centimeters, it was possible to set an upper limit of 20 degrees Kelvin with equipment available in 1946.[193] Hey points out that the report of

GALACTIC GAMMA-RAY EMISSION

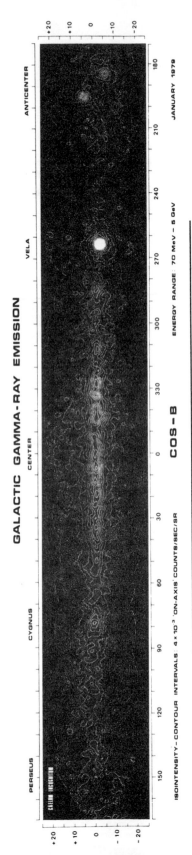

Figure 2.36 *Gamma-ray Map of the Galaxy*

In 1975 the European Space Agency's COS-B satellite was launched with a payload sensitive to gamma rays between about 50 MeV and several GeV, corresponding to wavelengths ranging from around 2×10^{-12} to below 10^{-13} centimeters. Since then the spacecraft has produced a map showing the diffuse gamma emission received from the galactic plane. Several regions of enhanced emission are seen, some of which are due to compact sources. Gamma-ray emission from the Vela pulsar is particularly pronounced. Other concentrations of gamma rays seem to come from cold dense clouds of gas in the plane of the Galaxy, where the gamma rays may be generated through the interaction of cosmic-ray particles with matter forming the clouds.

We do not yet know whether all gamma rays received from the universe can be attributed to individual sources or whether there also exists an isotropic background radiation at these wavelengths, similar in its uniformity to the microwave background.

Courtesy of the Caravane Collaboration for the COS-B Satellite[202] these observations and Gamow's original article on early cosmic evolution appeared only 232 pages apart in the same volume of the *Physical Review*.[194] But Dicke and Gamow apparently remained unaware of each other's work. In any case Gamow's ideas at that time were not as fully developed as necessary. The current background temperature was not yet explicitly mentioned. Three years later, however, the work of Alpher and Herman did explicitly mention a background temperature of a few degrees Kelvin.[195]

There matters stood until the early 1960s when the *Echo* and *Telstar* communications satellites were being engineered. Using the horn that later was to be used by Penzias and Wilson at shorter wavelengths, E. A. Ohm at Bell Telephone Laboratories reported his system's noise observations at 11-centimeters wavelength. As Penzias and Wilson pointed out in their own paper, Ohm's results were much the same as their own. He measured an excess noise of 2.1 degrees Kelvin but attributed it to a probable underestimate of the several sources of noise he knew existed in his system. These calculable sources of noise gave a predicted noise temperature of 18.9 ± 3 degrees Kelvin while Ohm's measured temperature was 21 ± 1 degrees Kelvin, a small enough difference perhaps, but one that was crucial in this case.[196] The 2.1 degree Kelvin difference seemed consistent with the 3 degree Kelvin uncertainties in his apparatus performance. A series of measurements conducted at Bell Laboratories two years later obtained similar results.[197]

It appears therefore that the microwave background radiation might have barely been discovered as early as 1960 if instrumental factors had been better understood. Dicke, however, feels that an attempt to measure the background radiation might have succeeded as early as 1946 if there had been sufficient motivation.[198] But the effort required would surely have been horrendous, and perhaps no one could have become sufficiently motivated when there were so many other, simpler observations to be made in astronomy.

41. X-ray Background

The microwave radio background currently is our strongest piece of evidence for the explosive origin of the universe some fifteen billion years ago. It also represents most of the energy residing in electromagnetic radiation criss-crossing the universe. The energy content of all the visible light in the universe is considerably lower. We also observe a cosmic X-ray background radiation bath whose energy content is lower and may just represent the total radiation received from X-ray stars and X-ray galaxies throughout the universe. The X-ray background was originally established on the same rocket flight that led Giacconi and Rossi to the discovery of X-ray stars.[199] Six years earlier, however, during a series of flights carried out by Rockoons (rockets launched from balloons), Friedman and Kupperian had already had substantial evidence for extraterrestrial X-radiation that did not come from the sun.[200] Therefore, 1956 or 1957 appears to be the earliest that instrumentation existed with which the X-ray background might have been discovered. The actual discovery took place in 1962.

42. Gamma-ray Background

At the very highest energies there also appears to exist an isotropic background bath of gamma rays—radiation at a wavelength less than ~10^{-12} centimeters where quanta are ~10^8 times more energetic than visible photons. This radiation was first detected by instruments placed aboard NASA's Third Orbiting Solar Observatory satellite by George Clark, Gordon Garmire, and William Kraushaar of the Massachusetts Institute of Technology.[201] A strong flux of gamma rays reaches us from the plane of the Milky Way, particularly from the central portions of our galaxy (figure 2.36), but there is an additional weaker component from the universe beyond. Future observations at higher spatial or spectral resolving power may explain the origin of this apparently isotropic flux of gamma rays.

43. Sources of Cosmic Rays

We have seen how the Austrian scientist Viktor Hess was led to the discovery of cosmic rays. But Hess had not been the first to suspect a cosmic contribution to the atmospheric ionization measured.

C.T.R. Wilson at Sydney Sussex College in Cambridge, England, had noticed earlier that even in a well-shielded ionization chamber a certain electrical leakage always occurred due to the production of ions in the chamber. Radioactive substances and X rays would have been stopped by the shields placed around the chamber. Therefore, Wilson theorized, some source of residual ionization must exist that could penetrate a great thickness of material.

Wilson suspected that the radiation might be cosmic. He set up his apparatus at Peebles in Scotland in a Caledonian Railway tunnel and found the same discharge rate outside and inside the tunnel. He concluded:

> There is thus no evidence of any falling off of the rate of production of ions in the vessel, although there were many feet of rock overhead. It is unlikely, therefore, that the ionisation is due to radiation which has traversed our atmosphere. . . .[203]

Although Wilson's conclusion was sensible for his time, it was incorrect. The ionization he observed is produced by highly penetrating secondary cosmic ray particles. Some of these persist even in the deepest mines a few thousand feet below the earth's surface where their intensity declines slowly. Wilson, however, was not as unfortunate in all of the studies he attempted: He was the inventor of the cloud chamber which played a fundamental role in the early days of experimental nuclear physics and won him the Nobel prize for physics in 1927.

Much of the information we do have on cosmic rays comes from a study of their interaction with the atmosphere. In addition, we have remote observational evidence for the existence of these particles throughout our own galaxy and in external galaxies as well.

The polarized radio waves emitted in supernova remnants and radio galaxies exhibit all the traits of synchrotron radiation emitted by highly energetic particles traveling through a magnetic field at speeds close to the speed of light. Some powerful emitters of radio waves and X rays, such as the Crab Nebula and the giant spherical galaxy M87, also exhibit structures that emit polarized visual radiation with a continuous spectrum. The Crab Nebula supernova remnant emits at least partially polarized radiation at X-ray frequencies. We know of no other mechanism except synchrotron emission from highly energetic particles that can produce these observed traits.

A third type of tentative but independent indication for the presence of cosmic-ray particles may be offered by the gamma-ray emission observed from the Galactic plane and predominantly from the direction of the central portions of the Galaxy. A statistical analysis shows a good correlation between the positions of gamma-emitting regions and of dark Galactic clouds containing dust and gas in molecular form. This suggests that cosmic rays passing through these dense clouds interact with matter and produce gamma rays which then signal the loss of energy by cosmic-ray particles that may have been formed in supernova explosions.

Repeatedly, discoveries of novel cosmic phenomena have been made by men and women who looked at the universe with new instruments and came to see it from a fresh perspective. Will this application of new methods continue forever to lead to further discoveries? How many more ways can we invent to view the universe in a new light? What is the scope of astronomy? How much more remains to be done?

These questions are tackled in the next two chapters.

CHAPTER

3

Observation

The Rate at Which Information Can Be Conveyed

The early part of this century was the heyday of radio communication. All of a sudden, messages could be beamed across oceans and voices from distant countries could be heard all over the earth. It seemed as though there could be no limits to the opportunities for communication provided by this new link. By dividing up the radio broadcast band into many sufficiently narrow channels, it seemed any number of people would be able to carry on any number of simultaneous radio conversations.

In 1922, however, John R. Carson of the American Telephone and Telegraph Corporation noticed a whole series of troublesome restrictions. In a paper he published on the theory of message transmission, he worried that all the ingenious schemes which promised increased rates of communication involved "a fundamental fallacy": He showed that in order to transmit certain amounts of information in a fixed time, it was necessary to use at least a minimum range of radio frequencies. The greater the rate at which information was to be transmitted, the greater was the bandwidth—the minimum range of frequencies that was required.[1]

Carson's ideas were far-reaching. Suppose that we wished to send a message across the Galaxy and that we decided to let a different color—frequency of light—stand for each word in the dictionary. How many different frequencies could we distinguish in going from red through orange, yellow, green, and so on, until we reached violet? Could we divide this range arbitrarily into many fine intervals, each corresponding to a different word?

Figure 3.1   *The Frequency of a Wave*

An electromagnetic wave moves past an observer with the speed of light. To measure its frequency the observer can count the number of crests passing by in a period of one second. This gives him a measure of the frequency; but the value obtained is uncertain by as much as one cycle per second. In the top two curves, one of which is drawn half a cycle out of phase with the other, there is a single cycle, and there is just one crest in both traces. In the second pair of curves, the solid curve shows two crests and the dotted trace shows just one. Both waves have a frequency of one and a half cycles, but there is an uncertainty of one cycle per second obtained if we merely count crests. In the third set of curves, the waves have a frequency of two cycles per second, and both traces show two crests.

We therefore see that a wave with a frequency of two cycles per second can be clearly distinguished from a wave whose frequency is one cycle per second, but neither wave is clearly distinguishable from a wave with a frequency of one and a half cycles each second. Such a distinction can be made, but only by counting for a longer interval than one second. When counting wave crests for a time interval, T, the count still remains uncertain by one cycle, but the total number of crests counted has increased by a factor of T, and the uncertainty in the frequency is now T times smaller.

No matter how long the counting interval T, two waves appear distinct only if one of them differs from the other by at least one full cycle over the interval T. If a wave of frequency f exhibits fT cycles, the next lower distinguishable frequency will be $(fT - 1)$. We can, therefore, distinguish no more than fT waves of frequency equal to f, or lower, if a time interval no longer than T is permitted for the observation. This restriction is one expression of the Heisenberg uncertainty principle.

To answer these questions we have to examine how we determine the frequency, f, of a wave, and how we distinguish two waves that have almost precisely the same frequency. One simple way to do that is to count the number of crests in each of the waves as it passes by. We see from figure 3.1 that the maximum number of distinct frequencies that can be sorted out in a time interval T is just fT.

Frequency, however, is not the only property of a light wave. Light also has intensity. If we take this into account as well, it is possible to prove that up to $2fT$ different waves—but no more—can be distinguished in a time interval lasting T seconds (appendix B).

Normally a transmission system does not allow us to transmit waves whose frequencies drop all the way to zero. Instead, the system only transmits a band that lies between a lower frequency, that we might label f_0 and a higher frequency, f_1. For a measurement lasting T seconds, we could therefore identify $2f_1 T$ distinct waves in the frequency range below frequency f_1, $2f_0 T$ distinct waves below f_0, and therefore a total of $2[f_1 - f_0]T$ waves in the frequency band of width $W = [f_1 - f_0]$. Put differently, the number of identifiably distinct waves that we can transmit in a time interval T is $2WT$—twice the bandwidth, W, of the communication link, multiplied by the time, T.

The impossibility of identifying a larger number of differing waves in the limited interval of time T, is one expression of Werner Heisenberg's uncertainty principle. The longer the available time, the finer the distinction we can make between adjacent frequencies. Conversely, the shorter the time available, the cruder is the distinction with which we will have to be satisfied. Curiously, the work of such men as Carson, Harry Nyquist[2] and others, which gave early expression to these ideas, was developed in the two years prior to 1924 and thus antedated Heisenberg's work. In fact, the precise connection between the two lines of argument remained undefined for twenty years. A clear mathematical demonstration of the relationship between quantum mechanics and information transmission was not provided until 1946 when Dennis Gabor, who later was to invent *holography*, first clarified the interdependence of physical and information theoretical restrictions.[3]

The result established by Gabor is this: If we wish to use different colors or intensities of waves in any combination whatsoever to transmit differing messages, we will be able to distinguish no more than $2WT$ different messages in a time interval T. There is no way to exceed that number no matter how ingeniously we try. The limitation lies in the nature of the waves and cannot be overcome. We have to recognize that no more than $2WT$ messages can be sent or received by any conceivable system of bandwidth W, if the total available transmission time is limited to an interval T.

Astronomical Information

Consider a telescope pointed in a specific direction in the sky. We can design this instrument to gather all the electromagnetic radiation—light, radio waves, X rays, and so on—in its field of view and to collect it onto a single detector.

We now ask how many different kinds of photons—quanta of light— might be clearly distinguished in this incoming stream of radiation. That question is interesting because we may consider each separate type of photon as a distinct carrier of information and a new channel through which we can observe the universe. Specifically, we might have equipment that is only able to distinguish light quanta from just two distinct groups, bluish light and light that appears yellow. Then we will only have two channels of information through which to observe. Nevertheless, with this simple instrument we would soon discover that some stars emit a great deal of blue light and very little yellow, though the vast majority of stars emits more yellow light than blue.

A capability for distinguishing increasing numbers of colors (wavelengths, spectral frequencies) permits astronomical observations through increasing numbers of spectral channels and construction of complete sets of maps of the sky, one map corresponding to each distinct color.

But we just saw that the light waves cannot be distinguished with arbitrarily fine color resolution, and we therefore cannot construct an arbitrarily comprehensive set of maps of the sky with each map in a distinct color. We can at best obtain $2WT$ differently colored maps.

In a similar vein, Heisenberg's principle imposes restrictions on the available angular, or pictorial, resolution. In order to draw increasingly detailed maps of the sky we need to know with ever-increasing accuracy the directions from which radiation is reaching us. By resolving two closely spaced directions in the sky, we may be able to distinguish whether a certain quantum of light has reached us from the bright star Sirius or from its faint white dwarf companion, less than 10 arc seconds removed in the sky. With poor angular resolution such a distinction cannot be made.

The uncertainty principle limits the smallest resolvable angle to a quantity which may be written as λ/D, the ratio of the wavelength, λ, divided by the dimension, D, of the available observing apparatus. If the apparatus consists of a pair of radio telescopes—one in Green Bank, West Virginia, linked to another roughly 3,000 miles distant at Owens Valley in California—then D is the separation between telescopes. More frequently, only a single telescope is employed, and then $D = A$ represents its aperture (figure 3.2).

Either way, if we are given observing equipment having a maximum dimension D, the best we can hope to do is to resolve the sky into roughly D/λ distinct elements along a north-south direction and a similar number along an east-west direction. The total number of distinguishable elements of area across the sky therefore is proportional to $(D/\lambda)^2$, and has an actual value of $4\pi(D/\lambda)^2$. Since the wavelength of radiation is just the speed of light, c, divided by the spectral frequency, f, we see that the total number of distinguishable spatial elements observed at spectral frequency, f, is just $4\pi(Df/c)^2$. The higher the spectral frequency and the larger the observing apparatus, the more sharply defined are the pictures of the sky we can obtain. That is why astronomers build increasingly large telescopes or arrays of telescopes to improve our maps of the sky.

One final characteristic for distinguishing two otherwise identical photons is polarization. For electromagnetic waves there are two—and only two—states of polarization. Each photon analyzed for polarization must be in one or the other of the two states. One way of describing

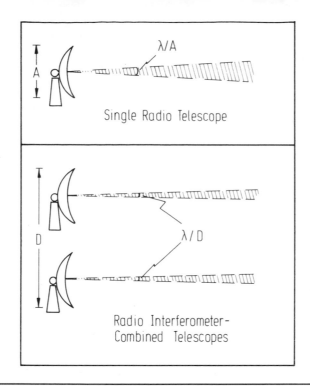

Figure 3.2 *Angular Resolution*

Any telescope or system of coupled telescopes has a limited capacity for determining whether the radiation received arrives from a single source or from two or more closely spaced sources. If observations are undertaken at a wavelength λ, and the telescope (top panel) has an aperture A across which light is gathered, then the smallest angle of separation that can be measured between two comparably bright sources is λ/A. When two or more telescopes (bottom panel) form an interferometer, none of whose members are separated by a distance greater than D, then the smallest resolvable angle is λ/D. A larger telescope, or a larger array, therefore helps us to determine small angular separations and permits us to construct finer maps of the sky.

polarization is in terms of the photon's spin. Every photon spins either in a right- or left-handed sense about its direction of propagation. Appendix B provides a quantitative discussion of the maximum rates at which photons from the universe can convey information to us.

Groups of Photons

In current practice we never even approach the limiting rates at which we might gather new astronomical information. We just are not prepared to look at each distinguishable wave individually, because far too many of them are available. Instead, waves are almost invariably observed in groups. Such a group may consist of a rather narrow range of spectral frequencies specified by two parameters, an average frequency and a bandwidth.

Two factors now come into play. Both result from dealing with

groups of waves rather than individual quanta, and both become apparent as soon as two or more photons are observed in combination: If
we add any two electromagnetic waves of differing frequency, they beat
against each other much in the way two pure tones of slightly different
pitch (frequency) produce sound beats (periodic changes in intensity)
as described in figure 3.3. The beats occur rapidly or slowly, depending
only on the difference in frequency between the two waves. Some observing equipment, however, may only respond to slow variations, unable
to follow rapid beats that occur on a scale of milli- or microseconds.

During much of the history of astronomy, rapid changes in astronomical phenomena were never known to exist because we lacked the
apparatus that could register fast variations. Equipment capable of
milli-second time resolution was only introduced into astronomy in the
mid-1960s; and the discovery of pulsars through rapid response radio
observations was a direct result of this new, expanded capability. On
a time scale of seconds or minutes, a pulsar's sharp pulses lose their
striking character and meld into a noisy hum indistinguishable from
the noise of other radio astronomical sources. For this reason, new equipment capable of rapid time resolution was the key to the discovery of
pulsars.

All time variations can be thought of as beats between different
waves, but in technical language, we speak of a *modulation* of the beam
rather than about beats. If an instrument is sensitive to modulation
on some time scale, *t,* it will normally act like a filter that is sensitive
also on somewhat higher and lower time scales, perhaps over a range

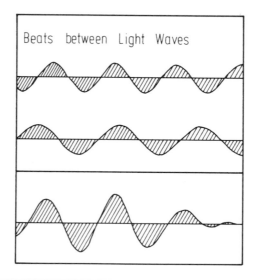

Figure 3.3   *Light Beats and Modulation*

When two light waves of slightly different frequency (top panel) reach an observer, they
add to produce beats—periodic increases and decreases in the amplitude of the combined
waves. We see these beats in the bottom panel as a slow increase and subsequent decrease
in the amplitude of the combined wave. We say that a wave of a given frequency is modulated,
by the addition to it of a wave of somewhat different frequency. This modulation produces
changes both in the frequency and the amplitude of the original wave. By modulating a
wave to alter its amplitude or frequency, we can use it to transmit messages. This is the
way in which radio broadcasting takes place. Effectively, it also is the means by which a
pulsar sends out its modulated—pulsed—signals.

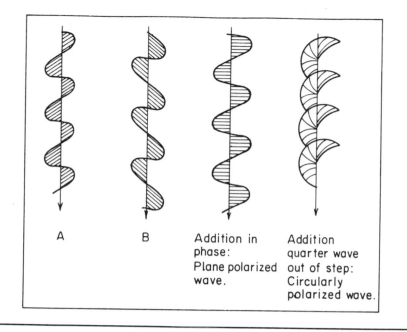

| A | B | Addition in phase: Plane polarized wave. | Addition quarter wave out of step: Circularly polarized wave. |

Figure 3.4 *Polarization*

Electromagnetic waves are transverse waves; they oscillate along a direction perpendicular to the direction along which the wave travels. If the wave is incident downward on the equator and the oscillations are north-south, we have a wave represented by trace A. If the wave travels downward, but its oscillations are along the east-west direction, we have a wave shown by trace B. These two waves can be combined in phase, so that a peak in the north-south direction is synchronous with a peak in the east-west direction. The result (third trace) is a plane polarized wave. Alternatively, the waves can be combined a quarter of a wave out of step. The result (fourth trace) is a wave that spirals down on the equator, its peak gyrating from north through east, south, and west before repeating the cycle. This is a circularly polarized wave. A further phase shift of half a cycle reverses the direction of gyration. By combining waves of the same frequency with different phase shifts, we can obtain either purely plane polarized waves, purely circularly polarized waves, or mixtures of circular and plane polarization that are said to be elliptically polarized. When a large number of waves are added with random phase shifts, no clear polarization state emerges, and we say that the wave is unpolarized.

from $t/2$ to $3t/2$. When we speak of equipment capable of detecting modulations on a time scale, t, which might be milliseconds for pulsars or hundreds of days for slowly pulsating red giant stars, we implicitly also have in mind a range of response times roughly comparable to $(3t/2) - (t/2) = t$. All this is tacitly assumed when we say that the time variations which can be observed with a given instrument correspond to the parameter, t.

The parallelism of two beams reaching us from a source in the sky can be defined in a similar fashion, by means of the spatial beats the two beams produce, and we can estimate the size of an astronomical source by measuring the separation between beats. This is the basis of stellar interferometry.

When light waves of vastly differing frequencies reach our apparatus, their beats are so fast we can no longer measure them as time variations. To define the spread in frequencies we may then wish to make use of a different instrument, a spectrometer.

Another type of beat phenomenon involves the polarization of groups of simultaneously observed waves (figure 3.4). If we take two plane polarized light waves of identical amplitude and combine them in phase, they will appear plane polarized and will either pass through a piece of Polaroid material or else be blocked by it, depending only on the orientation of the Polaroid. But if the phase of one wave lags the other by one-quarter of a wavelength, this blocking no longer occurs. The piece of Polaroid, no matter what its orientation, will pass just half the incident radiation. The electromagnetic wave is now said to be circularly polarized instead of plane polarized. The small change in phase has not particularly affected the mathematical character of the wave, but its interaction with the piece of Polaroid now drastically differs.

Classification of Observations

Any actually performed astronomical observation may be described in terms of the five parameters just discussed. We can

1. Report the spectral wavelength at which the observations were made
2. Specify the angular resolution obtained
3. Give the spectral resolution
4. State the time resolution
5. If the equipment also is sensitive to polarization, specify whether our observations tested for plane or for circular polarization of the received light

These parameters define the five orthogonal coordinate-axes of a five-dimensional space as listed in table 3.1. Any point contained in this space corresponds to a particular observation we can carry out, and conversely, any observation we might possibly conceive corresponds to one or more points that can be unambiguously located in the volume. We can call this space the *phase space of observations*. Sometimes we will also refer to a sixth dimension, intensity, shown in parentheses in the table.

TABLE 3.1
Phase Space of Electromagnetic Observations

Dimensions: Spectral frequency or wavelength
Spectral resolution
Time resolution
Angular resolution
Polarization
(Intensity)

Let us now see what a phase space represents.

The Phase Space of Observations

If we take a picture of the sky tonight, our photograph might include an unpretentious looking star about to explode as a supernova tomorrow, next year, or perhaps in the year 2119. At that time we would see an immensely bright source where an ordinary looking star had just been. One hundred years later, the bright source would have vanished, and a pulsar might be blinking in its place.

Each of these three phases of the star is best observed by quite distinct techniques. The pulsar phase is most readily detected by radio receivers capable of sensing pulses that last not much more than a few milliseconds. The supernova is most readily noted if we use optical techniques capable of showing changes occurring on a scale of a few days. The presupernova stage is least well understood but is likely to be comparatively quiet. It might best be observed by optical techniques that are sensitive to very slow changes occurring over centuries or millennia.

The detailed appearance of the sky seen from Earth changes from one millennium to the next because of the gradual motion of stars, appearance of new supernovae and pulsars, and new novae or new comets; but the average appearance of the sky stays the same. If there were five or six bright supernovae visible by eye in the last thousand years, there should be another five or six in the next millennium bright enough to be seen with the naked eye. If present-day methods lead us to find one or two new comets each year reaching Earth from the most distant parts of the solar system, then the same methods would lead to precisely the same rates of discovery a million years from now. The Galaxy and the solar system do not change very much over spans as short as a million years.

We can now think of a description of the universe that suppresses the positional changes of individual sources in the sky and instead culls the constant features that remain unchanged from one phase in the evolution of the universe to the next. The nineteenth-century American theorist J. Willard Gibbs first introduced such a description in chemical physics to show the conditions which exist in a vessel containing billions upon billions of gas molecules. The positions of individual molecules and their velocities will vary from one instant to another. The fraction of the molecules in any particular portion of the vessel, however, remains constant with remarkable precision and so does the distribution of velocities among the molecules. These traits do not change from what Gibbs called one *phase* in the evolution of the gas to another, even though all the individual molecules in the vessel may bounce from wall to wall changing directions and speeds as often as a hundred times a second. In a six-dimensional phase space, three coordinates can specify a molecule's position, the other three its instantaneous velocity, its speed and direction of motion. A point representing an individual molecule will move through this space from one phase in time to another. But the density of points in the space, corresponding essentially to what Gibbs termed the *density in phase* remains quite accurately constant, though the individual molecules involved in one measurement may be long gone by the time the next measurement is made.[4]

A similar constancy characterizes astronomical observations. If we set up radio telescopes to count the number of rapidly pulsing pulsars in the sky, we will discover about the same number, whether we search today or several million years from now. The density of pulsars observed will essentially remain constant from one phase in the evolution of the universe to another. We can therefore construct a phase space for astronomical observations in which we suppress detailed information about the absolute time and location of individual events and consider, instead, the density of different phenomena and their chief spectral, temporal, and spatial traits. The location or distribution of members of a class of phenomena in this phase space will be dictated by their characteristic structure, emission energy, spectral features, polarization patterns, and time variations. The density of the sources or events in the phase space will remain unchanged from one epoch to the next and will be the same for any observer. Only as the universe ages by billions of years, may long-term changes become apparent. For now we may neglect them.

Completeness

Once we are given the frequency of a light wave and have noted its location, its direction of travel, and its sense of polarization at a given time, we have completely specified all its attributes. If these traits are identical for any two photons, quantum mechanics tells us that the quanta have to be considered identical, for there are no other properties through which they might be distinguished. A photon is completely described when we specify a time, three positional and two directional parameters, one frequency, and one polarization parameter. An observation registering these eight traits yields all the information the observed photon can convey (figure 3.5).

In the five-parameter description provided by our phase space of observations, the three spatial coordinates of the photon are ignored. But these three coordinates simply tell the location of our observing equipment at the time the photon is registered; we do not need that information if we take seriously the idea that our astronomical results should largely remain unchanged no matter where we place the observing equipment.

In the phase space description the five remaining photon parameters translate into five phase space coordinates. The summation of individual frequencies for groups of photons give rise to new combinations (beats) that we observe as time variations when the beat frequencies are low, and observe as widening of spectral features when beat frequencies are high. The single parameter designating time therefore translates into two nearly orthogonal dimensions—time resolution and spectral resolution—in the phase space of observations (table 3.2). On the other hand, because we are not greatly interested in absolute positions in the sky, one angular or spatial resolution parameter suffices for the phase space description and replaces the two angular coordinates required

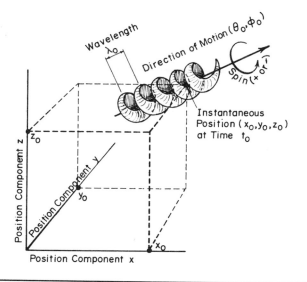

Figure 3.5 *Properties of a Photon*

A photon, or quantum of radiation, is completely specified if we provide eight pieces of information. Four of these are the photon's positional coordinates, (x_0, y_0, z_0) at the time t_0. Two more specify the direction (θ_0, ϕ_0) in which the quantum is heading at that time. A seventh property is the wavelength λ_0. The final trait which completes the characterization is the sense of the photon's spin—clockwise or anticlockwise. Photons have no additional traits by means of which they could be further distinguished. Two photons that are characterized by identical values of $x, y, z, t, \theta, \phi, \lambda$ and spin are indistinguishable from each other.

TABLE 3.2
Number of Orthogonal Dimensions for Electromagnetic Radiation

Properties of Photons	Phase Space of Electromagnetic Observations
1 Frequency coordinate	A. Spectral frequency (or wavelength)
1 Time coordinate	B. Time resolution C. Spectral resolution
2 Angular coordinates	D. Angular or spatial resolution
3 Spatial coordinates	
1 Spin parameter	E. Plane or circular polarization
Count of indistinguishable photons	F. Intensity

to describe the properties of a photon. Just as in table 3.1, a sixth dimension, intensity, can be added to the phase space. In chapter 4 we will compare the merits of these five- and six-dimensional versions of the phase space.

In the next few sections we will see that the phase space of observations is bounded for electromagnetic and, probably also, for all other types of radiation. The number of observations needed for a complete description of the radiation reaching us from space during a given epoch, therefore, can be expected to be finite, although large.

Energy Limits on Cosmic Photons

Most radio waves can pass through the spaces between stars almost unhindered, but extremely low-frequency radio waves are absorbed by the tenuous ionized gas which pervades the Milky Way plane. The absorption is essentially total at all frequencies lower than 100 kHz. At these frequencies the wavelengths exceed 3 kilometers and are longer than the radio waves in the long wavelength radio broadcast band. So strong is the absorption that the waves could never be transmitted even to the nearest stars (figure 3.6).

At extreme ultraviolet wavelengths between $\sim 10^{-6}$ and 10^{-5} centimeters, radiation also is strongly absorbed, this time by interstellar hydrogen atoms: The transmission within this band is wavelength dependent, and we may expect to see no further than 0.1 to 10 percent of the distance to the Galactic center, the precise distance depending on the frequency chosen.[5]

A similar limitation holds for photons at extremely short wavelengths—photons having the highest energies. At wavelengths shorter than 3×10^{-19} centimeters, a photon readily collides with a quantum of microwave radiation from the cosmic 3 degree Kelvin blackbody background. The two colliding photons give rise to an electron-position pair, and the original photon ceases to exist. This destruction of the high-energy photon is so likely to take place that a typical photon could hardly reach us from the center of the Milky Way, only 30,000 light-years away. The nearest galaxy comparable to our own lies 100 times further away than the Galactic center, and we certainly would not expect highly energetic radiation to reach us from that great a distance.

If the cosmic microwave background were the only source of background radiation, the prohibition against our detecting high-energy photons would not be quite complete. A positron-electron pair created by photons with significantly shorter wavelengths than 10^{-20} centimeters could effectively penetrate large distances through the blackbody background radiation, because the electron-positron pair can recombine to form a daughter photon that travels along the direction of the original photon and is only slightly less energetic.[6] The loss of energy for each transformation between photon and electron-positron pair is so small that photons of energies above 10^{18} eV—wavelengths below 10^{-22} centimeters—would readily penetrate distances comparable to those of the nearest large galaxy. At these high energies, however, a second more important source of destruction faces photons entering the Galaxy.

At radio wavelengths longer than 1 meter, a rapid increase in radio noise is perceived. It is strongest if we point our radio telescopes along the Galactic plane but diminishes only by a factor of 3 at higher galactic latitudes. This is the nonthermal radiation that Karl Jansky first discovered coming from the Galactic plane. At long wavelengths this background is far more powerful than the cosmic microwave radiation. On approaching our galaxy a photon with 10^{-21} centimeter wavelength would be met by a powerful blast of this radiation which would give rise to low-energy electron-positron pairs that soon dissipated the photon's energy. Even in intergalactic space, the nonthermal background radiation is sufficiently high to cause quite rapid destruction, and we can therefore be confident that we shall never detect extragalactic photons with energies exceeding 10^{15} eV and wavelengths below 10^{-19} centimeters. We will see later that similar limitations also hold for cosmic ray particles.

Spectral Resolution

A number of different factors limit the range of useful spectral resolution (figure 3.6). On the low resolution end of the scale we can usefully make photometric—intensity—measurements that cover a wavelength band $\delta\lambda$ as large as the wavelength λ itself. The resolving power $\lambda/\delta\lambda$ is therefore about unity. At the high resolution end of the scale we run into a variety of differing limitations. One or the other usually acts to keep the useful resolving power below 10^8—a resolution of one part in a hundred million. First, there is the natural line width, a broadening of any spectral line that is a fundamental consequence of Heisenberg's uncertainty principle. In the visible part of the spectrum, this limits the useful resolving power to $\sim 10^8$. However, thermal and turbulent Doppler broadening due to motion of the radiating matter within a cosmic source, or collisional broadening produced by encounters between radiating atoms, usually diffuse spectral features to such an extent that observations of much lower resolving power already suffice. For radio astronomical lines these secondary factors generally appear dominant.

Time Resolution

In the time domain, we can expect a range of variations stretching all the way from the apparent age of the universe—some 10 billion years or about 3×10^{17} seconds—down to the length of time it takes light to traverse the dimensions of an emitting source (figure 3.7). Faster varia-

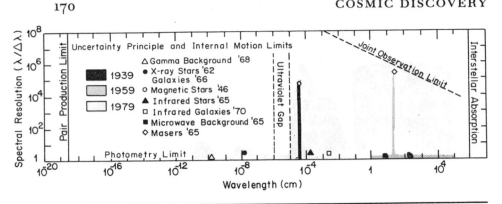

Figure 3.6   *Spectral Resolution Available in Different Years*

In this figure, and in the three figures to follow, we see a display of instrumental capabilities over a wavelength range stretching from 10^{-20} centimeters at the extreme gamma-ray end of the electromagnetic spectrum, to 10^6 centimeters (10 kilometers) long radio waves at the opposite extreme.

A wavelengths short of 10^{-19} centimeters, photons are destroyed through collisions with the cosmic microwave background radiation which produce electron-positron pairs. This defines a *pair-production limit*—the left boundary of the diagram. Beyond 10^6 centimeters, radio waves are so thoroughly absorbed by the interstellar ionized gases that we have no chance of seeing sources even as close as the nearest stars. This is the *interstellar absorption* limit that defines the right boundary. In the extreme ultraviolet portion of the spectrum, bounded by the two dashed vertical lines at 10^{-6} to 10^{-5} centimeters, interstellar absorption is similarly strong. This is the wavelength range labeled *ultraviolet gap*.

Instrumental capabilities available to us in 1939 are shaded dark. They deal mainly with the visible and near infrared parts of the spectrum, but there is a small shaded area at radio wavelengths that represents Karl Jansky's early radio efforts. By 1959 (lighter shading) instrumental capabilities had stretched over considerable portions of the radio spectrum, and by 1979 (lightest shading) infrared, X-ray, and gamma-ray observations had become possible. In the wake of these instrumental advances, new cosmic phenomena crystalized. The capabilities that made their discoveries possible are shown by specific symbols explained in the key. The dates that appear in the key show the prime historical trend: Methods for observing the gamma-ray background radiation, discovered in 1968, simply did not exist nine years earlier; nor did we then have the capabilities for detecting bright infrared galaxies discovered around 1970. Most of the phenomena discovered in the 1960s and 1970s appear in areas outside the shaded portions representing the capabilities for 1959. Infrared stars, discovered in 1965, could have been found with the spectral resolution available in 1939, but until well after World War II, we lacked detectors that had sufficient sensitivity.

Unshaded portions of the diagram show regions in which no instrumental capability exists. The diagram represents our competence in detecting cosmic signals; sometimes higher capabilities exist for solar observations made relatively easy by the sun's proximity.

The lower border is the limit of photometry, a limit in which we measure the energy received from sources in the sky, without knowing more than rough wavelength characteristics of the radiation. The upper border is approximately defined by two constraints. One is the natural width of spectral features defined by the Heisenberg uncertainty principle. The other is the width that is induced by relatively small motions within the emitting source. Both of these features tend to make a resolving power of 10^8 adequate for most sources. A second upper bound, related to the uncertainty principle, appears as the sloping dashed line at the upper righthand corner. Above that line it becomes impossible to obtain observations, unless a substantial sacrifice is made in time resolution. This limitation arises because spectral frequency can only be determined with limited accuracy in a finite observing time. We encounter a similar limitation in figure 3.7.

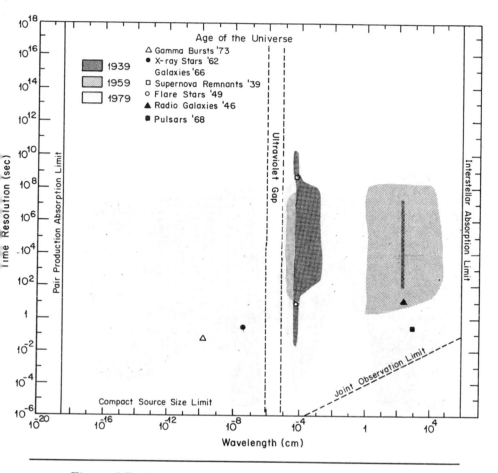

Figure 3.7   *Time Resolution Available in Different Years*

This plot is similar to figure 3.6, but deals with our ability to resolve astronomical signals on different time scales.

At the top of the figure, we are bounded by the age of the universe. Any evolution or change that has occurred must have taken place on a time scale that cannot well exceed this age. At the bottom, there is a limit of a microsecond. This might correspond to signals we could receive from the most compact sources in the sky—sources that are not more than a fraction of a kilometer in size and, perhaps, rotate rapidly. There is a further limit previously mentioned in the caption to figure 3.6. It precludes high spectral resolution on short time scales. Observations below the dashed line in the lower right portion of this figure are incompatible with observations above the dashed line of figure 3.6.

The lightly shaded areas show the rapid advances made between 1959 and 1979. Further progress may, however, turn out to be much slower. Observations lasting several thousand years may be needed just to fill in the portion of the diagram, ranging up to the level of 10^{11} seconds (3,000 years). Slow changes can only be observed by carrying out measurements for long periods. The upper portions of the diagram may therefore remain blank for millennia or longer.

tions than those can usually be ruled out because if an entire source is to start or stop radiating, some message to effect this change has to be transmitted across the source; and that cannot happen at a speed faster than the speed of light. The smallest sources we need to consider are about 1 kilometer in diameter, so that the fastest time variations of concern are about 3 microseconds. An exception to this general rule might be found in rapidly rotating sources that emit a narrow beam of radiation. Such a searchlight beam could sweep across our telescope more rapidly. A second, more stringent limitation may then be invoked: The observed signal modulation frequency never can exceed the frequency bandwidth of the radiation, or ultimately the radiant frequency itself.

Angular Resolution

At least one source of radiation covers the entire sky. The microwave background radiation appears to be completely isotropic and is detected when we are able to suppress our sensitivity to all other radiating sources detected at higher angular resolution.

At the other end of the scale, a number of distinct high resolution limitations can be enumerated. Angular resolution is limited by the ratio of wavelength, λ, to instrumental size, or to separation, D, between receivers. With instruments located on Earth, which are the only kind that currently can be obtained in practice, the smallest angle that can be resolved is of order $\lambda/(3 \times 10^9)$ radians, where the wavelength, λ, is measured in centimeters. By this we mean that an angle of one radian—roughly one-sixth of a full circle—can be resolved into $(3 \times 10^9/\lambda)$ distinguishable parts. This restriction is imposed by the maximum distances $D \sim 10,000$ kilometers over which earth-bound apparatus can be deployed. If a baseline of the order of the solar system is used instead, at some time in the future, the smallest resolved angle will be thirty thousand times smaller—$(\lambda/10^{14})$ radians (figure 3.8).

Whether angular resolution so refined can be fully utilized will depend on a number of practical factors. For example, current theory indicates that luminous Galactic objects are unlikely to be smaller than the Schwarzschild radius of a star having the sun's mass. If the sun were to be compressed and became smaller and smaller, it would ultimately reach this radius—roughly 1 kilometer, or a few city blocks. At this stage its self-gravitation would be so great that it would irrevocably collapse into itself and no particles or radiation could escape its surface. At distances of the order of the Galaxy's diameter, a kilometer-sized object would subtend an angle of $\sim 10^{-18}$ radians, and at greater distances we probably would not detect these objects at all because they would be so faint.

At extragalactic distances, we might expect collapsed objects with the masses of galaxies and Schwarzschild radii of the order of 10^{16} centimeters. Located at the furthest reaches of the universe, such an object would subtend an angle of the order of 10^{-12} radians.

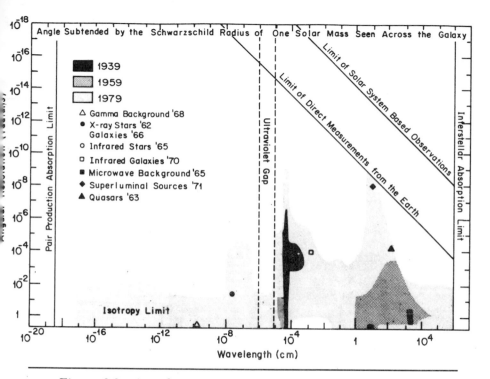

Figure 3.8 *Angular Resolution Available in Different Years*

This diagram is similar to figures 3.6 and 3.7 but illustrates the bounds placed on angular resolution in the electromagnetic wave domain.

The angular resolution to which we have access is limited by isotropy—insensitivity to any angular structure—at the bottom of the diagram. At the top of the figure, a variety of limits compete. The smallest observable source might be a black hole in distant parts of our galaxy. Two diagonal lines show the limits, respectively, for observations with instruments tied to the earth and with apparatus confined to the dimensions of the solar system. To improve on these capabilities, we would have to send instruments into interstellar space. Particularly at short wavelengths, such drastic measures may not be worth while, since the smallest sources we might expect to detect are likely to be resolved with equipment that might be placed in Earth orbit.

This plot, also, shows the rise in observing capability in the four decades between 1939 and 1979 and the rapid discovery of novel phenomena in the wake of instrumental advances.

Polarization Analysis

An electromagnetic wave may exhibit circular or plane polarization, a mixture of these states, or no systematic polarization at all. Capabilities for observing polarization at different stages in the development of astronomical techniques are shown in figure 3.9.

Sensitivity Improvement

In most portions of the electromagnetic spectrum, enormous improvements in sensitivity have come about since 1959. The early 1960s

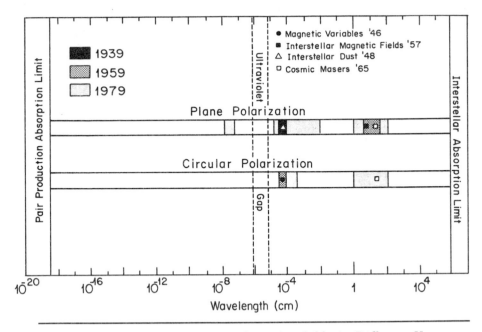

Figure 3.9   *Polarization Analysis Available in Different Years*

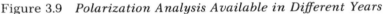

This figure uses the same format as figures 3.6 to 3.8.

Plane polarization and circular polarization usually are detected by rather different instrumental means, as indicated by the historical trends indicated by the shadings. Circular polarization has been more difficult to sense and has lagged in development. Although some attempts were made to observe polarization before World War II, success was achieved only in solar work where magnetic field strengths were measured with equipment sensitive to circular polarization. Not until after the war, however, were cosmic observations truly successful. Then, a series of discoveries involving interstellar dust, interstellar magnetic fields, magnetic stars, and masers all became possible as a result of increasing ability to detect polarization.

Figure 3.10   *Sensitivity Improvement Obtained Through the Use of Telescopes, Photographic Plates, and Electronic Devices*

The first technical improvements in astronomy beyond visual observations with the unaided eye were independently introduced by Harriot in England, Simon Marius in Germany, and Galileo in Italy. All three started observations in 1609, but Galileo's instrument was the most powerful. The second half of the seventeenth century saw the design of larger lenses, an improved eyepiece by Huygens, and the use of long-focal length (aerial) telescopes that provided higher spatial resolution by avoiding chromatic aberration. By the mid-eighteenth century, reflecting telescopes started to show improvements. James Short in England pioneered the construction of sizable speculum mirrors. His largest reflector had a mirror whose diameter measured 21.5 inches. It was built for Lord Spencer in 1742, and like most of Short's other instruments never appears to have been seriously used for astronomy: A 1768 prospectus advertises an 18-inch reflector for sale at 800 guineas, a price that only rich customers could afford. By 1789, William Herschel had built his 4-foot diameter speculum reflector fitted into a 40-foot-long reflecting telescope. This was by far the largest aperture telescope that had ever been built and was to be exceeded in size only when Lord Rosse installed a 72-inch reflector at Parsonstown in Northern Ireland in 1845.

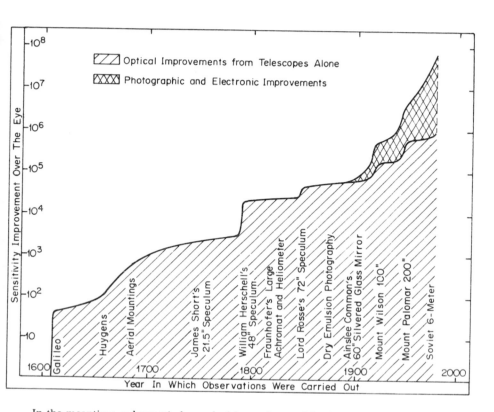

In the graph:
- Y-axis: Sensitivity Improvement Over The Eye (10^8, 10^7, 10^6, 10^5, 10^4, 10^3, 10^2, 10)
- X-axis: Year In Which Observations Were Carried Out (1600, 1700, 1800, 1900, 2000)

Legend:
- ▨ Optical Improvements from Telescopes Alone
- ▨ Photographic and Electronic Improvements

Labels: Galileo, Huygens, Aerial Mountings, James Short's 21.5" Speculum, William Herschell's 48" Speculum, Fraunhofer's Large Achromat and Heliometer, Lord Rosse's 72" Speculum, Dry Emulsion Photography, Ainslee Common's 60" Silvered Glass Mirror, Mount Wilson 100", Mount Palomar 200", Soviet 6-Meter

In the meantime, achromatic lenses had been pioneered by John Dollond and his son, Peter, at their optical works in London. By the early nineteenth century, large enough blanks could be made to permit Fraunhofer the construction of a large refractor for the astronomer Struve at the Dorpat Observatory and a large heliometer for Bessel at Königsberg. These refractors had smaller apertures than the speculum reflectors, but their upkeep was not as difficult, and their light losses were smaller. One of the first, large silvered glass mirrors was introduced into astronomy in 1870 by Henry Draper, a medical doctor and physiologist in New York. The reflectivity of these mirrors was much higher than that of speculum, and they were well suited for photographic work which was limited by absorption in refractors. Ainslee Common's 60-inch mirror was the largest of this kind built in the nineteenth century, and all large telescopes subsequently built were fashioned along these lines.

Although a daguerreotype of the sun had been obtained by the French physicist-astronomers Foucault and Fizeau as early as 1845, astronomical photography was not widely accepted until the silver-bromide-gelatin-emulsion dry plates were introduced in 1871.

Since World War II, highly improved emulsions have had to compete with photosensitive electronic devices. The diagram shows improvement in sensitivity in observations of individual stars. The best current combination of telescope and photosensitive electronic device can see a star 100 million times fainter than the unaided eye. Not shown in this diagram is the improved information flow obtained because photographic plates simultaneously record images of thousands of stars, while the eye tends to register the magnitudes of individual stars, one at a time.[7]

saw the start of X-ray and far-infrared astronomy from the most modest foundations. Gamma-ray astronomy became possible in the late 1960s. Radio astronomy, which had experienced a hesitant start in the 1930s, came into its own during the years immediately following World War II, and increases in sensitivity continue. For optical astronomy, in contrast, a gradual improvement in sensitivity has taken place over a period exceeding three centuries. This development is shown in figure 3.10.

Summary of Limitations on Photons

While some of the limits just cited might be challenged, let us assume that they are at least approximately correct and that the phase space of possible observations is finite for electromagnetic waves and confined to:

Wavelength range:	3×10^{-19} centimeters to 3×10^5 centimeters
Spectral resolution:	1 to 10^8
Time resolution:	3×10^{-6} second to 3×10^{17} second
Angular resolution:	2π to 10^{-18} radians except at the longer wavelengths where Earth's diameter or Earth's orbital diameter limit the resolution respectively to about $\lambda/(3 \times 10^9)$ or $\lambda/10^{13}$, and λ is measured in centimeters.

Figures 3.6 to 3.8 show these limitations. They also give an idea of the capabilities of currently available apparatus and of the development that has taken place in observing techniques in past decades. These drawings have been compiled by gathering results from the widest variety of astronomical observations. Many of the capabilities might be in some dispute, primarily because technological advances have been rapid, particularly in the gamma-ray, X-ray, infrared, and radio domains, and not all observers would be in complete agreement on the practical accomplishments of different techniques. Nevertheless, the figures probably provide data which are sufficiently reliable for present purposes.

The capabilities shown refer to Galactic and extragalactic observations. Solar astronomers who deal with a bright nearby source often have been able to extend techniques substantially further; but the discussion here concentrates on the universe beyond the solar system.

The figures show a number of limiting bounds which correspond to the natural or instrumental constraints just discussed. Figures 3.6 and 3.7 also contain lines labelled "Joint Observation Limit." This line is drawn to emphasize the impossibility of a search for high frequency time variations whenever we begin to approach an instrument's bandwidth, W, or ultimately the spectral frequency, f. This can happen most readily in the radio domain where the spectral frequencies are low. An attempt to obtain both high-spectral and high time resolution simultaneously violates Heisenberg's uncertainty principle. If we simultaneously took the highest spectral resolving power, say 10^8, and the shortest time resolution, say ~3 × 10^{-6} seconds, then we would be attempting conflicting observations if we were to carry the measurements out at wavelengths longer than 10^{-3} centimeters. These limitations are roughly taken into account by the dashed lines in figures 3.6 and 3.7. We may undertake only those time resolution studies above the dashed line in figure 3.7 and spectral resolution studies below the dashed line in figure 3.6; otherwise we run into trouble with the uncertainty principle.

Cosmic-ray Particles as Information Carriers

Highly energetic protons, neutrons, electrons, positrons, and a wide variety of heavier nucleons constitute the cloud of Galactic cosmic-ray particles. Antinucleons, consisting of antimatter, have been sought, and recently antiprotons have been found. Unstable subnuclear particles are not expected to figure significantly as carriers of cosmic information because their short lifetimes preclude travel over long distances in the universe, unless the most extreme energies are assumed.

If neutrons exist at energies of the order of 10^{20} eV, they must suffer such a strong relativistic time dilation—slowing down of the flow of time—that they can travel a distance of three million light-years before suffering decay. As seen by the neutron, its life is still restricted to about a thousand seconds, but to a stationary observer on Earth the lifetime of the energetic neutron would appear to be three million years. Highly energetic neutrons could therefore reach us from as far away as the nearest galaxies. But lower energy neutrons or shorter-lived particles could not reach us because they would decay too rapidly. Neutrons of higher energy also could not be transmitted over greater cosmic distances even though at 10^{21} eV the neutrons left to themselves could travel a distance of 30 million light-years without decaying. At these high energies, however, the probability of destructive interaction with individual photons of the microwave cosmic background radiation becomes high, and 30 million light-years becomes the maximum distance any neutron could cover even at the highest energies. With a few conceivable exceptions, neutrons therefore are not likely to be useful as carriers of cosmological information.

Protons at their highest energies also cannot reach us from distances greater than 30 million light-years. The protons and other charged particles, however, also have other limitations as information carriers. At lower energies they are strongly bent by passage through cosmic magnetic fields. Their directions of arrival or time of arrival, therefore, tell us little about the sources of origin. The particles approach us essentially isotropically and with no recognized time variation (except for a variation produced by activity of the sun). The energy of the particles also is altered by passage through cosmic electromagnetic fields; consequently we cannot expect to uncover any finely structured energy features either. As carriers of cosmic information, the cosmic rays, therefore, are more restricted than electromagnetic radiation. However, they can carry some nucleo-chemical information, and they can carry coarse energy information. Both of these are interesting data.

At energies about 10^{11} eV, cosmic-ray particles have some difficulty diffusing into the solar system to reach Earth, since the solar wind carries outflowing magnetic fields in which the low energy cosmic-ray particles can become enmeshed and swept away. Particles below this energy, therefore, also are not especially useful carriers of information to Earth, although they may well provide interesting information once we have spacecraft that leave the solar system and make measurements from interstellar space.

To summarize the limitations on cosmic-ray particles as information carriers: The energy range of Galactic, or possibly extragalactic cosmic-ray particles that we can hope to use as carriers of information lies between about 10^{11} eV and 10^{21} eV. But most of the stable cosmic-ray particles are charged, and therefore studies at high angular resolution, at high spectral (energy) resolution, or at high time resolution are not likely to yield a great deal of information because Galactic magnetic fields scramble all memory of where the particles originated and make uncertain the time of origin, as well as the initial energy of the particle. Elemental and isotope abundance studies for the more massive nuclei may, on the other hand, tell us about the processes that gave rise to the cosmic rays in the first place and may also elucidate the subsequent interactions they suffer with matter in interstellar space.

Other Carriers of Information

We still know too little about neutrino astrophysics and gravitational waves to set up any significant bounds on the phase space of observations for these two carriers. A variety of theoretical discussions do suggest limitations, but since neither type of carrier has ever been detected in astronomical observations, we do not even have a good understanding of the expected magnitude of an eventually observed flux. One interesting data point that we do have is an upper limit to an observable intensity or actually to the energy density in space that these carriers could be contributing. The general theory of relativity predicts that if there were a great deal of neutrino or gravitational radiation, the cosmos would be appreciably curved. However, we find that light waves do not significantly deviate from straight-line trajectories in our universe, and this observation yields a weak upper bound to the energy density of unseen carriers of information. Right now, the energy density of gravitons and neutrinos required to cause an observable curvature of space is much higher than the energy density of all electromagnetic radiation known to occupy the universe; so this upper limit does not tell us very much. But in the future, we may be able to cite more significant bounds.

For meteorites the problem is somewhat different. Here the limitations on information that can be derived are somewhat similar to those that hold in laboratory experiments and are more difficult to define. We should, however, note that meteorites appear to bring us information solely about the solar system. If similar particles or chunks of matter occasionally reach us from interstellar space as well, we have not yet learned to recognize them.

The Logarithmic Phase Space

Why should the phase space of observations constructed here exhibit logarithmic scales—scales labeled in powers of ten—as shown in figures 3.6 to 3.9? Would a linear scale not have done just as well?

Let us look at a spectrum which exhibits a spectral line whose width, W, is to be measured. We may choose to analyze the line by means of a number of different spectral filters (figure 3.11).

The best possible filter for this particular observation would have had a width, W, because it transmits all the light in the line and no extraneous radiation. But in general an observer's collection of filters may not include one with precisely the bandwidth, W, and he must then choose that filter which comes closest to optimum performance—in figure 3.11 either the filter with width $W/2$ or the filter with width $3W/2$, depending on whether there is strong added continuum radiation or not.

In astronomy we never know in advance what line widths might characterize a novel phenomenon. In making new searches, therefore, we may simply resign ourselves to a choice of a standard filter set in which no one filter is necessarily of optimum bandwidth. If these filters have bandwidths that systematically increase by factors of 3, as in the example above, we may do well enough. At least one filter can then always be found that performs at a level better than 57 percent of optimum, and we may decide that that is good enough. If, on the other hand, we require better performance, we can choose a set of filters whose widths are still more closely matched. Their bandwidths might be 0.75 W, 0.83 W, 0.91 W, W, 1.1 W, 1.2 W, and so on, each filter having a width 10 percent wider than the previous one. In that case there will always be a filter that comes within at least 95 percent of optimum performance, no matter what the width of our spectral feature might be.

Looked at in this way any given criterion for expected performance relative to an optimum requires observations made through filters whose widths are related by constant ratios. Such a series of bandwidths when plotted on a logarithmic scale defines evenly spaced intervals. To probe for new phenomena exhibiting as yet unknown bandwidths, with the use of a set of filters that guarantees coming within a certain range of optimum performance, we therefore require probings at a series of resolving powers that are evenly spaced on a logarithmic plot. The space of all possible observations with predefined quality of performance therefore is properly defined by a logarithmic scale. The higher the required performance, the finer the spacings between adjacent points at which observations must be carried out to cover the entire volume.

We have only discussed spectral resolving power here, but the same argument also applies to angular and time resolution.

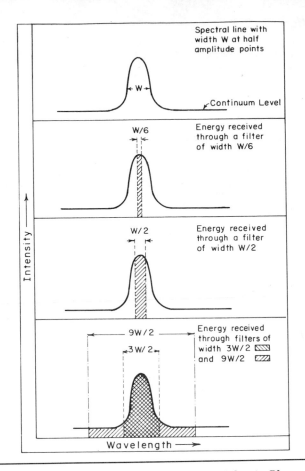

Figure 3.11 *Matching Filters and the Logarithmic Phase Space of Observations*

A spectrum is a plot of light intensity at different wavelengths. In a given part of the spectrum the intensity usually consists of a broad continuum on top of which one or more spectral lines may be superposed. A well-matched filter may permit us to readily detect a weak spectral line against a strong continuum base. The width of the spectral line, W, is the separation, in wavelength, between points at which the intensity of light given off in the spectral line falls to half the value seen at peak intensity (panel 1). If we look at a line having width W through a narrow filter that transmits only that portion of the light which lies in a wavelength range of width $W/6$ (panel 2), we receive only a small part of the radiation in the line. If we increase this width threefold (panel 3) and choose a filter of width $W/2$, we receive about half the light seen in the spectral line, and we also triple the amount of continuum seen. By tripling the filter width in successive steps, we receive increasing fractions of the total intensity in the spectral line, but we also see a successive tripling of the continuum radiation seen through the filter. The filter having width $9W/2$ is no longer optimum because it transmits only marginally more line radiation than a filter of width $3W/2$ and does not reject the continuum well enough. An optimum choice is a filter with a width somewhere between $W/2$ and $3W/2$. For these widths, a good fraction of the line radiation is transmitted, but most of the continuum is rejected, and the line is more readily seen against the continuum.

Looked at in this way, we see that performance criteria, defined relative to some optimum, require observations through filters whose widths bear a specific ratio to the wavelength of the radiation actually observed. Since it is the ratio of filter width to wavelength that is ultimately important, a logarithmic scale is the most appropriate choice for plotting the phase space diagrams (figures 3.6 to 3.9). This choice also corresponds to the information content of an observation which depends on the logarithm of the resolving power, as explained in connection with table 3.3 below.

The wavelength scale is also logarithmic in the plots we show and for similar reasons. The minimum spectral resolution we define at any given wavelength in our plots is unity. By this we mean that we simply take all of the radiation available and measure its intensity. If we choose the bandwidth, W, to equal the central wavelength, λ, in the band, then the band will cover all wavelengths from $\lambda/2$ to $3\lambda/2$.

With $W = \lambda$ we can now define a series of nonoverlapping wavelength bands centered on $\lambda/9$, $\lambda/3$, λ, 3λ, 9λ, and so on, each of which has a width equal to its central wavelength. When plotted on a logarithmic scale, these central wavelengths are evenly spaced. If we choose central wavelengths defined by this series of values and carry out observations at logarithmically spaced spectral, spatial, or angular resolution values over each wavelength band, our coverage will include all conceivable observations, and the quality of the data will be uniform throughout.

A final question that may arise concerns the selection of wavelength, λ, to define the horizontal coordinate scale in our diagrams. How is the relationship between wavelength and spectral resolution different from the relationship between the actual time at which we observe and the time resolution? The answer to this again derives from what we already know about the universe. Our universe appears to be isotropic, homogeneous, and slowly evolving. Statistically this means that we observe pretty much the same kind of behavior whether we look north or south in the sky, and we would observe phenomena rather similar to those seen locally if we went to a galaxy five hundred million light-years away. We also see rather similar behavior this week as last. The absolute time, and the absolute direction in which our telescope is pointed, therefore, is not terribly important. We see the same phenomena, statistically speaking, wherever and whenever we look. In the wavelength domain, however, this is no longer true. We see quite different phenomena when we go from one wavelength to another that is several thousand times longer or shorter. Different physical processes are observed in these widely spaced wavelength ranges, and the wavelength therefore appears as a dimension in our phase space of observations, whereas absolute time does not.

The Range of Intensities

The intensity of radiation that we can expect to observe in a new range of spatial, spectral, or time resolution is limited both by an upper and a lower bound.

The upper bound is quite simply obtained because by now we have measured the total amount of radiation impinging on Earth from the cosmos in virtually every spectral range. This energy is measured through the use of receivers that gather radiation incident from large portions of the sky. No attention is paid to the question of whether the radiation comes from stars, clouds of gas, or giant galaxies, whether it comes in short bursts or continuously, and whether it is monochromatic or covers a broad part of the spectrum. We simply measure the total amount of radiation that we receive; and, if we assume that the flux is fairly constant in time, this measurement provides an upper limit to the radiation we may expect in any chosen range of wavelength, angular size, or time duration.

Similarly we can place a lower bound on the amount of radiation we can expect to measure as we embark on new varieties of observations. This limitation arises from two quite distinct factors, *noise* and *confusion*.

There are many different kinds of noise that can hinder a given observation. Every radiation detector produces at least an occasional spurious signal, even if it is in a completely darkened box where we know that no radiation is incident. This *dark noise* arises because the atoms, ions, or electrons that make up the detector are in continuous thermal motion. Occasionally there is a temperature fluctuation in a submicroscopic portion of the detector, and this may have all the earmarks of having been produced by an incident quantum of radiation. It may produce a measurable current through the detector, and the electronic discriminator may be unable to distinguish whether the current was produced by a thermal fluctuation or by an incident photon.

By cooling the detector we can lower the thermal noise, but other sources of noise then begin to dominate: The detector may have some sensitivity to cosmic rays, since these highly energetic particles rain down on our atmosphere in a random patter, each particle giving rise to a shower of secondary electrons, nuclei, and mesons as it traverses the atmosphere. When such a secondary particle hits the detector, all kinds of spurious signals may result. This source of noise can be minimized by shielding the detector with absorbing material, but some of the secondary particles have strong penetrating power, and shielding beyond a certain limit becomes impractical.

Before an observation is started, the dark noise of the detector can be measured, and this specifies the minimum signal measurable in a given observation. It defines a lower bound on the brightness of novel phenomena we may expect to detect during the available observing time.

Not all sources of noise are related to the detector. The universe itself also adds to the noise. Contributions arise in differing ways in different parts of the spectrum. In the visible range the performance of a telescope mounted in an earth-orbiting satellite is impeded by the zodiacal glow. The zodiacal light is not perfectly steady but varies slightly as different grains of dust enter or leave the field of view or turn this face or that toward the sun. When our detector registers a signal suggesting the discovery of a faint new source, we therefore must be careful to exclude the possibility that we have merely observed an undulation in the glow coming from the zodiacal dust cloud.

Even if noise is a negligible hindrance, we must still worry about confusion. Confusion arises because the universe is so rich in sources that a slight spillover from a source just outside our field of view may be mistaken for a signal from the source we are trying to observe. Because of the diffraction of light, again related to the Heisenberg uncertainty principle, we are unable to define with absolute precision the direction from which a photon approached just before it entered our telescope. We may therefore think we are observing a single star, but instead we might be gathering radiation also from several adjacent stars in the sky. In the radio domain, where wavelengths are very long, this effect is strongest, and radio astronomers must take the greatest pains to exclude false signals produced through confusion of neighboring sources in the sky. If our universe were more sparsely populated by sources, this limitation would drop off, and the complexity of equipment required to unravel the confusion would diminish. As it is, the cost of building equipment immune at least to the most blatant forms of confusion can be quite high since complex sets of measurements are required if false signals are to be excluded.

If we take the total received power to be an upper bound, and the expected noise and confusion to define a lower bound, we can estimate the range of intensities that might be expected to span all possible observations. The upper bound will be the same for observations at any resolution, since this upper limit is simply the total radiation received from the general portion of the sky we plan to observe. The lower bound may vary, depending on the resolution, provided the noise and confusion levels change for different resolving powers. Typically, however, the range of signals lying between the upper and lower observable bounds in any spectral domain and at any resolution will not exceed twelve powers of 10—a factor of 10^{12}. As elsewhere in this discussion, however, we exclude the sun, the earth, and the moon which are far brighter than typical stars, typical planets, or typical satellites only by virtue of their exceptional proximity.

The minimum signal cited in this context is taken to be some absolute minimum related to cosmically produced noise as measured from the earth or earth-orbit—foreground or background emission from sources in or near the field of view, penetrating cosmic ray bombardment, and so on. Some types of noise could be reduced by traveling outside the solar system where noise from solar cosmic-ray particles or from zodiacal glow fluctuations is cut down. New lower bounds could then be achieved, but in many cases the gains are only small. Galactic cosmic-ray fluxes persist even after we leave the solar system, and a faint glow from distant stars and galaxies and from light scattered by interstellar dust replaces the zodiacal dust glow as a new limiting source of noise.

In visible light, the brightest star in the night sky is Sirius—a minus 1st magnitude star. The faintest stars or galaxies currently observed are about 26th magnitude objects. Not far beyond this faintness we reach the point at which distant galaxies start to overlap because there are so many of them. Then it no longer is clear how foreground objects are to be distinguished from this matrix of galaxies near the cosmic horizon. Luminous matter at these distances melds into a patchy blur.

The brightest planet, Venus, at maximum angle from the sun can be 10 times brighter than Sirius. The light we receive is then close to the total brightness of all the sources in the night sky put together, about 10^{12} times brighter than the light received from the faintest galaxies we can observe.

Precision, Resolving Power, Stray Light Rejection, Noise

Friedrich Wilhelm Bessel determined the parallax of the star 61 Cygni in 1838 and six years later published his determination of the orbital motions of Procyon and Sirius. To arrive at these results he had to observe the positions of stars with a precision measured in fractions

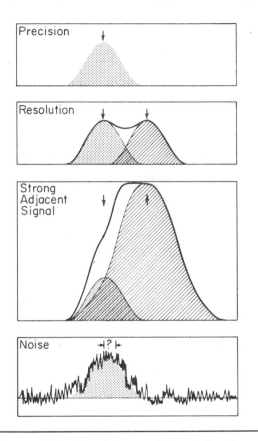

Figure 3.12   *Precision and Resolving Power*

In spectroscopy, precision represents the accuracy with which we can name the exact wavelength of a spectral line (top panel). Spectral resolution, in contrast, represents the capability of measuring a separation between two closely lying spectral lines of comparable intensity (second panel). Similarly, in astrometry, precision reflects our ability to accurately measure the position of a star, while angular resolution specifies our capability for distinguishing two closely spaced stars of comparable brightness from each other. When a faint star lies very close to a brighter companion, or a faint spectral line lies near a far brighter feature (third panel), the two may not be separable with a given resolving power, even though a pair of equally bright stars or spectral lines would readily be resolved at comparable separation. The brighter companion, in this case, simply appears to merge with the fainter, as shown by the heavy solid curve, and a higher resolving power is required to separate the two.

The precision of which an instrument is capable may be substantially greater than its resolving power. If the source of radiation is known to be symmetric, its exact center can often be discerned even though the source by no means is resolved. The effective limit on precision usually is dictated by noise that interferes with the observation. If a scan of the brightness of a source ideally has the shape shown in the top panel, noise in the apparatus, or from the sky, may actually give a result approximated by the jagged trace shown in the last panel. Now the center of the trace no longer is well defined. The uncertainty is of the order of the separation between the arrows embracing the question mark. This is the limitation on precision produced by noise in the observation.

of a second of arc. The annual parallax that Bessel announced for 61 Cygni—half the star's annual displacement in the sky—amounted to 0.3136 seconds of arc. Two years later Bessel published a revised value of 0.3483 seconds, and these values may be contrasted with today's accepted value of 0.294 seconds of arc.[8] Bessel's results, therefore, appear to have had a precision of about $\frac{1}{20}$ of an arc second.

We know today that Bessel's telescope could rarely have offered him a stellar image less than 1 second of arc in diameter, and his resolving power generally must have also been poorer than 1 second of arc. However, as explained in figure 3.12, the precision of a measurement often exceeds resolving power by one or two orders of magnitude; and that is consistent with Bessel's results.

Phase Space Filters

Whenever we uncover a new way of viewing the universe, we effectively are observing through a brand new kind of filter (figure 3.13). This filter may not just be a device which isolates different colors. Instead it may isolate a certain portion of the phase space and specify a range of resolving capabilities that characterize the observation.

We will call a device a *phase space filter* if it has the following properties: Within a certain spectral frequency range it passes all the radiation that could be detected for a given state of polarization; and from this radiation it selects for transmission only those features that lie within a given range of time resolution, spectral resolution, and angular resolution or source size. These five traits correspond to the five dimensions of the phase space of observations. We might add, as a sixth dimension, a range of source intensities which the filter selectively rec-

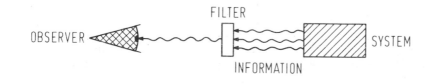

Figure 3.13   *Filter*

A filter is a device that systematically sorts radiation into two components. One component is blocked; the other is transmitted, ideally without distortions of any kind, from a system that is being studied to the observer seeking information.

The transmitted component usually consists of a range of wavelengths or frequencies. The width of that range is the bandwidth. While filters that pass a certain band of electromagnetic frequencies are the most often used devices, there also exist spatial filters that are tuned to pass only those spatial components in a scene that have a certain spatial frequency band. The action of such a filter is somewhat like that of a sieve. It transmits those components of a scene that have a selected coarseness of structure. A third kind of filter is a polarization analyzer. It can be used to highlight features in a scene that are plane or circularly polarized. Finally, we can construct filters that are tuned to transmit only those time variations in a signal that are particularly interesting or block other components that are undesirable.

The action of filters of this type is demonstrated in figure 3.14. Figure 3.15 describes a phase space filter, a device which combines spectral, spatial, time, and polarization filters in one.

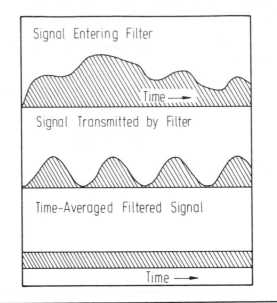

Figure 3.14 *Action of a Filter*

A filter that selects for transmission only a certain set of spectral frequency components or time variations from a signal works as illustrated here. The top panel shows the variation in the signal's amplitude as time elapses. The filter selects only that portion of the time variation that has a certain frequency (second panel). This is a frequency in a range around four cycles in the time frame shown. The filtered portion of the signal is transmitted to the observer, while the rest is blocked.

Normally, the observer wants to know the amplitude of the transmitted component. He can obtain that by allowing the filtered signal to pass through some kind of an averaging device, usually an electronic instrument, which averages over the crests and troughs in the wave (bottom panel) and provides the mean amplitude of the selected frequency component in the original signal.

ognizes. Figure 3.14 shows a trace which represents the signal strength at different times. It also shows the frequency to which the filter responds and the output of the filter as the signal passes through it.

A spectral frequency filter as used here is similar to an electronic filter in a radio. It seeks out and detects specific structural components in the spectrum transmitted through it. It is not what is usually called a spectral frequency filter in optics: In optics a filter is tuned to a specific wavelength. In our phase space application, it selects features at any wavelength we choose to examine that have a given spectral structure— a coarseness or regularity of spaced features at wavelength intervals specified by the filter width.

In general, a phase space filter appropriate for ready detection of a cosmic signal must be matched to the angular, spectral, and time variations of the source at the wavelength of maximum emission. Such a filter appropriate for the detection of the magnetic variable stars mentioned on page 112 picks out point sources against a background of diffuse galactic radiation. It is sensitive to circular polarization since the observed effect is discerned only if we can discriminate between left- and right-circularly polarized light; it is sensitive to spectral changes since we are looking for a wavelength shift of left-circularly polarized light relative to right-circularly polarized light; and it is sensitive to time variations since we are searching for a wavelength displacement of circularly polarized components as the star's magnetic field rotates with the star to point a magnetic pole toward the observer or away from him.

To detect this phenomenon, we need a phase space filter with angular resolving power of a few seconds of arc, to isolate the star from its neighbors in the sky; spectral resolving power of about 10^4; the capability of resolving light into circularly polarized components; and a time resolution of perhaps ten days—comparable to the rotation period of the star.

A schematic projection of a phase space filter is shown in figure 3.15.

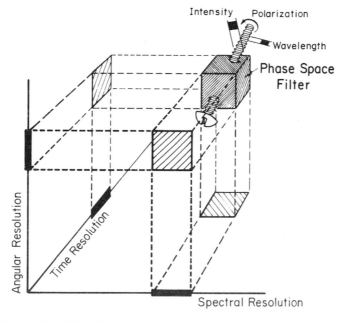

Figure 3.15 *Phase Space Filter*

A phase space filter combines the properties of several different kinds of filtering devices. Three of the properties that characterize it are its angular resolution, time resolution, and spectral resolution. These traits specify the band of spatial frequencies and time variations transmitted to the observer. They also specify the coarseness of spectral features transmitted by the filter. If the spectral resolution is high, the filter will selectively emphasize spectral lines of high color purity. If it is low, monochromatic lines become de-emphasized at the expense of any broad spectral continuum features that might be present in the incident radiation.

The angular, time, and spectral resolution are shown here as three dimensions of a space within which any phase space filter can be located. The filter, however, is characterized by three further properties. Since I could not draw a six-dimensional space, these three other properties are represented by the characteristics of the screw shown in the diagram. The wavelength band transmitted by the filter is represented by the pitch of the screw—the separation between adjacent threads. The diameter of the screw represents intensity of radiation and the sense of the screw—left- or right-handed—represents the sense of polarization selected by the filter.

The properties of any phase space filter bear a one-to-one correspondence to individual points in a six-dimensional space. Any point in that space corresponds to a basic observation that can be undertaken, and conversely, any conceivable observation corresponds to one point or a set of such points in the six-dimensional space. The six properties, or dimensions, shown in the figure correspond to the six traits listed in the right-hand column of table 3.2.

Instrumental Sensitivity

There are usually some twelve decimal ranges of apparent bright-ness separating the strongest signals that can be observed in our universe from the faintest signals detectable in the presence of cosmically pro-duced noise. Moreover, we can take as a fairly useful rule of thumb that the sensitivity of most instruments in active use cannot be improved by factors of more than 1,000 before some cosmic noise limitation is encountered: We generally find that an improvement in sensitivity by many orders of magnitude pays off, not in improved performance in previously possible observations; rather, the main gains appear in terms of higher spectral or time resolution that the improved sensitivity makes possible. In those regions of phase space in which observations are possi-ble at all, we are frequently able to cover most of the ultimately accessi-ble intensity range, and the efficiency of our observations appears to be rather high, though our methods may be clumsy and our speed in gathering data on large numbers of sources may be low.

We might at first be surprised at this, particularly since observers generally strive for the greatest sensitivity in their equipment and since certain discoveries have largely been made possible through improve-ments in sensitivity, rather than in resolution. However, once a few astronomical sources have been detected through a new phase space filter—let us label it X—improvements in sensitivity by many orders of magnitude almost automatically show up so many further sources all of the same generic type that the observer soon becomes limited by confusing sources, unless he simultaneously improves his resolution in at least one of the phase space dimensions. That procedure then will normally carry his observations into a new phase space domain, and it is there that further discoveries may be made.

If confusion by extraneous sources does not produce serious prob-lems, the higher sensitivity may nevertheless not fully pay off if natural fluctuations in background radiation or some other new fundamental limitations are encountered. Such limitations again may sometimes be minimized by improvements in resolving power of one kind or another. Sensitivity increases therefore tend to lead to improved observations, primarily in permitting the observer access to new phase space domains previously beyond his reach. Sensitivity improvements, of course, also increase the speed at which analytical work can be carried out routinely on large numbers of sources; but, taken alone, they do not seem to favor discovery of new phenomena.

There is at least circumstantial evidence that this view is correct. Despite a continuing increase in the sensitivity of optical detectors, and despite continued construction of new, larger optical telescopes, no new astronomical phenomena have been discovered in the past several de-cades by optical techniques alone, except in those instances involving substantial advances into new domains of the phase space. Thus the introduction of polarization techniques led in rapid succession to the discovery of polarization by interstellar dust and synchrotron emission from interstellar relativistic particles. Similarly, the implementation of circular polarization techniques combined with high spectral resolution and sensitivity led to the discovery of magnetic variables. But we have no instances of recently discovered major new phenomena in any of the parts of the phase space that had been accessible with pre-World War II techniques.

Radio astronomy also is now a sufficiently old discipline that we might have expected some discovery of a new phenomenon with techniques and capabilities known for more than twenty-five years. But that has not happened. Despite significant advances in instrumental sensitivity, most of the discoveries that have been made in recent years have been the direct result of improved resolving capabilities introduced shortly before.

All in all, we might estimate that a typical intensity coverage amounting to half the available logarithmic range of intensities describes our capabilities in those domains of the phase space that appear shaded in figures 3.6 to 3.9.

The Information Content of an Observation

As already discussed in chapter 1, we can specify the result of an electromagnetic observation by naming the phase space filter used and the intensity recorded. This view will be substantiated later; for the moment, however, let us accept it as correct. Then, if each filter bandwidth amounts to 3 decades—a factor of 1,000—we have the following options for covering the entire range of observations feasible in the electromagnetic domain: We have eight choices of wavelength bands, ranging from 10^{-18} to 10^6 centimeters. Any particular choice of one of these 8 filters can be specified by a three-digit binary number since $2^3 = 8$. A choice among 16 filters would require a four-digit number.

We have 18 decades, or 6 filter bandwidths, in angular resolution. We have 8 decades, or roughly 3 filter bandwidths, that we can specify in the spectral resolution domain, and another 24 decades, or 8 filter bandwidths, along the time resolution axis. The polarization of a source may be either circular, linear, unpolarized or some combination of these, and we need two binary digits—bits—of information to define which of these choices is being tested. With this variety of parameters any arbitrary filter we can conceive can be labeled with a 13-digit binary number constructed in accordance with table 3.3

TABLE 3.3
Binary Digits Needed to Specify Phase Space Filters

Wavelength	Angular Resolution	Spectral Resolution	Time Resolution	Polarization
3 bits	3 bits	2 bits	3 bits	2 bits

We therefore need a 13-bit number to specify the filter. And if our total range of intensities never exceeds 12 decades, or 4 factors of 1,000, we only need 2 more binary digits to specify the result of the observation made through that filter. For any particular field of view observed at some given time, we can, therefore, describe an observation and its results by a 15-digit binary number.

Our current capabilities are limited to about a twentieth of the entire available set of filters covering the totality of observables for the electromagnetic spectrum. We therefore could currently get by with something like 4 bits of information less per observation. An 11-bit number would do to specify all the observational results we could obtain by viewing a given portion of the sky.

If we widened our scope to include all five carriers of information, we would probably have to add another 2 bits to our 15-bit number, to finally arrive at a total word length of 17 bits, to describe the filter used in any conceivable observation of a particular field of view, where today we would need only some 11 bits. With an 11-bit word we are capable of enumerating roughly 2 percent of all the results we could describe with a 17-bit word.

Actually, our intensity measuring capabilities under most observing conditions do not cover more than 1 or 2 factors of 1,000, out of the possible 4 that we have mentioned. Our estimate of current capabilities might, therefore, need to be revised downward by another factor of 2 to 4. This would reduce the current capability to a range of complexity that can be described by just 9 or 10 bits.

We are now ready to decide on the gain in information that we are likely to achieve if we expand our technical expertise to the point where we can make use of all conceivable phase space filters rather than just those we now have at our disposal.

In the theory of information first developed by Claude Shannon, the information content of a vocabulary is proportional to the number of symbols or words the vocabulary contains. In the astronomical context the vocabulary consists of the names of phase space filters through which we can observe the sky. We can see that directly if we consider how our observations might be carried out. We can point our observing apparatus at some preselected region in the sky and look at the same region successively through each phase space filter we have. For each filter, we record the name or designating number of the filter in one column, and in a column next to it we write down either "Y" for "Yes" if we detect a signal, or "N" for "No" if we do not. Alternatively, we could simply write down "1" or "0" to designate these two possible results.

An entire observation then can be recorded as a word consisting of a filter designation followed by a "0" or "1," and if the filter designation is written in binary digits, our entry in the observing log might read 110010111001000101. In this 18-bit word the first 17 entries designate the filter through which we are observing and the last digit tells us that a signal was recorded through this filter when we pointed it in the preselected direction in the sky.

If we currently only have 1 percent of all eventually usable filters at our disposal, our vocabulary now has only 1 percent of its full scope, and we therefore are capable of obtaining just 1 percent of the information that will ultimately be at our disposal, provided that the words we construct in this way are truly independent.

That independence, however, does not exist. If all our observations yielded truly independent, unpredictable results, we would have to conclude that there were no scientific laws governing the behavior of cosmic events. We know, on the contrary, that such laws do exist, at least for many of the phenomena we now recognize.

Every language has its syntax. Words in a sentence do not appear arbitrarily; their order is governed by rules of grammar. Similarly, the sequence of letters in a word is not completely random but follows certain rules. In roughly the same way, the grammar that tells us ahead of time whether a given 18-bit observational result will end with the digit "1" or "0" is an expression of the laws of astrophysics. Often we predict the outcome of a new observation on purely theoretical grounds, and a correct prediction of the observation strengthens our confidence in the theory. However, even if our theory proves to yield incorrect predictions, there presumably exist correct theories—whether we are aware of them or not—that do make one observational result a consequence of another. And in that sense the words in our vocabulary of observations are not truly independent.

Currently we can make observations through about 1 percent of all ultimately usable phase space filters. We therefore can look forward to 100 times this number of observations to be carried out in the future. The information remaining to be uncovered can at best be 100 times the size of our current store of cosmic information. This number 100 is an upper limit, and the additional information to be uncovered through further observations may very well amount to a factor far lower than that. We will argue in chapter 4 that we may ultimately only triple our recognition of major cosmic phenomena: More specifically, we may already have observed roughly one-third of the phenomena we finally might hope to uncover. It is the astrophysical syntax that lowers the expected wealth of discoveries from 100, down to perhaps just 3 times the presently known number of phenomena.

An important point still to be noted is that our choice of filter bandwidth—a bandwidth amounting to a factor of 1,000 in the particular choice of phase space filters made here—is not strongly influential in determining how much more information we might expect to gather in the future. Had our choice of filter bandwidth been narrower, we would have found that the ultimately usable number of filters would have increased; but the number in current use would have increased in the same proportion, and the net fraction of currently-used to ultimately-usable filters would have remained approximately 1 percent. With that ratio fixed, the added information we may ultimately hope to gain also should remain approximately fixed.

Information Rates

By increasing the sensitivity of the detectors he uses, the astronomer achieves two aims: He can reach out to observe ever-fainter sources, or else he can reliably register signals from bright sources at an ever-increasing rate. The first of these capabilities allows him to observe intrinsically faint stars or galaxies and to see more luminous galaxies at greater distances across the universe. Our coming to grips with the structure of our universe depends on such observations. The second capability permits the study of increasing numbers of sources: The greater the sensitivity of the equipment, the sooner reliable results are gained; the astronomer is able to study the sources on his observing list in quicker succession; the rate at which he gathers information is increased.

The rate at which information can be collected may, however, also be increased without increased sensitivity. If we were to use a telescope to gather light from ten different portions of the sky onto ten distinct detectors, then each detector could be receiving radiation from a different galaxy, and the rate at which the telescope yielded results would increase tenfold.

As the number of detectors used with a telescope is increased, the rate at which information can be gathered also increases, provided we make sure that each detector collects useful information. The photographic plate is an extreme example of an array of many millions of tiny detectors. Each grain in the emulsion essentially acts like an independent detector. Thousands upon thousands of individual stars can simultaneously be registered on a photographic plate, and the rate at which information can be gathered increases in direct proportion to the number of stars viewed.

This feature of photographic plates—their ability to increase the rate at which information is gathered—was the essential element that led to the discovery of cluster variables toward the end of the nineteenth century. S. I. Bailey, working at 8,000 feet in the Peruvian Andes, simultaneously followed the pulsations of some eighty-five variable stars in the globular cluster Messier 5. Without photography, individual portions of the cluster would have had to be observed sequentially, the variable stars picked out among the far larger number that shone steadily, and each variable then pursued through several periods of its pulsation—a truly immense amount of work.

Improved sensitivity increases the rate at which information is received by greater or lesser amounts, depending on the nature of the observations: The information theorist Claude E. Shannon first showed that the capacity of a channel or detection system to convey information, measured in bits per second, is

$$C \sim W \log_2\left[1 + \frac{S}{N}\right]$$

where C is the capacity, W is the bandwidth of the detector, S is the received signal power, and N is the noise power assumed to be random and equal at all frequencies.[9] Appendix B discusses this equation. In the extreme limit of small signal-to-noise ratios the capacity reduces to the simple expression

$$C \sim WS/N$$

and the rate at which information is conveyed is directly proportional to the signal-to-noise ratio. For very large transmitted signals, the rate at which information can be received increases only logarithmically with signal-to-noise ratio. When $S/N = 1,000$, halving the noise in order to make $S/N = 2,000$ only increases the maximum rate at which we can receive information by 10 percent. If $S/N = 10^6$, a further doubling of the ratio increases the capacity to receive information by only 5 percent. This must be contrasted to the situation that holds for very small ratios, when S/N is much less than unity. In that case doubling the ratio S/N does double the rate at which information is received.

CHAPTER

4

Detection, Recognition, and Classification of Cosmic Phenomena

Classification of Phenomena

Each major cosmic phenomenon recognized today has a quite distinct appearance. Some phenomena differ from others in sheer scale—neutron stars, white dwarfs, and red giants differ by many orders of magnitude in their radii. Other sources differ strikingly in the wavelengths at which they radiate the bulk of their energy—X-ray binaries, ordinary stars, infrared galaxies, and pulsars emit in widely differing portions of the electromagnetic spectrum. Even within the radio spectrum alone there is a great difference in color purity between masers, which emit within the narrowest spectral confines, and radio galaxies, which radiate over a wide band of frequencies.

Huge distinctions can also be seen in the explosive energy liberated in the outbursts of flare stars, novae, and supernovae. Similarly, pulsars pulse many orders of magnitude faster than Cepheid variables.

Some of the phenomena differ so drastically from anything else we have observed that they amaze us even at first glance. The gamma-ray bursts fall into that class. They compare to nothing else we observe. In contrast, X-ray galaxies become startling only when we recognize their distance and infer the true scale of their X-ray emission, thousands of times greater than the emission from X-ray stars.

Most astronomers would agree that whenever we detect some strikingly new pattern of events in our observations, we sense that we are dealing with an entirely new process—a new phenomenon. What we mean by *striking* in this context must, of course, be settled; but most individuals would agree that differences spanning many orders of magnitude—differences amounting perhaps to at least a factor of 1,000 in some trait—would label such a newly observed object or set of events as strikingly different from anything previously discovered.

A corollary to this view is that observing techniques must be capable of recognizing strikingly novel traits if new phenomena are to be discovered: We have seen that all conceivable observations can be defined in terms of a finite phase space that specifies the wavelength, polarization, and spectral, spatial, and time resolution of any given measurement we might undertake. If we are to discover features that differ by many orders of magnitude from known behavior, then our observing techniques must be capable of recognizing these new traits. It follows that we facilitate the discovery of a new phenomenon whenever we find ways of making observations in previously unexplored portions of the phase space: Pulsars were discovered soon after astronomers first applied techniques capable of resolving outbursts that last less than 1 second. Quasars were only recognized as highly unusual radio sources when they were found to be strikingly compact—when the angular resolving power of radio astronomical techniques was upgraded so that features as fine as seconds of arc could be resolved. The earliest radio astronomical techniques lacked sufficient reliability for faint absolute flux measurements: Recognition of a completely isotropic radiation bath, however, depended on just that reliability, and the discovery of the cosmic microwave background radiation, therefore, had to await the construction of equipment sufficiently reliable to succeed.

All this suggests that recognition, classification, and distinction of major cosmic phenomena might center on the strikingly disparate appearance of different phenomena. Theorists may feel uncomfortable with such a superficial approach; but in many areas of science, classification based simply on superficial features has had significant success.

Distinctions Between Major Phenomena

Let us now return to the notion of a phase space of observations to see whether it can help in classifying new phenomena. We can think of taking the entire volume V of this bounded space and dividing it into N different compartments each of volume $u = V/N$. Figure 4.1 shows one way of visualizing such a compartment. Here u is a measure of the range of observations we are able to make through a phase space filter corresponding to one of these compartments. As the number N into which we divide the phase space is increased, the volume u diminishes and the observations possible with any given filter become more specific and more restricted.

To understand what an observation through a phase space filter entails, let us suppose that we are viewing a particular section of the sky through the filter, and there is no transmitted radiation at all. We then log a signal which, as well as we can judge, has zero intensity. The uncertainty in our judgment about whether the signal is truly zero, or just very small, is the noise associated with the measurement. We

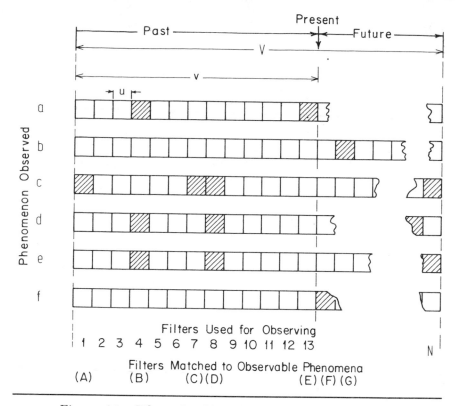

Figure 4.1 *Filters for Observing Different Phenomena*

A selection of different cosmic phenomena, here labeled *a, b, c, . . . f,* can be sought with the aid of a wide variety of observing techniques. Each of these techniques can be thought of as a phase space filter which provides the observer with a specific range of observing capabilities. We can label the complete set of all conceivable filters with numbers 1, 2, 3, . . . *N.* If the entire phase space corresponding to all possible observations has a volume *V,* and the *N* filters divide *V* into equal subvolumes, then each filter exhibits a range of capabilities that fills a volume $u = V/N$ of the space. We label those filters that pass observable signals with letters (A), (B), (C). . . .

In this figure we examine a time in the history of astronomical observations when only 13 phase space filters have come into use. At this stage the volume of capabilities attained corresponds to $v = 13u.$ The competence ratio $r = v/V$ is a measure of instrumental achievement during the era labeled *Past.* As shown here past observations have led to the discovery of phenomenon *a* through the use of (shaded) filters 4 and 13. Phenomena *b* and *f* have yet to be recognized; phenomenon *c* is recognized through the use of filters 1, 7, and 8, and the phenomena *e* and *d,* while known through observations with filters 4 and 8, remain indistinguishable from each other and are considered as one. The existence of two separate phenomena *d* and *e* may not be recognized until instrumental competence is almost complete and the final filter *N* comes into use. At that stage phenomenon *e* will be recognized through traits exhibited in observations with this last filter; it will be recognized to be different from phenomenon *d,* undetected through filter *N.*

With the introduction of filters 14 and 15, phenomena *f* and *b* will become recognized.

Phenomenon *c* is unique; it differs from other phenomena in two entirely distinct ways: It exhibits a signal through filter 1 and is detectable through filter 7. It can therefore be independently recognized through either filter.

Looked at in this way, the history of astronomy is this: With just filter 1 available, only phenomenon *c* was known. Introduction of filters 2 and 3 did not lead to the recognition of further phenomena, but filter 4 led to the recognition of what appeared to be a single phenomenon, although we would discover that three different phenomena, *a, d,* and *e,* were involved. . . .

The thesis presented here is that cosmic phenomena can be adequately distinguished by an enumeration of the phase space filters through which the phenomena are detected. Table 4.3 shows that this view is largely correct and that a list of phenomena culled by means of distinguishing filter monograms closely approximates the list of currently recognized astronomical phenomena.

will take it for granted here that an astronomical signal can only be detected if it is strong enough to be noticed in the presence of noise. With this in mind, let us look at a specific discovery.

Johannes Hartmann's finding of interstellar gas along the line of sight to the binary δ Orionis was based on his recognition that some of the spectral lines appeared to move in time—in synchronism with the emitting stars—while one of the calcium absorption lines remained stationary. He inferred that the calcium must be interstellar.

In terms of our phase space description we would say that Hartmann had made use of two separate filters. They both had to provide high enough spectral resolution to isolate spectral lines and to note the small displacements that Hartmann was analyzing. The unusual feature of Hartmann's observation was that many of the lines in the spectrum moved with time but that one of the lines did not vary at all. The discovery, therefore, required one filter capable of time resolution fast enough to follow the displacement of the moving lines over the orbital period of the stellar binary. This filter, however, would not convey the presence of the stationary calcium line because that line showed no time variations at the specified periodicity. A second filter, with slower time resolution, was therefore needed to show that the stationary calcium line existed as well. With this combination of a faster and a slower time resolution, both filters transmitted appreciable signals. Hartmann would not have thought of it in just this way, but he did recognize that this behavior was strikingly different from previous reports on stars and implied an important discovery.

In this particular description the system observed by Hartmann yielded discernible signals when viewed through two distinct phase space filters whose time resolution differed by orders of magnitude. There might be other phenomena which could be recognized, not through the difference in frequency filters which pass recognizable signals but rather by the total strengths of these signals. Supernovae, for example, can be distinguished from novae in the strengths of the signals observed through filters. The brightness changes that occur in supernovae are many orders of magnitude greater than in novae.

All this suggests that differences in the phase space filters or combinations of filters that pass detectable amounts of radiation may suffice to distinguish between the phenomena known to us today. Clearly we have far more available information on any of these phenomena; but if the additional data are not needed, the phase space filter description alone can provide sufficient information, and may permit us to arrive at an objective definition for what we mean by a cosmic phenomenon.

Objective Classification

The simplest way to classify astronomical phenomena might be according to the names given to events or objects by common agreement among astronomers. Such a consensus approach might even do rather well for just the reasons that Polanyi mentions.* It did not take astronomers long to agree that pulsars deserved a name of their own: Pulsars simply differed too much from any other known phenomenon.

What we need is an objective prescription that will, to a greater or lesser extent, tell us when an observed set of events deserves a novel name and when it does not. At what point can a nova outburst no longer be considered a nova but must instead be recognized as a supernova— far, far brighter than an ordinary nova and presumably powered by mechanisms of a completely different character?

Details appear to play but a minor role in the classification of astronomical phenomena. And if only major, substantial differences are to be taken into account, we should be able to define a cosmic phenomenon solely in terms of particular phase space filters that pass discernible signals.

Figure 4.1 illustrates the approach. When different objects or events in the sky are viewed through phase space filters 1, 2, 3, . . . N, . . . some will provide a signal while others will not. We can therefore construct a table in which we list each observed event and those filters that transmit an observable signal. If we chose to relabel these particular filters with letters A, B, C, . . . as shown, phenomenon c in figure 4.1 could be designated *(ACD)*, phenomena d and e, *(BD)*, and phenomenon a, *(BE)*. Viewed in this way *the classification of a phenomenon observed in the universe becomes a matter of constructing monograms, such as (ABFY) designating the group of filters A, B, F, and Y through which a set of events or group of objects viewed in the sky yield observable signals. Each distinct phenomenon is then characterized by a unique monogram or set of monograms.*

This way of designating cosmic phenomena has many elements that we would like to see in an objective definition. But there are two different ways we could choose to view the universe, and we must first decide which one to select. The first is that of an observer located on Earth. While this is the normal view of the universe that we gain from raw data, we seldom restrict ourselves for very long before converting these data into the perspective that a cosmic traveler would have if he could translate all observations into scale models of observed events.

This second viewpoint compensates for the large differences in distance at which various sources are normally observed. Remote galaxies no longer appear tiny as they do in the sky; instead they take their rightful places with our Milky Way as phenomena on a far grander scale than individual stars or groups of stars.

Each of these vantage points—the local and the cosmic—has its strengths. The earthbound view will lead to a definition of phenomena that we call *Class I.* The cosmic view, which scales all events to proper perspective, will be called *Class II.*[1]

Before discussing these two prescriptions for classification more carefully we might briefly compare their advantages and disadvantages.

---

* See page 36.

The difference between the two schemes relates to the alternate descriptions of cosmic phenomena discussed earlier.* The Class II selection closely corresponds to the conceptual models astronomers employ. It incorporates such features as actual size and actual luminosity of a source. It provides a sense of proportion and, in that respect, reflects a more faithful view of the universe. However, this classification requires a number of astrophysical assumptions, particularly about the distance at which events occur. These assumptions frequently are uncertain; in that respect, the Class I description, though more primitive, is less speculative.

Class I and Class II Phenomena

When we make an observation of a spiral galaxy through phase space filters, we are actually obtaining a mixed set of data about the source. The spectral characteristics, the time variations, and the polarization give direct information about the properties of the source, provided the radiation reaches us in unaltered form after its long trek through space. The surface brightness of the galaxy—its brightness divided by the solid angle it covers in the sky—also is an intrinsic characteristic independent of distance.

Purely angular measurements differ in being extrinsic. Without knowing the distance of the galaxy the angular dimensions provide only partial information. They can tell us whether the galaxy is round or whether it is long and filamentary. Similarly intensity observations yield restricted information. If a galaxy's brightness changes as it did when the supernova of 1885 exploded in the Andromeda Nebula, we can tell the ratio of maximum to minimum emission; but the absolute brightness of the source—its intrinsic luminosity—eludes us unless we know its distance.

The Class I description is restricted solely to directly observed traits, mainly because these traits provide the most generally available information about any observed source when no distance measure is available. For most astronomical sources we do, however, have a fairly good measure of distance. We can therefore sharpen our description somewhat by adopting the Class II depiction, which replaces angular coordinates by actual spatial size and apparent brightness by the intrinsic luminosity of a source. We are then closer to an accurate description of the most important source traits, having removed those characteristics that are introduced by the chance location of the observer—the type of trait that makes the planet Jupiter look brighter than the star Betelgeuse, when actually that star is a 5×10^{12} times more luminous, but lies ten million times further away.

Each Class I phenomenon can be identified by a set of phase space filters through which detectable energy is received. Each filter is assigned a designating code consisting of four numbers and a letter, in

* See pages 55 ff.

accordance with the parameters shown in table 4.1. The first number represents the wavelength band in which the phenomenon is primarily observed. The second number gives the angular resolution, the third the spectral resolution, the fourth the time resolution, and the letter indicates the polarization of the source. A dash implies uncertainty about the characteristics. An italicized number, for example, *3*, means that this represents a usual value but that a neighboring value, for example 2 or 4, might sometimes also be representative.

TABLE 4.1

Key to Filter Designation Code for Class I Phenomena

Band	Wavelength (or Frequency, Energy)*	Angular Resolution	Spectral Resolution	Time Resolution	Ellipticity of Polarization†
1	10 meter to 10 km (3×10^7 Hz – 3×10^4 Hz)	isotropy to 20′	photometry – 10^3	10^{-5} sec – 10^{-2} sec	L
2	1 cm to 10 meter (3×10^{10} Hz – 3×10^7 Hz)	1″ to 20′	$10^3 – 10^6$	10^{-2} sec – 10 sec	C
3	10 μ to 1 cm	10^{-3}″ to 1″		10 sec – 3 hrs	E
4	100 Å to 10 μ	unresolved		3 hrs – 4 months	U
5	10^{-9} cm to 10^{-6} cm (100 keV – 100 eV)			4 months – 300 y	
6	10^{-12} cm to 10^{-9} cm (100 MeV – 100 keV)			> 300 y (or unresolved)	
7	10^{-15} cm to 10^{-12} cm (100 GeV – 100 MeV)				

* See glossary tables G.1, G.2 for conversion of units.
† L = Linear, C = Circular, E = Elliptical, and U = Unpolarized.

In these terms we list one of the identifying traits of comets as (4114U), since comets are observed primarily as visible objects (band 4), apparent at low angular and spectral resolution and with time variations predominantly in the three-hour to four-month range. Their radiation is unpolarized (U). Table 4.3 lists the filters associated with different cosmic phenomena, including filter (4114U) for comets. In the table, meteorites and other entities that are directly collected carry the designation "direct collection."

In the Class II description the angular resolution phase space filters are replaced by a set of physical size filters—a set of sieves. Each filter selects a size range spanning three factors of 10—a factor of 1,000. In this way we can use some eight filters to cover a range of source sizes going from 10^5 to 10^{28} centimeters—a range that includes essentially all astronomical phenomena from asteroids and neutron stars on the smallest scale, to clusters of galaxies and the scale of the entire universe at the other extreme.

Similarly we can specify sets of filters that will only pass radiation within a given range of intrinsic source luminosity. Here we might concentrate on the range from about 10^{29} erg per second—somewhat less than the energy emitted from the faintest stars observed—to more than the 10^{47} erg per second emitted from large clusters of galaxies. The luminosity range spanned by these sources, $10^{18}:1$, may be compared to the brightness range accessible through noise-restricted observations. That range is restricted to $\sim 10^{12}:1$.*

A Class II filter can now be specified by a sequence of digits and one letter, thus: (4324C4). The first digit specifies the wavelength band of observations; the second gives the size range of the region under investigation; the third digit shows the spectral resolution range in which the observation is made; the fourth indicates the time resolution required; and the last provides the luminosity of the event. Between the fourth and last digits, a letter U, L, C, or E is inserted to show unpolarized, linearly, circularly, or elliptically polarized radiation. A more complete description of these filters is given in table 4.2.

There is only one scale parameter that we are free to choose in attempting to fit the Class I and Class II characterization to actual cosmic phenomena. It is the bandwidth of the filters, or equivalently the size of the phase space cells probed by these filters. We had previously labeled the cell volume, u. If we now choose the dimension of u to be three decades (a factor of 10^3) in each of the continuous phase space coordinates, a selection of Class I and Class II phenomena is obtained that closely matches our previously compiled list of forty-three cosmic phenomena. This fit appears to be good for two reasons: First, a factor of 1,000 in signal strength or a factor of 1,000 in spectral frequency or structural scale appears to distinguish the various listed phenomena. Second, many cosmic phenomena occur with a range of traits that varies over a somewhat narrower bandwidth than the factor of 10^3; this means that relatively few phenomena would appear strong when viewed through two adjacent filters in the phase space of observations.

The 23rd listing in table 4.3 deals with magnetic stars and shows the filter designation (4324C4) cited above. This tells us that these stars are visible objects in the size range from 10^{11} to 10^{14} centimeters. Their chief characteristic, the spectral lines whose positions shift with circular polarization and vary with time, are best seen with a circular polarization filter sensitive to variations on a time scale of three hours to four months. The indicated luminosity lies in the range of 10^{35} to 10^{38} erg per second. While these ranges are extremely wide, covering a factor of 1,000, they nevertheless provide a sufficiently specific description to uniquely select magnetic stars.

By and large our list of rather subjectively selected phenomena is well duplicated by the listing of Class I and Class II phenomena as shown in table 4.3. However, there are exceptions. Ionized hydrogen regions and planetary nebulae merge into a single phenomenon since there is no sufficiently pronounced distinction between these objects to show up in the signals transmitted through the relatively coarse Class II filters. Similarly, superluminal sources cannot be distinguished from quasars, or rings from the planets they encircle. These phenomena therefore are designated (R0) because they are not separately recognized

* See page 183.

TABLE 4.2
Key to Filter Designation Code for Class II Phenomena

Band	Wavelength (or Frequency, Energy)*	Size	Spectral Resolution	Time Resolution	Ellipticity of Polarization†	Intensity
1	10 m to <10 km (3×10^7 Hz to $>3 \times 10^4$ Hz)	$10^5 - 10^8$ cm	photometry to 10^3	10^{-5} to 10^{-2} sec	L	$<10^{29}$ erg/sec
2	1 cm to 10 m (3×10^{10} to 3×10^7 Hz)	$10^8 - 10^{11}$	$10^3 - 10^6$	10^{-2} to 10 sec	C	$10^{29} - 10^{32}$
3	10 μ to 1 cm	$10^{11} - 10^{14}$		10 sec to 3 hrs	E	$10^{32} - 10^{35}$
4	100 Å to 10 μ	$10^{14} - 10^{17}$		3 hrs to 4 months	U	$10^{35} - 10^{38}$
5	10^{-9} to 10^{-6} cm (100 keV – 100 eV)	$10^{17} - 10^{20}$		4 months to 300 y		$10^{38} - 10^{41}$
6	10^{-12} to 10^{-9} cm (100 MeV – 100 keV)	$10^{20} - 10^{23}$		<300 y (or unresolved)		$10^{41} - 10^{44}$
7	10^{-15} to 10^{-12} cm (100 GeV – 100 MeV)	$10^{23} - 10^{26}$				$10^{44} - 10^{47}$
8		$>10^{26}$				$>10^{47}$

* See glossary tables G.1, G.2 for conversion of units.
† L = Linear, C = Circular, E = Elliptical, and U = Unpolarized.

TABLE 4.3

Identification of Class I and Class II Phenomena

Class I descriptions consist of four digits and a letter, Class II phenomena of five digits and a letter (see tables 4.1 and 4.2). The number of independent ways in which a phenomenon is recognized is also shown. (R0) and (R1) respectively designate unrecognized and singly recognized phenomena, etc.

1.	Zodiacal cloud of interplanetary matter		
	Zodiacal glow	(4115U)	(4315U1)
	Meteorites	Direct collection	Direct collection
	Meteors		
	Visual	(4112U)	
	Radar	(2112U)	
		(R3)	(R1)
2.	Planets		
	Orbits		
	Visual	(4115U)	(4315U1)
	Radio	(2115U)	(2315U1)
	Size		
	Angular	(4215U)	Visual (4215U1)
			Radio (2215U1)
		(R2)	(R2)
3.	Asteroids		
	Orbits	(4115U)	(4315U1)
	Size and brightness variations	(4414U)	(4114U1)
		(R1)	(R1)
4.	Moons		
	Solar orbits	(4115U)	(4315U1)
	Planetary orbits	(4214U)	(4214U1)
	Size	(4416U)	(4116U1)
		(R1)	(R1)
5.	Rings	Same as for planets	
		(R0)	(R0)
6.	Comets		
	Orbit	(4115U)	(4415U1)
	Brightness variations	(4114U)	(4314U1) same
			(4315U1) object
		(R1)	(R1)
7.	Main sequence stars		
	a. Faint stars		(4216U2)
		(R0)	(R1)
	b. Intermediate	(4416U)	(4216U3)
		(R1)	(R1)
	c. Bright stars		(4216U4)
		(R0)	(R1)
8.	Subgiants and red giants		
	Visually resolved broadband	(4315U)	(4315U4)
	variable	(R1)	(R1)

9. Pulsating variable stars
   a. Cepheids and intermediate
      period stars

Brightness variation	(4414U)	(4314U4)
Spectral shift	(4424U)	(4324U4)
	(R1)	(R1)

   b. Long period stars

Brightness variation	(4415U)	(4315U3)
	(R1)	(R1)

   c. Dwarf cepheids and short
      period stars

Brightness variation	(4413U)	(4213U3)
	(R1)	(R1)

10. Multiple stars
    a. Close binaries

Eclipses	(4414U)	
Doppler variation	(4424U)	
	(R0)	(R0)

    b. Intermediate separation

Spectral shift	(4424U)	
Orbital motion		(4314U3)
	(R1)	(R1)

    c. Visual and astrometric

Orbital motion	(4215U)	(4415U3)
	(R0)	(R1)

11. White dwarfs

Motion across sky	(4215*U*)	
Size		(4216*U*2)
	(R1)	(R0)

12. Galactic clusters

Grouping	(4216U)	(4516U5)
Individual stars	(4416U)	(4316U4)
	(R1)	(R1)

13. Globular clusters

Grouping	(4216U)	(4516U5)
Individual stars	(4416U)	(4316U3)
Variable star content	(4414U)	(4314U4)
	(R1)	(R1)

14. Planetary nebulae

	Same as for ionized hydrogen regions	
	(R0)	(R0)

15. Ionized hydrogen regions

Visual lines	(4226U)	(4526U5)
Infrared continuum	(321--)	(351--5)
Radio recombination lines	(2226U)	(2526U2)
Radio thermal continuum	(2216U)	(2516U3)
	(R2)	(R3)

16. Cold gas clouds

Stellar line variation	(4424U)	
Interstellar lines	(4226U)	
Visible interstellar absorption cloud		(4526U*1*)
Radio emission and absorption lines	(2226U)	(2426U1)
	(R1)	(R2)

TABLE 4.3—*Continued*

17. Interstellar dust and reflection nebulae		
Polarized continuum	(4216L)	(4516L3)
Polarized lines	(4226L)	(4526L3)
	(R1)	(R1)
18. Supernovae		
Brightness change	(4414U)	(4314U6)
	(R1)	(R1)
19. Eruptive variables		
a. Novae		
Brightness changes	(4413U)	(4313U5)
	(4414U)	(4314U5) same
Recurrence	(4415U)	(4315U5) object
	(R1)	(R2)
b. Symbiotic stars	(4413U)	(4313U3)
	(R0)	(R1)
c. R Coronae Borealis stars	(4415U)	(4315U4)
	(R0)	(R0)
20. Variable stars associated with nebulosity		
Nebulosity variations	(4215L)	(4515L2)
Stellar variability	(4414U)	(4314U3)
	(R1)	(R1)
21. Infrared stars—circumstellar dust clouds		
Slow variable continuum	(3215–)	(3315–4)
	(R1)	(R1)
22. Flare Stars		
Flaring	(4412U)	(4212U2)
Binary star	(4424U)	(4314U3)
Radio variability	(2314U)	(2414U1)
	(R1)	(R1)
23. Magnetic stars		
Variable circularly	(4424C)	(4324C4)
polarized lines	(R1)	(R1)
24. Cosmic masers		
a. Circularly polarized	(2424C)	(2424C2)
	(R1)	(R1)
b. Linearly polarized	(2424L)	(2424L2)
	(R1)	(R1)
25. Pulsars		
Radio pulses	(2412L)	(2112L2)
	(R1)	(R1)
26. X-ray stars		
Variable continuum	(5–13–)	(5213–4) same
	(5–14–)	(5214–4) object
	(R2)	(R2)

27. Supernova remnants
    Polarized continuum   (4215L)   (4525U4)
    Unpolarized filament lines   (4225U)   (4515L4)
    Polarized radio continuum   (2215L)   (2515L3)
    X-ray continuum   (521--)   (551--4)
                                       (R2)   (R2)

28. Interstellar magnetic fields
    Zeeman splitting   (2226C)   (2526C-)
    Faraday rotation   (2212L)   (2612L-)
    Faraday rotation
       of pulsar radiation   (2412L)   (2112L2)
                                       (R2)   (R2)

29. Galaxies containing gas
    dust polarized continuum   (4216L)   (4616L6)
    Visual Doppler shift   (4216U)   (4616U6)
    Radio continuum   (2216U)   (2616U4)
    Radio Doppler shift   (2216U)   (2616U3)
                                       (R2)   (R2)

30. Galaxies devoid of gas
    Continuum appearance   (4216U)   (4616U6)
    Doppler shift   (4216U)   (4616U6)
                                       (R1)   (R1)

31. Clusters of galaxies
    Distribution   (4116U)   (4716U7)
    Doppler membership   (4216U)   (4616U6)
                                       (R1)   (R1)

32. Radio galaxies
    a. Extended sources
       Visual Doppler appearance   (4216U)   (4616U6)
       Polarized radio continuum   (2216L)   (2616L6)
                                       (R1)   (R1)

    b. Compact Sources
       Visual Doppler appearance   (4216U)   (4516U6)
       Radio scintillation   (2315L)   (2515L5)
                                       (R1)   (R1)

33. Unidentified radio sources
    Radio scintillation   (2312L)   (2-12L-)
                                       (R1)   (R1)

34. Expansion of the universe
    Red-shift distance relation   (4826U6)
                                       (R0)   (R1)

35. Quasars
    Visual time variations   (4415-)   (4515L7)
    Radio polarization and
       variation   (2315L)   (2515L7)
    X-ray emission   (521--)   (551--7)
                                       (R1)   (R2)

36. Superluminal radio sources   Same as for quasars
                                       (R0)   (R0)

37. X-ray galaxies and clusters
    a. X-ray galaxies
       Visual appearance   (4216U)   (4616U6)
       X-ray emission   (521--)   (561--6)
                                       (R1)   (R1)

TABLE 4.3—*Continued*

b.	X-rays from clusters		
	Visual appearance of	⎧ (4116U)	(4716U7)t316*U*)
	cluster of galaxies	⎨ (42*16U*)	
	X-ray emission	(521--)	(571--6)
		(R1)	(R1)
38.	Infrared galaxies		
	Visual galaxy	(42*16*U)	(46*16*U*6*)
	Infrared emission	(321--)	(361--*6*)
	Polarized dust continuum	(4216L)	(4616L6)
		(R1)	(R1)
39.	Gamma-ray bursts	(6-12-)	(6-12--)
		(R1)	(R1)
40.	Microwave background radiation	(31*16*U)	(38*16*U-)
		(R1)	(R1)
41.	X-ray background	(511--)	(581---)
		(R1)	(R1)
42.	Gamma-ray background		
	a. Cosmic	(611--)	(681---)
		(R1)	(R1)
	b. Galactic	(611--)	(661---)
		(R0)	(R1)
43.	Sources of cosmic rays		
	Visible polarized continuum	(42*15*L)	(45*15L4*)
	Radio polarized continuum	(22*15*L)	(25*15L3*)
	Collection at Earth	Direct	Direct
		(R2)	(R2)

through the use of these filters. In contrast, phenomena that are singly or doubly recognized, respectively, are labeled (R1) and (R2) in table 4.3. At any rate, table 4.3 compares the forty-three phenomena cited in chapter 2 to lists of phenomena that come out of a blindly applied Class I or a Class II selection. For many phenomena the table lists several identification regions in the phase space of observation. This set of regions is not meant to be exhaustive. Most of the phenomena listed can be detected through other filters not included. When such a filter has been omitted, it is either because it was too common to many other phenomena to serve as a useful label or because its use depended on observations through some other filter which might be needed first to locate the source.

Tables 4.4 and 4.5 complement table 4.3 by listing the various phenomena respectively recognized through different Class I and Class II filters.

A useful feature of both the Class I and Class II prescriptions is their selection of three different types of pulsating variables; indeed, as we have already seen, some three, largely distinct groups might exist.

Table 4.3 distinguishes forty-nine different Class II phenomena while the listing of chapter 2 described forty-three phenomena; some forty-three, however, were presented as possibly comprising two or three actually distinct phenomena: The descriptions given in chapter 2 provided no way of deciding whether all main sequence stars should be viewed as a single phenomenon or whether their great range in luminosity should lead us to consider faint dwarfs, stars of the sun's luminosity, and bright supergiants as distinct phenomena. Under the Class II prescription these are listed as three separate phenomena. However, the faintest group has the same characteristics as white dwarfs. White dwarfs and faint main sequence stars cannot be distinguished through the Class II filters. In contrast, Class I filters only identify two types of main sequence stars, but do distinguish white dwarfs.

There is an interesting aspect to the relatively close relationship between phenomena selected through phase space filters and our earlier list of forty-three phenomena. That listing essentially represented a glossary of astronomical sources and events, a compilation of the classes

TABLE 4.4

Class I Phenomena and Their Filters

Filter*	Phenomena†	Filter*	Phenomena†
2 1 1 2 U	1	4 2 1 5 L	20, 27, 43
2 1 1 5 U	2	4 2 1 6 U	12, 13, 29, 30, 31, 32a,
2 2 1 2 L	28		32b, 37a, 37b, 38
2 2 1 5 L	27, 43	4 2 1 6 L	17, 29, 38
2 2 1 6 U	14, 15, 29	4 2 2 5 U	27
2 2 1 6 L	32a	4 2 2 6 U	14, 15, 16
2 2 2 6 U	14, 15, 16	4 2 2 6 L	17
2 2 2 6 C	28	4 3 1 5 U	8
2 3 1 2 L	33	4 4 1 2 U	22
2 3 1 4 U	22′	4 4 1 3 U	9c, 19a, 19b
2 3 1 5 L	32b, 35	4 4 1 4 U	3, 9a, 10a, 13, 18, 19a, 20
2 4 1 2 L	25, 28	4 4 1 5 –	9b, 19a, 19c, 35
2 4 2 4 C	24a	4 4 1 6 U	4, 7b, 12, 13
2 4 2 4 L	24b	4 4 2 4 U	9a, 10a, 10b, 16, 22
3 1 1 6 U	40	4 4 2 4 C	23
3 2 1 – –	(15), (14), (38) . . . (2)	5 – 1 3 –	26
3 2 1 5 –	(21) . . . (14), (15), (38)	5 – 1 4 –	26
4 1 1 2 U	1	5 1 1 – –	41
4 1 1 4 U	6	5 2 1 – –	28, 35, 37a, 37b
4 1 1 5 U	1, 2, 3, 4, 6	6 1 1 – –	42a, 42b
4 1 1 6 U	31, 37b	6 – 1 2 –	39
4 1 1 4 U	4′	Direct A	1
4 2 1 5 *U*	2, 10c, 11	Direct B	43

Notes: Parentheses, (15), indicate that the appearance overlaps that of another phenomenon listed for an immediately adjacent filter and noted at the end of the same line.

A prime (′) appended to a number indicates that independent recognition is not possible through that filter alone. Another filter must first be used to locate the source.

* The numbers denoting the filters are explained in table 4.1.
† The numbers denoting the phenomena are given in table 4.3.

TABLE 4.5
Class II Phenomena and Their Filters

Filter*	Phenomena†	Filter*	Phenomena†
$2\,1\,1\,2\,L\,2$	25, 28	$4\,3\,1\,5\,U\,1$ ⎫	1, 2, 3, 4', 6
$2\,2\,1\,5\,U\,1$	2	$4\,3\,1\,5\,U\,3$ ⎪	9b
$2\,3\,1\,5\,U\,1$	2'	$4\,3\,1\,5\,U\,4$ ⎬	8, 19c
$2-1\,2\,L\,-$	33	$4\,3\,1\,5\,U\,5$ ⎭	19a'
$2\,4\,1\,4\,U\,1$	22'	$4\,3\,1\,6\,U\,3$ ⎫	13
$2\,4\,2\,4\,L\,2$	24b	$4\,3\,1\,6\,U\,4$ ⎬	12
$2\,4\,2\,4\,C\,2$	24a	$4\,3\,2\,4\,U\,4$	9a'
$2\,4\,2\,6\,U\,1$	16	$4\,3\,2\,4\,C\,4$	23
$2\,5\,1\,5\,L\,3$ ⎫	27, 43	$4\,4\,1\,5\,U\,1$ ⎫	6
$2\,5\,1\,5\,L\,5$ ⎬	32b	$4\,4\,1\,5\,U\,3$ ⎬	10c
$2\,5\,1\,5\,L\,7$ ⎭	35	$4\,5\,1\,5\,L\,2$ ⎫	20
$2\,5\,1\,6\,U\,3$	(15) . . . (14)	$4\,5\,1\,5\,L\,4$ ⎬	27, 43 . . . (17)
$2\,5\,2\,6\,U\,2$	(15)' . . . (14)'	$4\,5\,1\,5\,L\,7$ ⎭	35'
$2\,5\,2\,6\,C\,-$	28	$4\,5\,1\,6\,L\,3$	(17) . . . (27)
$2\,6\,1\,2\,L\,-$	28	$4\,5\,1\,6\,U\,5$ ⎫	12, 13
$2\,6\,1\,6\,U\,3$	29'	$4\,5\,1\,6\,U\,6$ ⎬	(32b) . . . (29)
$2\,6\,1\,6\,U\,4$	29	$4\,5\,2\,5\,U\,4$	27
$2\,6\,1\,6\,L\,6$	32a	$4\,5\,2\,6\,U\,1$ ⎫	16
$3\,3\,1\,5\,-\,4$	21	$4\,5\,2\,6\,U\,5$ ⎬	(15) . . . (14)
$3\,5\,1\,-\,-\,5$	(15) . . . (14)	$4\,5\,2\,6\,L\,3$	17
$3\,6\,1\,-\,-\,6$	38	$4\,6\,1\,6\,L\,6$	29, 38
$3\,8\,1\,6\,U\,-$	40	$4\,6\,1\,6\,U\,6$	(29), 30, 31, 32a,
$4\,1\,1\,4\,U\,1$	3		37a, 38 . . . (32b)
$4\,1\,1\,6\,U\,1$	4	$4\,7\,1\,6\,U\,7$	31, 37b
$4\,2\,1\,2\,U\,2$	22	$4\,8\,2\,6\,U\,6$	34
$4\,2\,1\,3\,U\,3$	9c	$5\,2\,1\,3\,-\,4$	26
$4\,2\,1\,4\,U\,1$	4'	$5\,2\,1\,4\,-\,4$	26
$4\,2\,1\,5\,U\,1$	2	$5\,5\,1\,-\,-\,4$ ⎫	27
$4\,2\,1\,6\,U\,2$ ⎫	7a, 11	$5\,5\,1\,-\,-\,7$ ⎬	35
$4\,2\,1\,6\,U\,3$ ⎬	7b	$5\,6\,1\,-\,-\,6$	37a
$4\,2\,1\,6\,U\,4$ ⎭	7c	$5\,7\,1\,-\,-\,6$	37b
$4\,3\,1\,3\,U\,3$ ⎫	19b	$5\,8\,1\,-\,-\,-$	41
$4\,3\,1\,3\,U\,5$ ⎬	19a	$6-1\,2\,-\,-$	39
$4\,3\,1\,4\,U\,1$ ⎭	6	$6\,6\,1\,-\,-\,-$	42b
$4\,3\,1\,4\,U\,3$ ⎫	10b, 20, 22	$6\,8\,1\,-\,-\,-$	42a
$4\,3\,1\,4\,U\,4$ ⎪	9a, 13	Direct A	1
$4\,3\,1\,4\,U\,5$ ⎬	19a	Direct B	43
$4\,3\,1\,4\,U\,6$ ⎭	18		

Notes: Braces indicate filters that would combine into a single filter if absolute luminosity were not measured.

A prime (') appended to the number indicates that independent recognition is not possible through that filter alone. Another filter must first be used to locate the source.

Parentheses, (15), indicate that the appearance overlaps that of another phenomenon listed with an immediately adjacent filter and noted at the end of the same line.

* The numbers denoting the filters are explained in table 4.2.
†The numbers denoting the phenomena are given in table 4.3.

of behavior considered sufficiently important to be given distinct names. It is true that there is no perfect correlation between all the various eruptive and pulsating types of variables for which we have names and the number of related phenomena listed in table 4.3. Nevertheless where there are several different types of named subgroupings of stars or nebulosities, a correspondingly larger number of phenomena often is listed in table 4.3. This suggests that astronomers tend to give new names only to events that differ radically—by factors of 1,000 or more—from previously known classes of behavior. The remarkable point, perhaps, is that by suitably selecting the one arbitrary parameter of the classification—the factor of 1,000—we find a close correspondence between the purely superficial properties of phenomena—their brightness, size, polarization, and spectral and time characteristics—and the subjective distinctions which previously had been made on what appeared to be more detailed information. Either our penetration into astronomy and astrophysics has been equally superficial to date, or else the phase space classification criterion provides useful means for distinguishing classes of behavior that genuinely differ in important ways. Either way, when we come to a calculation of how many more such cosmic phenomena await future discovery, we will be able to make use of the phase space criteria just described.

We might think that the loss of absolute size and absolute luminosity information in the Class I description would be an important hindrance to setting up a list of cosmic phenomena. However, there are a number of factors which counteract this: Jupiter and Betelgeuse appear about equally bright in the sky but actually are quite different types of objects. The distinction between a planet like Jupiter and a star like Betelgeuse, nevertheless, is simple even though the Class I description lacks an absolute sense of distance: Betelgeuse, for all purposes, appears fixed in the heavens, while Jupiter circles the sun once in eleven years. This indirect distance criterion is present for all of the solar system phenomena we observe, and in a very real way provides a direct observational parameter for labeling these phenomena. In the Class I description planets do not so much differ from stars in angular size or apparent brightness as in the rapid excursions they make across the sky. This, in fact, is just the distinction that Copernicus emphasized more than four centuries ago when he argued that stars must be enormously distant, since they exhibited no discernible motion in the sky.

When we look beyond the solar system, similar distance criteria are not quite as apparent but to some extent still are there. We can only detect white dwarfs and flare stars that happen to be quite close to the solar system. When much further away they become too faint to be seen. Because they are so close to us, white dwarfs and flare stars tend to have substantial angular motions across the sky, often amounting to seconds of arc in a century. These apparent motions are due partly to the stars' displacements and partly to the sun's travels through the Milky Way. In the Class I description such stars are therefore distinguished not by their low luminosities but rather by their small but measurable angular excursions—an equally good criterion for distinction.

One more factor should be mentioned. The relative intensities observed through individual filters also may be used as a Class I criterion for distinction, provided the intensity ratios are sufficiently great. Consider the two events in figure 4.1 that are detectable through filter combinations *BD*. Suppose that for event *d* the radiation received through filter *B* is just as intense as the radiation received through filter *D,* while for event *e* filter *B* transmits a million times more radiation than filter *D.* The events *e* and *d* would then represent distinct phenomena because the intensity ratios differ by more than a factor of 1,000.

Our Class I designation might therefore have carried an added intensity label, much like the last digit in the Class II filter designation. In fact, such a list was initially set up, but proved to be almost identical with the Class I listing shown in table 4.3. The relative intensity marking was therefore dropped because for now it appears to provide only redundant information. In time, however, this may change. As observing techniques and sensitivity keep improving, we may ultimately find a larger number of phenomena, all of which can be detected through identical sets of phase space filters. At that stage the order of magnitude distinction in intensity ratios will begin to become important. When we reach that stage of development in astronomical technique, it will be worth remembering that a relative intensity label actually should accompany the Class I filter characterization but has not been included for now.

Independent Recognition

We should now take a more careful look at how cosmic phenomena are actually recognized. Our earlier emphasis on differences by orders of magnitude is not quite enough to qualify a new pattern of events as a phenomenon. We require in addition a positional identification—except in the case of isotropic phenomena. The positional criterion allows us to differentiate between the rediscovery of a phenomenon that we already recognize and the genuine discovery of a new phenomenon: When we discover a new variety of signal coming from certain parts of the sky, we must first check whether there are other signals that can be detected through different phase space filters viewing the same parts of the sky (figure 4.2). If we find that a well-known class of object invariably occupies the same field of view as the new source, then we may wish to check whether each member of the previously known class also exhibits the new variety of signal (figure 4.3). If we find a one-to-one-spatial correlation between sources of the new signals and sources representing the previously known class, we are able to conclude that there has been no new discovery—simply a recognition of a new trait characterizing the previously known phenomenon.

In some instances this new characteristic, by itself, can uniquely identify the previously known class of objects. We then find that we are able to distinguish the phenomenon through either of two sets of totally independent filters. In other words, such a phenomenon is independently recognized through observations carried out in two completely unrelated portions of the phase space of observations. It is as though

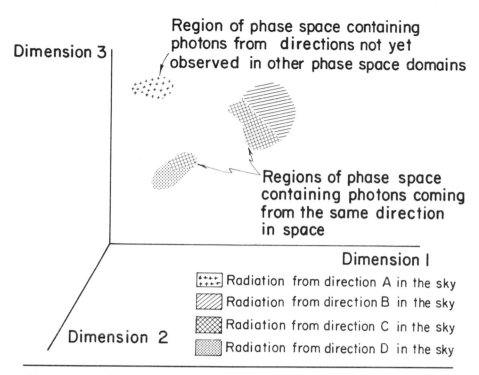

Dimension 3

Region of phase space containing photons from directions not yet observed in other phase space domains

Regions of phase space containing photons coming from the same direction in space

Dimension I

Dimension 2

Radiation from direction A in the sky
Radiation from direction B in the sky
Radiation from direction C in the sky
Radiation from direction D in the sky

Figure 4.2 *Phase Space Filters Used to Observe Different Portions of the Sky*

Suppose we have been searching the sky with a new phase space filter—perhaps a radio frequency detector that can respond to changes in signal strength occurring on a time scale of a few days. We find a portion of the sky from which such signals can be detected, but this is a blank portion of the sky as far as any other phase space filter is concerned. This filter fits into a phase space region shown in the upper left part of the diagram. We can contrast it with a pair of filters shown by crosshatched shading. Each of the corresponding regions represents a separate observing technique; but whenever signals coming from a certain part of the sky can be detected through one of the filters, signals are also obtained through the other.

Unidentified radio sources exhibit the first described behavior. The second type of source might correspond to spiral galaxies, which can be detected both through radio observations of steady signals and observations at visual wavelengths.

Each of the crosshatched regions is associated with a second kind of shading—closely spaced dots on the lower left and parallel lines on the right. These other shadings indicate the existence of sources other than spiral galaxies that emit a steady radio signal though no visual radiation, as well as sources such as stars that emit at optical wavelengths, but not at radio frequencies.

Since it is difficult to represent a space with more than three dimensions, this figure acts as though the phase space only had three dimensions instead of the six shown in figure 3.15.

we were recognizing a friend at an appreciable distance by his peculiar walk, or alternatively from a close-up photograph of his face. The first form of recognition effectively uses a filter that permits us to see time variations but no detail; the second involves spatial detail but no time variations.

Let us again label each phase space filter by means of a single letter, *(A)*, *(B)*, *(C)*, . . . as we had done in figure 4.1. We can then characterize a phenomenon by a monogram constructed from the letters representing the filters through which the phenomenon is recognized. A phenomenon

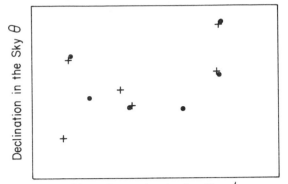

- Sources having properties $(\nu_1, \Delta\nu_1, \Delta t_1, \Delta\alpha_1, \pi_1)$
+ Sources having properties $(\nu_2, \Delta\nu_2, \Delta t_2, \Delta\alpha_2, \pi_2)$

ν Specifies the spectral frequency at which we observe
$\Delta\nu$ Specifies the spectral resolution
Δt Specifies the time resolution
$\Delta\alpha$ Specifies the angular resolution
π Specifies the polarization

Figure 4.3   *Phase Space Filters Through Which Signals Can Be Detected from Different Portions of the Sky*

In this diagram, signals obtained through two different phase space filters, labeled with dots and crosses respectively, are shown at given source locations (θ,ϕ) in the sky. For several positions signals are received through both phase space filters, even though the spectral frequency at which observations are conducted ν, the spectral resolution $\Delta\nu$, the time resolution Δt, the angular resolution $\Delta\alpha$, and the polarization to which our observing equipment is sensitive π, all may differ by many factors of 1,000.

If we find a one-to-one correspondence between those parts of the sky in which we see a signal through one filter and those in which we see signals through the other, then we say that we recognize a given phenomenon through two different filters. If, on the other hand, there exist portions of the sky labeled by dots but not by crosses, or portions labeled by crosses but not by dots, then we say that we distinguish two, or possibly three, different phenomena, potentially those labeled by dots, those labeled by crosses, and those labeled by both dots and crosses.

For a phenomenon to be recognized in two truly independent ways, the two filters or two sets of filters through which the phenomenon invariable provides signals must be able to function independently and must, in addition, provide a unique identifying monogram for the phenomenon.* Not all filters currently in use can be employed independently. Most far-infrared observations can only be carried out if we know the position of a source from X-ray, optical, or radio source catalogues. Far-infrared spectral observations, in that sense, cannot be carried out independently from other sightings.

* See Figure 4.4.

recognized in two different ways, therefore, is one whose monogram can be split into two nonoverlapping parts, each of which independently identifies the phenomenon and differentiates it from all others. Independent recognition of this type will play a prominent role when we come to estimate the total number of phenomena that characterize the universe.

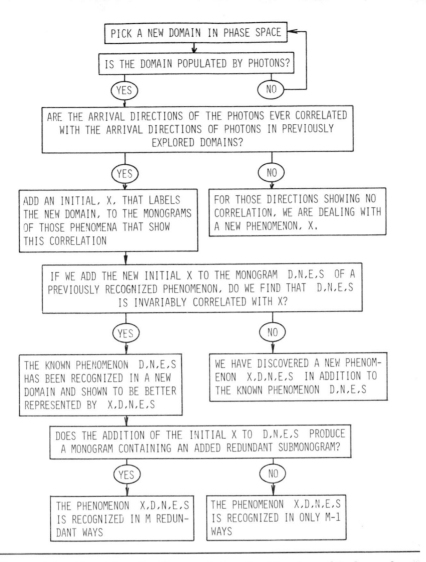

Figure 4.4 *Algorithm for Determining the Number of Independent Ways in Which a Phenomenon is Recognized*

An objective approach to defining what is meant by a novel phenomenon or by a phenomenon recognized in several independent ways, requires a prescription—a set of procedures to be followed. Only in that way can different researchers agree on what constitutes a cosmic phenomenon and what does not. The algorithm presented here is a procedure of this sort. It prescribes tests to determine whether a new set of observations has discovered a novel phenomenon. In principle, computerized machinery could be set up to survey the sky in the manner prescribed. In practice, we have not yet approached such a degree of automation in astronomy. Tables 4.3 to 4.5 provide lists of filters through which different phenomena are singly or redundantly recognized. The procedure followed in establishing those tables were based on the methods advocated in the algorithm shown here.

Figure 4.4 shows an *algorithm*—a procedure—for deciding whether a newly observed event represents a new phenomenon, adds to our information about known phenomena, or leads to a novel, independent recognition of a previously known phenomenon.

Finer Structural Detail

To some extent, the positional criterion for identification of a new phenomenon discriminates against certain finer features characterizing a source. For example, the filamentary structure of a supernova remnant is not recognized as a phenomenon in its own right. The filaments merely act as substructure. Neither are rings around planets recognized in this classification scheme.

If this feature of the Class I or Class II definition proved to be objectionable, one could perhaps consider different classes of phenomena—call them *Class III* or *Class IV*—that would give greater prominence to substructure. We would then obtain a list of phenomena, rather different from ours, appreciably longer, and peculiarly concerned with detail which we have passed over as insufficiently important. The statistical conclusions that we will reach in the next few sections would also have to be modified for such smaller scale phenomena; but that would not be disturbing because the statistics would in fact govern quite different groups of objects and events. We would be considering a different class of entities.

This re-emphasizes the twofold nature of the study we have made. First, there is the formalism we have set up to discuss cosmic phenomena in some reasonably objective manner. And second, there is the fit we have established between the size of our phase space filters and the set of phenomena. It is this fit that can be altered to suit a variety of classes of phenomena of potential interest. The underlying formalism, however, can remain unaltered for these differing classes and therefore may prove to have a wider range of applications.

Changing Lists of Phenomena

A phenomenon that falls into the Class I or Class II mold today may later turn out to be a composite in the sense that actually several different phenomena have been mistaken for a single one. A good example of this kind of splitting is given by the listings in Messier's catalogue of diffuse sources. When it was published in 1781, the catalogue listed diffuse patches stationary in the sky and therefore different from comets. Comets orbit the sun, but Messier's nebulosities remain fixed in the sky. We now know that Messier's list comprises galaxies, ionized hydrogen regions, galactic and globular clusters, planetary nebulae, supernova remnants, and several other distinct phenomena. This successive splitting is particularly noticeable when we compare the discoveries in figure 1.1 to the list of forty-three phenomena enumerated in chapter 2.

Splitting, however, is not the only change that can occur. There can equally well be an amalgamation of phenomena once considered separate into a single more variegated phenomenon. A phenomenon such as the gamma bursts might eventually turn out to be part of some other well-known class of events. The astrophysicist Stirling Colgate, for example, has argued for a supernova origin for gamma bursts.[2] If the bursts proved to be consistently emitted in supernova explosions, we would say that the bursts do not constitute a separate phenomenon but rather are one more manifestation of supernovae. If, on the other hand, only certain types of supernovae exhibit gamma bursts, then we would still be dealing with a new phenomenon, namely supernovae that emit gamma bursts as distinct, by orders of mangitude in this trait, from supernovae which do not.

Suppose that a certain object we see in the sky exhibits behavior characteristic of some phenomenon a. At a later epoch, however, it starts to exhibit properties characteristic of the phenomenon c. We might find, if we wait long enough, that this is a quite common occurrence and that all the various sources representing the phenomenon a eventually become members of the class c. In that case we might conclude that there is only a single phenomenon showing a certain type of variability. If we must wait a long time for this variability to become evident, we may have to wait until astronomy has become old enough to give us access to some very slow time scale filters in our phase space of observations.

A concrete example of this type of transition involves supernovae and supernova remnants. These are listed separately in table 4.3. At the moment we do not know whether all supernovae give rise to supernova remnants, nor whether all the remnants we currently observe were produced by supernovae. The evolutionary time scale of these remnants is about a thousand years; we may therefore have to wait that long to answer this question to our complete assurance. In the meantime, however, we have listed remnants and supernovae as separate phenomena because observationally their one-to-one correspondence has not been established.

Such a changing list of phenomena is somewhat inconvenient. We would prefer to deal with phenomena that are identified once and for all, with no ambiguities about how they are to be classified and no possible mistakes in classification. But this shifting base of phenomena is inevitable as long as our knowledge of the grand features of our universe remains incomplete. A stable listing will only emerge after all major cosmic phenomena have been correctly identified.

Statistics of Cosmic Phenomena

We can return to a point of view examined earlier, rephrasing it now in terms of the phase space volume, V, containing all possible phase space filters. Let us consider that the universe may be characterized by n distinct phenomena. If all possible observations we might undertake can be arranged to fill a space of volume, V, and if the observations that we have actually made occupy a subvolume, v, of this space, what fraction of the n phenomena will we have recognized thus far? To answer this question we can proceed in the following way:

We assume that those observations needed to recognize a given phenomenon are distributed across the phase space volume, V, in a manner independent of the sequence of observations actually undertaken. In astronomical observations this random historical approach appears, in fact, to have been taken. Novel techniques implemented in astronomy frequently make use of discarded military components and have little to do with expected astronomical source strength: The postwar advances in radio, radar, and rocket ultraviolet astronomy were brought about largely in this way. They took place in spite of theories previously held by optical astronomers. More recent examples are the U.S. Air Force mid-infrared sky survey from rockets and the discovery of gamma bursts by the Vela military satellites.

Such examples imply that there is little correlation between the observations actually made and those that would ideally lead to the largest payoff in new discoveries.

Let us define two parameters:

1. An efficiency parameter ϵ which represents the probability of discovering a phenomenon when observations are carried out in a portion of the space where the phenomenon can be recognized
2. A *modality* parameter m which specifies the number of independent regions in volume V, where recognition of a given phenomenon is possible

Both these parameters depend on the nature of cosmic phenomena that we have not yet discovered; in addition, our proficiency, represented by ϵ, varies from one region of V to another. Nevertheless, we will assume that there exists some average value of m that would be characteristic of all phenomena, and we can also assume that there is some mean value of ϵ that holds for the subvolume v which our present astronomical techniques have allowed us to search. For our argument, we will not need to know the actual values of ϵ, m, and v; we will need to know only that such average values exist. What these values actually are, fortunately, is unimportant.

Having stated all these provisos, we can go ahead to estimate the total number of phenomena, n, in the universe. We first write down an expression for the number of phenomena, A, discovered without redundancy thus far. Our assumption of statistical independence allows us to state (appendix A) that

$$A = n \, (m\epsilon v/V)[1 - \epsilon v/V]^{m-1} \tag{1}$$

provided v is much smaller than V, and provided the value of m is appreciably larger than one. We know from our previous discussion in chapter 3* that v currently is only 1 percent of the volume V. We do not know much about the value of m and will have to discuss a variety of possibilities later. For now, however, we can also write down an expression for the number of phenomena that could have been recognized in two independent ways.

* See page 191.

This independence involves recognition of a phenomenon in two distinct sets of domains of v. For a random distribution, appendix A shows the number of doubly recognized phenomena to be

$$B = \frac{nm(m-1)}{2} (\epsilon v/V)^2 (1 - ev/V)^{m-2} \tag{2}$$

The combination of equations (1) and (2) now allows us to estimate n as

$$n \approx A[A + 2B]/2B \tag{3}$$

where A is the number of phenomena singly recognized thus far and B is the number of phenomena recognized in two independent ways. Since the values A and B can be observationally determined, we can obtain a direct evaluation of the number of cosmic phenomena n.

In general, equation (3) holds only when A is much larger than B. But since by far the largest fraction of the phenomena known to us is clearly recognized only through the full set of data that we possess, the number of redundantly observed phenomena B indeed is small compared to A. This condition, therefore, is reasonably well satisfied.

Let us now turn to the question of possible variations in m. We can estimate this effect realistically if we suppose we are dealing with two classes of phenomena. There are n_1 phenomena with a modality m_1 and another n_2 phenomena with modality m_2. The true number of phenomena then is $n_1 + n_2$, but use of equation (3) will always lead to a value that is less than or, at best, equal to the true value. We treat one extreme example, $m_1/m_2 = 16$, in appendix A, and show that it does not alter our estimate of the true value of $n_1 + n_2$ by much more than a factor of 2, for particular values $A = 37$, $B = 7$, which approximately correspond to values for A and B cited for various classes of phenomena listed in table 4.3 and in table 4.6 below. In appendix A we also note how the number of triply recognized phenomena can be used as a guide to check whether m appears to vary widely. It does not.

It is important to note just how general the result summarized by equation (3) really is. It does not strongly depend on the number of recognition domains m, or on the efficiency ϵ of our technique, or on an accurate estimate of the total volume V—as long as it is finite—or on the portion v in which we have carried out observations. It depends only on a proper accounting of the phenomena thus far observed and recognized: How many have we recognized? How many have we recognized in two independent ways?

In table 4.3, we provided a list of forty-four phenomena chosen following as faithfully as possible the Class I criterion. About seven of these are recognized in two ways, and one, interplanetary matter, in three ways. We can make use of this result and set $A = 36$, $B = 7$, and $C = 1$.

If we substitute the values $A = 36$, $B = 7$ in equation (3), we see that the universe should contain a total number of phenomena n

amounting to about 129, of which the phenomena listed in table 4.3 constitute about 34 percent.

Very few phenomena are really understood in two redundant ways: Planets are just barely recognized redundantly. To be sure, the radio emission from Jupiter has been known for some years, but it would have taken about eleven years to establish the closed elliptical orbit about the sun. The existence of planets other than our own would therefore be a very recent discovery if all optical data were somehow kept from us, as they would be if we were inhabitants of the ever-cloudy planet Venus.

Gas-containing galaxies now also are redundantly recognized (figure 4.5). Within a few years supernova remnants should also become redundantly recognized by virtue of their X-ray emission.

Interestingly, it is not until B starts approaching the value of $A/2$ that we will be close to approaching completeness of cosmic observations. At present we are quite far from that goal, but when we approach it, we will need to make use of the number of triply and quadruply identified phenomena to arrive at a best estimate for n. This is illustrated in figure 1.9 which refers both to unimodal and multimodal phenomena. In chapter 1 we estimated the maximum number of currently known unimodal phenomena—phenomena for which $m = 1$—to be $A' = 4$. The four phenomena referred to were the poorly understood gamma bursts and gamma-ray background, unidentified radio sources, and X-ray background. If we take $A' = 4$ and $V/v = 100$, we see that the number of unimodal Class I phenomena we could eventually discover is at most of the order 400, *at the present efficiency level ϵ.* The number of unimodal phenomena in our universe can therefore not be orders of magnitude higher than the number ~130 we estimate for multimodal events. An expression that gives upper and lower bounds on n when both unimodal and multimodal phenomena exist (and $B \ll A-A' \ll n$) is

$$A[A + 2B]/2B < n < (A - A')[A - A' + 2B]/2B + A'V/\epsilon v \qquad (4)$$

Figure 4.5   *The Spiral Galaxy M81 Seen at a Wavelength of 21 Centimeters*

At a wavelength of 21 centimeters, atoms of hydrogen emit a spectral line that can be used to uniquely identify this atom. Radio observations at these wavelengths therefore permit us to map the distribution of hydrogen within a source. Here we see that the interstellar hydrogen in the galaxy M 81 is distributed largely in the spiral arms. The central portion of the galaxy appears devoid of these atoms.

Doppler shifts can be measured in the 21-centimeter line just as readily as at optical wavelengths, and the rotational velocity becomes a clear identifying radio feature for the gas-containing galaxies. From this picture it is apparent that gas-containing galaxies are now identified as readily at radio wavelengths as in the visual part of the spectrum, and we can consider these galaxies to be independently recognized through two quite distinct sets of observational capabilities.

Composite by A. H. Rots and W. W. Shane, courtesy of *Astronomy and Astrophysics*[9]

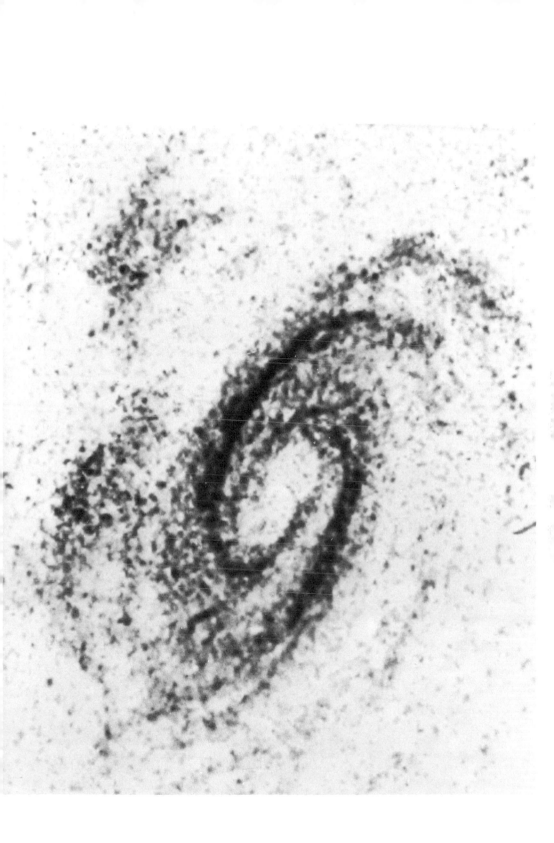

Stability of the Results

We chose to define the concept of an astronomical phenomenon in one particular way and called phenomena fitting into this mold Class I phenomena. The calculations we have made have concerned these, and we might therefore wonder what would have happened had we chosen to define phenomena in a different way.

We can at least make an estimate of the stability of our results—their insensitivity to changes in definition—by comparing the results obtained for Class I phenomena with those that we would obtain if we took the Class II description. In table 4.3 we compare the number of singly and doubly recognized members of the set to arrive at a projected value for the total number of phenomena described in this way.

The Class II phenomena can be divided into two subclasses, those in table 4.3 and a similar tabulation that would be obtained if we omitted the intensity characterization of the phase space filters and only retained the absolute size designation. We might call these classes respectively *Class IIa* and *Class IIb*. The Class IIb filters have characteristics that appear to lie between those of the *Class I* and *Class IIa* filters.

When we compare these listings, we find the results shown in table 4.6.

TABLE 4.6
Number of Phenomena of Different Classes

	Singly Recognized	Doubly Recognized	Triply Recognized	Expected Total
	A	B	C	n
Class I	36	7	1	129
Class IIa	39	9	1	124
Class IIb	37	7	0	135

The variations in these results do not appear to be excessive and suggest that any objective definition of phenomena that came out with a similarly good fit to the forty-three phenomena we had originally selected as archetypes (figure 1.1) is likely to come up with roughly similar results. However, as astronomy develops, and our capacity to observe improves, we may find that Class I phenomena no longer bear a close relationship to what we would instinctively call an astronomical phenomenon today, and a different classification could then be defined. At any epoch, however, we may expect that the general approach outlined here will permit us to make a quantitative estimate of the discovery rate and the total number of additional discoveries to be expected in further observations. It should also permit selection of the kind of observations that promise the highest discovery rate for the immediate future. At present our best strategy lies in the utilization of novel carriers or of techniques involving new wavelength ranges and improved resolving power.

New Phenomena We Might Expect to Detect

We may expect to discover three particular phenomena in the future: disk stars, black holes, and intelligent life in the universe.

Disk Stars

We are on the verge of discovering disks around stars. A picture of a disk star would look somewhat like the planet Saturn with its rings. A number of hints for the existence of such objects are seen in young stars of the T Tauri type which exhibit evidence of such a surrounding disk.

Disks also may surround X-ray stars: Here the emission of X rays occurs when matter from an extended star leaves its surface and is gravitationally funnelled onto a more compact companion star. This transferred gas falls onto the compact star with such energy that X rays are emitted when the gas finally is slowed down near the star's surface. As it falls into the compact star, however, the gas approaches along a spiral path, forming a disk in the same plane in which the two stars orbit each other. Such a disk might eventually be detected directly when X-ray observations attain sufficiently high spatial resolution.

Other types of disks might also be found in the universe.

Black Holes

Black holes have been discussed with much enthusiasm in the past few years, but none have yet been unambiguously identified. There are a number of ways in which they might be observed: One is to see an ordinary star orbiting about a massive hole. The hole itself emits no light, and we would observe the ordinary star orbiting around essentially nothing we could see.

Matter falling into a black hole can also produce X rays just before it crosses the star's Schwarzschild radius, the radius from within which no matter or radiation can escape. X radiation, however, also is seen from neutron stars, which would be difficult to distinguish from black holes through X-ray emission alone. But a black hole may rotate much faster than neutron stars might be expected to spin without flying apart; and if an extremely short period rotation, measured in fractions of milliseconds, were observed modulating the X-ray or gamma-ray emission, then a black hole would have to be involved. High time resolution and extremely high spatial resolution X- and gamma-ray observations would probably be needed for this type of recognition.

Intelligent Life in the Universe

If life prevails throughout the universe—not just on our planet—it must be counted as an important cosmic phenomenon. We may doubt that life exists elsewhere, but that is simply a matter of taste, not of fact. Many astronomers argue that life should exist somewhere in those hundred billion galaxies, each consisting of a hundred billion stars, a good fraction of which may support planetary systems.

Let us be persuaded that life might exist elsewhere in the universe and that we should search for it. Intelligent life willing to communicate with other civilizations is the most readily detected species. None of us knows, of course, how such a civilization might choose to communicate, but we can be sure that there would be attempts to cut costs and to communicate as economically and as effectively as possible. Bernard Oliver, an avid enthusiast in the search for life, who also is vice president for research and development at Hewlett-Packard Corporation, has argued that the most likely carriers of communications from such civilizations lie in the 1.4 to 1.7 GHz frequency band of the radio spectrum.[4] Here the interstellar gases in our galaxy produce very little emission that would compete with a beamed message; the dispersion, which causes waves of different frequencies to be transmitted at different speeds and tends to diffuse pulsed signals, is low; and sensitive receivers, as well as efficient transmitters, are readily constructed.

If we take the suggestions of Oliver and his colleagues in the Search for Extraterrestrial Intelligence (SETI) study seriously, we should construct equipment to detect radio emission from distant planets—point sources in the sky. The signals would be transmitted in a narrow frequency band in order to catch our attention, and this radio frequency would be encoded with some time varying pattern to further call attention to itself and deliver a message. To detect such a signal we need antennas with high spatial resolution to single out point sources, with high spectral resolution to detect narrow lines, and with precise (high) time resolution to cull the encoded regularities—the time of arrival of photons—in the signal.[5]

Observation, Exploration, and Experiments

Many branches of science can be pursued by means of a mixture of observation and experimentation. Medicine is one of these, and astronomy is another. Thus far, we have dealt with the observational side of the discipline, but some kinds of astronomical information are better obtained through local studies: Analyses of meteoritic material falling onto Earth from interplanetary space provide one example of laboratory experiments which, in fact, play an essential role in unraveling the chemistry of the early solar system.

The main difference between an experiment and pure observation is just this: An observer receives information by means of carriers such as photons, cosmic-ray particles, meteors, and so on. The influx of these carriers is beyond control. The collection of data is passive, except insofar as the observer may make a decision to select only limited portions of the available data. An experimenter, in contrast, actively influences the system he studies. He first transmits information to the system and in return receives information conveying the system's response. These ideas are not new. As long as a century ago, in his *Introduction to the Study of Experimental Medicine,* the French physiologist Claude Bernard was writing about these distinctions, as applied to medicine.[6]

We may wonder whether the limitations we find in observational astronomy have any analogue in the experimental and exploratory sciences. How do the limits of astronomy, which is largely an observational science, compare to those of physics or biology, where experimentation is the main method of advance? In these sciences the phase space needed to describe all possible experiments is substantially more complex than the phase space dealing solely with observations.

We can subject a system to experimental conditions defined by a number of different parameters—the temperature at which the system is kept, the magnetic field strength in the system's vicinity, the electromagnetic radiation, if any, incident on the system during the observations, and so on. We may then assign an order of complexity to this experiment. An experiment of zero order is one in which no parameter can be varied at the observer's will. Such an experiment is purely an observation. A first order experiment in contrast permits the experimenter to control a single parameter, perhaps the temperature or, possibly, the magnetic field. A second order experiment has two such parameters that may be varied independently: the magnetic field, for example, and the temperature. Experiments of third, fourth, and higher order can be thought of in similar ways (figure P.1 of preface).

Many of the experiments we perform today are first, second, and third order experiments. An experiment of fifth order becomes quite complex; the data to be gathered have, in some sense, to be displayed in a five-dimensional space, and the costs of the experiment increase quite rapidly.

These costs can be estimated by imagining a simplified example. Consider an experiment that has 5 different parameters, all of which can be independently varied. We would like to vary each parameter over a range of some 10 different values. The observed system response to each such combination of imposed parameters is recorded. For 5 parameters each of which assumes 10 values, we will need to record the results observed for 10^5 different combinations of values. If the cost of each independent measurement is fixed, for example, by the time and effort required to record individual results, we see that the cost of a fifth order experiment is 10 times higher than the cost of the corresponding fourth order experiment. A sixth order experiment would be correspondingly more expensive than a fifth order investigation. Quantitatively expressed, the cost depends on the cost in dollars per individual operation, c, the number of parameters varied, P, and the number of values, s, that we are allowed to assign to a given parameter. Here we had taken $s = 10$, but it can take on whatever magnitude the experimenter deems necessary. The cost in dollars then becomes

$$\text{Cost} = cs^P$$

Information and Expense

The amount we may expect to learn about a system through observation or experimentation ultimately is limited by the amount of information we can gather. But, as we just saw, information can only be obtained at a cost.

Science already is costing the United States a small percentage of its annual national budget—if applied science and military research are included. At the present national wealth and current prices, we therefore cannot expect to significantly increase the amount of research undertaken. Conceivably the cost of research may be lowered in time, but there are fundamental limitations to the costs, which eventually will provide real bounds on the complexity of practical experiments.

These limitations can be exemplified by two types of experiments—those that involve very high energy particles produced in an accelerator and those that involve extremely low temperatures (energies). Both types of experiments exhibit energy limitations and ultimately cost limitations. High energy experiments eventually must be limited, if by no other factor, then by the finite energy resources available for the construction of accelerators and for the acceleration of individual particles. Low temperature efforts similarly depend on the energy that can be budgeted for running refrigerators that yield ever lower temperatures. Since the efficiency of refrigerators rapidly decreases at lower and lower temperatures, a finite amount of expendable energy will limit us to cooling a given amount of matter to a finite temperature above absolute zero. Cooling to even lower temperatures, by the laws of thermodynamics, would then require even greater expenditure of energy than we had budgeted. In both these situations, then, limits on information that can be gathered are set by available energy resources.

A related restriction concerns data banks used to store information. Each bit of information entered into such a bank is accompanied by a small but finite heat input. In appendix B, we show that the energy accompanying one bit of information brought in along a channel kept at a temperature T is $0.693\ kT$ ergs. If we add this information to a data bank, we impart heat to the bank; and in order to keep the temperature from rising excessively, this heat must be removed by refrigerators. The efficiency with which we can cool the bank is at best T/T_0, where T_0 roughly corresponds to the ambient temperature at Earth. Hence, we must invest an energy of at least $0.693\ kT_0$ for each stored bit, and

T_0 certainly is at least as high as 3 degrees Kelvin, the temperature of the cosmic microwave background. Data processing energy requirements take similar tolls in energy and ultimately in cost.

Expenditure Rates and Information Rates

While the total expense of a scientific project is an important limitation, we are frequently restricted even more severely by limited annual expenditure—the amount of money we can spend on a project each year. The rate of spending in this instance is more of a constraint than the total amount of money to be spent over the entire life of the project.

As science develops, we may ultimately face a way of doing research vastly different from the methods we now know. To carry out experiments of increasingly higher order, or cosmic observations on very faint or very slowly changing phenomena, we may have to allocate large sums of money over periods spanning many millennia. Science then would no longer be the exciting, highly personalized adventure it has become for many scientists in our time. It would be a project run according to a predetermined schedule over a span of centuries or millennia by a cult of specially trained, dedicated disciples.

Whether science carried out in that fashion can maintain its momentum is not clear. Perhaps projects requiring sustained efforts lasting many millennia cannot be contemplated. Perhaps mankind's attention span is too short, and scientific questions that can only be answered through such sustained efforts will ultimately not be considered worth answering. Science is a venture that does depend, more than we generally acknowledge, on individual style and flair; and if the style in which our society approaches problems always requires the quick solutions that science provides today then certain classes of questions may have to remain permanently unanswered.

If great projects requiring efforts and patience well beyond any we have known are to become realities, we will need to understand how societal factors influence scientific progress, how the education of young astronomers channels their efforts in later life, how the methods controlling the distribution of grants determine the resulting forms of research, and how the rewards we offer scientists affect the style in which they work. More than mere technical expertise, a vigorous future will require an enlightened approach to our cosmic search, an understanding of the scientific enterprise well beyond the limited insights we have gleaned to date.

The Fringes of Legitimacy — The Need for Enlightened Planning

The Ways of Discovery

Any attempt to chart the future of astronomy must be based on lessons learned from previous successes and failures. We saw in chapter 2 that no two astronomical searches follow quite the same course; the forty-three discoveries we examined exhibit the diversity of paths that can lead to the recognition of a new phenomenon. Yet even in this varied set of histories a few trends emerge, and these can serve as useful guides to the future. Table 5.1 summarizes the data and provides an overview.

The first column of the table names the phenomenon. The second gives the year in which the discovery was made. Frequently, several dates are given because discoveries often proceed in discrete steps. Globular clusters were first recognized only as nebulosity that clearly differed from ordinary stars. Successively these clusters also became distinguished from planetary nebulae, galactic clusters, galaxies, and so on. The different dates entered for globular clusters reflect these stages. To avoid excessive repetition, the differentiating steps are entered in the table for only one of the several phenomena distinguished through a key observation.

The third column presents the instrumental capabilities that were required for a phenomenon to become recognized. Planetary nebulae and globular clusters both can appear as perfectly round nebulae. Seen through a spectroscope, however, they are clearly distinct. The planetary nebula shows a few bright spectral lines, the globular cluster a diffuse spectrum. When William Huggins introduced the spectroscope to nebular astronomy in 1864, he saw this clear-cut distinction at a glance. Before the introduction of this instrument, a century had been spent

TABLE 5.1

Summary of Data on Forty-Three Astronomical Discoveries

Phenomenon	Year of Discovery or Rediscovery	Instrumental Requirements	Age of Technology*	Military (M) Communications (C)	Chance Discovery	Discoverers	Professions	Remarks
1. Interplanetary Matter								
a. Meteors	1798	Eye	>50y		No	H. W. Brandes	Physicist	Height and velocity
						J. F. Benzenberg	Physicist	
b. Meteorites	1803	Eye	>50y		No	J. B. Biot	Physicist	Examination of a fall
c. Zodiacal dust	1934	Coronal spectra	~25y		No	W. Grotrian	Phys./Astron.	Fraunhofer lines
d. Radar meteors	1946	Radar	<5y	M	Yes	J. S. Hey	Physicist	Echoes from trails
2. Planets	Antiquity	Eye	>50y		Yes			Orbital motion
	1610	Spyglass	<5y	M/2	Yes	Galileo	Physicist	Disk resolved
	1728	Zenith tube	~50y		Yes	J. Bradley	Astronomer	Earth's motion
	1955	Radio telescope	~10y	M/2	Yes	B. F. Burke	Physicist	Radio rediscovery of planets
						K. L. Franklin	Astronomer	
3. Asteroids	1801	Telescopic Astrometry, Daily observations	~50y		Yes	G. Piazzi	Theology/Math/Astron.	Orbit predicted by J. H. Bode
4. Moons	1610	Spyglass	<5y	M/2	Yes	Galileo	Physicist	Orbital motion
5. Rings	1655	Few arc second resolution	<5y		No	C. Huygens	Physicist	Saturn's rings
6. Comets	1577	Arc minute precision parallax observations	~10y		No	Tycho Brahe	Astronomer	Shown to be extraterrestrial
	1705	Century long orbital records	~30y		No	E. Halley	Astronomer	Periodic orbit

7. Main sequence stars	1717	Positional precsn: arc minutes/century	~50y	No	E. Halley	Astronomer	Proper motion
8. Subgiants and red giants	1838	≤1 arc sec/year	~10y	No	F. W. Bessel	Math/Astron.	Annual parallax
	1890	Objctv. prism. photogr.	~10y	No	E. C. Pickering	Physicist	Spectra and distances for large numbers of stars
	1910	Arc second/year	~20y	No	E. Hertzsprung / H. N. Russell	Chem. Eng. / Astronomer	
9. Pulsating variable stars	1596	Eye	>50y	Yes	D. Fabricius	Pastor/Astron.	Variability
	1912	Photog. photometry	~10y	No	H. S. Leavitt	Astronomer	Period/Luminosity
	1914	Precision spectra	~10y	No	H. Shapley	Astronomer	Expansion/Contr.
10. Multiple stars	1672	Eye	>50y	Yes	G. Montanari	Math/Phys./Astron.	Eclipsing var.
	1803	Arc second astrometry	~50y	Yes	W. Herschel	Musician	Visual binaries
11. White dwarfs	1834	Sub-arc second astrometry	~50y	No	F. W. Bessel	Astron./Math	Required selection of high proper-motion stars
	1862	Arc second resln. low scatter	~20y	Yes	A. G. Clark	Telescope maker Astronomer	⎫ Companion was very
	1915	Arc second resln; low scatter spectroscopy	~20y	No	W. S. Adams	Astronomer	⎭ faint compared to Sirius
12. Galactic clusters	1754	Small telescope	>50y	No	N. Lacaille	Abbé/Astronomer	Resolved stars
	1864	Spectroscope	~50y	No	W. Huggins / W. A. Miller	Gentleman Chemist	Found stellar continuum spectrum
13. Globular clusters	1781	Low magnification	~50y	No	W. Herschel	Musician	Globular appearance partially resolved into stars
	1864	Spectroscope	~50y	No	W. Huggins / W. A. Miller	Gentleman Chemist	Stellar continuum Spectrum
	1899	Photography	~25y	Yes	S. I. Bailey	Astronomer	Cluster variable stars

* Rounded off to 5-year intervals.

TABLE 5.1—*Continued*

Phenomenon	Year of Discovery or Rediscovery	Instrumental Requirements	Age of Technology*	Military (M) Communications (C)	Chance Discovery	Discoverers	Professions	Remarks
14. Planetary nebulae	1790	Arc sec. resolution, low scatter	~40y		Yes	W. Herschel	Musician	Star centered on nebula
	1864	Spectroscope	~50y		Yes	W. Huggins W. A. Miller	Gentleman Chemist	Isolated emission lines
	1961	Short wavelength radio sensitivity	<5y	M/2	No	C. R. Lynds	Astronomer	Radio rediscovery
15. Ionized hydrogen regions	1865	Spectroscope	~50y		Yes	W. Huggins W. A. Miller	Gentleman Chemist	Isolated emission lines
	1954	Short wavelength radio sensitivity	<5y	M/2	No	F. T. Haddock C. H. Mayer R. M. Sloanaker	Physicist Electr. Engr. Physicist	Thermal radio continuum emission (rediscovery)
16. Cold gas clouds	1903	High resolution spectroscope	<15y		Yes	J. Hartmann	Astron./Astrophys.	Doppler-shifted & unshifted lines
	1951	High spectral radio resolution	~5y	M/2	No	E. M. Purcell H. I. Ewen	Physicist Physicist	Theoretically predicted (rediscovery)
17. Interstellar dust and reflection nebulae	1914	Faint, medium spectral resln.	<25y		No	V. M. Slipher	Astronomer	Reflection nebulae
	1930	Photometry	~50y		No	R. J. Trumpler	Astronomer	Dust absorption
	1948	Polarization	~50y		Yes	W. A. Hiltner J. S. Hall	Astrophys. Astronomer	Polarization

Phenomenon	Date	Observational requirement	Duration	M/2	Yes/No	Discoverer	Profession	Nature of discovery
18. Supernovae	Antiquity	Eye	>50y	M/2	No	Chinese Imperial Astrologers	Astronomer	Systematic sky watch
19. Eruptive variables	1885	Small telescope	>50y		Yes	E. Hartwig	Astronomer	SN in Andromeda
	1917 to 1934	Large telescope, sensitive photography	~10y		Yes	G. W. Ritchey	Telescope designer, astronomer	Nova in Andromeda
						H. Shapley	Astronomer	Brightness estimates
						W. Baade	Astronomer	
						F. Zwicky	Physicist	
20. Nebular variables	~1861	Long-term observations	~50y	No	Yes	J. R. Hind	Civil Engr./Astronomer	Variable nebula
	1945	Faint object	<10y		No	A. H. Joy	Astronomer	Variability & emission line spectra; ambient clouds
	1951	Variability & spectroscopy	~50y		No	G. Herbig	Astronomer	
	1952		~40y		Yes	G. Haro	Astronomer	
21. Infrared stars, circumstellar dust clouds	1965	Sensitive detectors, sky emission subtraction	<5y	M/2	Yes	G. Neugebauer	Physicist	Infrared sky survey
						R. B. Leighton	Physicist	
22. Flare stars	1949	Fast response	~50y	No	Yes	W. J. Luyten	Astronomer	Flares recorded in earlier decades
23. Magnetic stars	1963	Large radio telescope	<5y	M/2	No	A. C. B. Lovell	Physicist	Radio flaring
	1947	High spectr. resln. & sens.; circ. polar.	<10y	No	No	H. W. Babcock	Astronomer	Zeeman split stellar lines
24. Cosmic masers	1965	High spectr. resln., ~10d time variations & polarization	~10y	M/2	Yes	H. Weaver	Astronomer	Highly polarized OH emission from compact sources
						S. Weinreb	Electr. Eng.	
						A. H. Barrett	Physicist	
25. Pulsars	1968	Sub-second time resln.	~5y	M/2	Yes	A. Hewish	Physicist	Radio pulsations
						Jocelyn Bell	Physicist	
26. X-ray stars	1962	X-ray sensitivity	<5y	M/2	Yes	Bruno Rossi	Physicist	Continuum X-ray emission
						R. Giacconi	Physicist	
27. Supernova remnants	1937	Low resln. nebular spectr.	~20y	M/2	Yes	N. U. Mayall	Astronomer	Doppler-split spectral lines
	1939	High ang. resln. photogr.	~20y		No	J. C. Duncan	Astronomer	Expanding size of nebula
	1942	Historical data	~50y		No	J. H. Oort	Astronomer	Connection with SN of 1054
						J. J. Duyvendak	Historian	

TABLE 5.1—Continued

Phenomenon	Year of Discovery or Rediscovery	Instrumental Requirements	Age of Technology*	Military (M) Communications (C)	Chance Discovery	Discoverers	Professions	Remarks
28. Interstellar magnetic fields	1957	Optical polarization and polarization at short radio wavelengths	5y	M/2	No	V. A. Dombrovsky, C. H. Mayer, T. P. McCullough, R. M. Sloanaker	Astronomer, Electr. Engr., Physicist, Physicist	Polarized as theoretically predicted
	1969	High spectrl resln. & circ. polarization	<5y	M/2	No	G. Verschuur	Radio astronomer	Zeeman split hydrogen lines (rediscovery of field)
	1972	Low spectrl resln. & lin. polarization	<10y	M/2	No	R. N. Manchester	Phys./Geophysicist	Interstellar Faraday rot. (Accurate field measure)
29. Galaxies containing gas	1845	High ang. resln.	<5y		Yes	William Parsons	Earl of Rosse	Spiral structure
	1917	Large telescope sens. photography	~10y		Yes	G. W. Ritchey	Astronomer	Extragalactic novae
30. Galaxies devoid of gas	1925	High sens., ~2-sec. ang. resln., 1 mon., vis. varbl.	~25y		No	E. P. Hubble	Astronomer	Cepheid-based distance
	1927	High spectrl. resln.	~10y		No	J. H. Oort	Astronomer	Galact. diff. rot. from Doppler vel.
	1939	Faint nebular ultra-violet spectra	~15y		No	N. U. Mayall	Astronomer	Intrastellar gas
	1954	High spectrl & spatl. resln. spectr.	<10y	M/2	No	F. J. Kerr, J. C. Hindman, B. J. Robinson	Radio Physicist, Radio Techncn., Physicist	21-cm rotation. (Rediscovery of galaxies with gas)

No. / Discovery	Year	Instrument / Technique	Lead time	Class	Predicted	Discoverer	Profession	Finding
32. Radio galaxies	1932	Radio telescope	<5y	C	Yes	K. Jansky	Physicist	Galactic radio emission
	1946	Radio telescope	<5y	M	Yes	J. S. Hey	Physicist	Cygnus A
33. Unidentified radio sources	1974	Powerful optical telescopes	~10y	M/2	No	J. Kristian, A. Sandage	Physicist, Astronomer	Nondetection of compact radio sources
	1972	Radio fast-time resln.	<10y	M/2	No	B. J. Harris	Astronomer	Radio scintillation
34. Cosmic expansion	1912	High sens. & med. spectrl. resln.	<5y	M/2	Yes	V. M. Slipher	Astronomer	Red shift of galaxies
	1929	High sens. & ang. resln.	<10y	M/2	No	E. Hubble	Astronomer	Distances of galaxies
35. Quasars	1960	Radio arc sec resln.	<5y	M/2	Yes	H. P. Palmer, T. A. Matthews	Meteorologist, Astronomer	Stellar size, high radio brightness; Precise position
	1963	Radio occulation med. opt. spectr. resln.	<5y	M/2	Yes	C. Hazard, M. Schmidt	Physicist, Astronomer	Red shift
36. Superluminal sources	1971	Very long baseline interferometry	<5y	M/2	Yes	I. I. Shapiro	Physicist	Rapidly separating compact radio sources
37. X-ray galaxies and clusters	1966	High X-ray sensitivity	<5y	M/2	Yes	H. Friedman, T. A. Chubb, E. T. Byram	Physicist, Physicist, Electr. Engr.	Discovery of M 87 and Cyg-A as X-ray emitters
38. Infrared galaxies	1966	High sens. PbS detectors	<5y	M/2	No	H. L. Johnson	Astronomer	3-micron emission
	1970	High sens. bolometers	<5y	M/2	No	F. J. Low	Physicist	20-micron emission
39. Gamma-ray bursts	1973	Gamma sens. to ~1 sec bursts; multiple detectors	<5y	M	Yes	R. W. Klebesadel, I. B. Strong, R. A. Olson	Engr./Phys., Physicist, Physicist	Nearly simultaneous bursts detected at several locations
40. Microwave background	1965	Sensitivity to isotropic radio flux	~5y	C	Yes	A. A. Penzias, R. W. Wilson	Physicist, Physicist	3° K microwave radiation
41. X-ray background	1962	X-ray sensitivity	~5y	M/2	Yes	B. Rossi, R. Giacconi	Physicist, Physicist	X-ray backgr. from beyond atmosphere
42. Gamma-ray background	1968	Gamma-ray sensitivity	<5y	M/2	Yes	G. W. Clark, G. P. Garmire, W. L. Kraushaar	Physicist, Physicist, Physicist	Gamma-ray flux detected from Earth-orbiting satellite
43. Sources of cosmic rays	1912	Balloon-borne electroscopes	<10y	M/2	Yes	V. Hess	Physicist	Extraterrestrial origin

debating which nebulae, if any, would ultimately be shown to resolve into stars. Huggins's spectrometer showed that the bright-line nebulae were gaseous and plainly had no stars to resolve.

The fourth column gives the length of time that the required instrumentation had been in use in astronomy before the discovery was made. Sometimes the introduction of a technique into astronomy also coincides with its invention. The high angular resolution obtained with astronomical instruments is unmatched in other branches of optics, radio physics, or X-ray technology and has been especially developed for astronomical use. On the other hand, radio astronomers for many decades used equipment that was slow in response. Later, when fast techniques, long available in ordinary radio communication, were introduced into radio astronomy for observing radio scintillation produced by the interplanetary plasma, the groundwork was laid for the discovery of pulsars. For entry 25, pulsars, therefore, the fourth column in table 5.1 reads $\sim$ 5y, since fast response times at the requisite wavelengths had only been introduced into astronomy some five years before the discovery of pulsars. While these techniques had already served as the basis for radio communications for more than half a century, their potential in astronomy had not been realized.

The entries in the fourth column of the table are rounded off to five-year intervals to avoid unnecessarily detailed debate. Even for Galileo's discoveries, which are known to have taken place within eighteen months after the invention of the spyglass and within nine months of the first introduction of the spyglass into astronomy, the entry simply reads < 5y—less than five years.

The fifth column devotes itself to the involvement of the military or the communications industry. Discoveries made directly as a result of military or communications industrial activity are labeled M or C. The discovery of gamma-ray bursts occurred as part of purely military surveillance and is labeled M. The discovery of the Galaxy's radio emission resulted from attempts to check a new communications link and is labeled C. In many instances, the military or communications influence is somewhat diluted and a listing M/2 or C/2 is used. Often this simply means that the detectors, which were essential for the observation, were available solely because they had been built for military or communications purposes. Most of the infrared detectors used in astronomy since World War II have been of this variety, though a highly sensitive bolometer developed by Frank Low, then at Texas Instruments Corporation apparently had no direct military ties.[1] Early postwar radio work also was done almost entirely with discarded military equipment, and that trend still persists today.[2] Where a blank has been left in column 5, I have no evidence for either communications or military involvement in the discovery. That does not, however, mean that such an involvement has necessarily been absent.

Earlier in this century Viktor Hess's cosmic-ray balloon flights of 1911 and 1912 were launched and piloted by military personnel, and military support may have been given to a wide variety of other astronomical developments. This question is further discussed below.

The sixth column attempts to show the extent to which chance has been involved in astronomical discoveries. A simple yes or no often is an oversimplified answer to the question, Was this discovery largely luck, accidental, surprising, serendipitous? The previous account of each discovery presents matters in greater detail and may permit the reader to judge for himself.

The rationale used in deciding whether or not a discovery was unexpected is somewhat subjective. Hey's discoveries of a radio galaxy and the existence of radar meteors are classed accidental because Hey was engaged in wartime efforts at the time of these discoveries. Astronomical and atmospheric effects of the kind he found were noticed not because he searched for them but because they made themselves evident as a nuisance in radar efforts. On the other hand, I have not classed Huygens's discovery of the rings of Saturn accidental because Huygens was actively trying to understand the odd appendages of Saturn that had puzzled astronomers since Galileo first thought he was seeing two moons orbiting Saturn very close to the planet's disk. Huygens did not expect to find a ring, and to that extent the discovery was surprising to him. But he was trying to study Saturn through use of more powerful techniques and, by means of these efforts, managed to solve the puzzle of the planet's appearance.

A judgment of this kind is never beyond debate. Hey's discovery of radar reflections from meteor trails has more of the elements of chance than his discovery of the radio galaxy Cygnus A, which he found shortly after the war was over. The first discovery resulted from an investigation into false alarms registered during radar watches for V2 missiles. The second discovery was more deliberate and arose from an attempt to further study the Galactic radio emission originally observed by Jansky and later by Reber. The rapidly varying signal detected from the Cygnus region of the sky, however, was totally unexpected, was unrelated to the emission from the Galaxy, and ultimately led to the surprising realization that enormously powerful radio galaxies exist in the universe. Since there was little relation between the source originally to be studied and the source actually found, the entry for radio galaxies indicates chance discovery.

Column 7 names the main observers involved. Where a discovery was made by several different researchers or groups, only the principal names are cited. Fortunately, there are not very many discoveries for which priority is under serious dispute.

Column 8 gives the profession for which the discoverer was trained. Over a number of centuries the names of professions change. Physicists did not exist by name in the seventeenth century. Yet, Galileo and Huygens are here called physicists because most of their activity corresponded to the work a physicist does today. Tycho Brahe, on the other hand, clearly worked in a style much more nearly related to the activities of present-day astronomers, even though he shared, with Galileo and Huygens, a love for the construction of new instruments and a preoccupation with novel astronomical techniques.

The last column in the table briefly remarks on the critical observation involved in each of the entries listed.

A number of results are readily recognized from the table:

1. Approximately half the researchers credited with making discoveries came from professions other than astronomy.
2. Instrumental power appears to play a key role in bringing about a discovery. Many discoveries, particularly those made in the post-World War II era, depended on equipment less than five years old. In the quarter century since 1955, few of the phenomena discovered could have been found with equipment available as long as ten to twenty-five years earlier—none with instruments that had been around longer.
3. About half the discoveries made have been serendipitous. At the time of discovery, the investigator often was exploring entirely unrelated effects. The discovery forced itself upon him.
4. Since World War II the large majority of discoveries was made with equipment originally built for the military or was made in the course of military or communications industrial activities. The extent to which progress in astronomy has rested on military and industrial factors in previous centuries has not been sufficiently analyzed here but may have been substantial.

Further columns might have been added to Table 5.1. The impact of ideas and theories on the course of observational discoveries and the contributions made by technical assistants have not been included. History is curiously silent on this last point, but the question should perhaps receive serious consideration: New phenomena are often uncovered within a few years after the introduction of the technical means required for their discovery. This suggests that a specific piece of equipment rather than a specific observer might well be credited with making a discovery; and this is quite at variance with historical tradition.

Because of the important role assumed by novel instrumentation, the part that a key engineer or skilled technician played in constructing a critical piece of observing equipment should be examined. Most astronomers have the good sense to appreciate improved performance of an instrument, but not all observers have the skills in design or the manual dexterity to produce an improved piece of apparatus. Frequently, design alone is not enough. The construction of equipment may also require craftsmanship well beyond ordinary capabilities. Yet scientific tradition encourages the astronomer alone to author a final astronomical report on observations obtained. The engineer and skilled technician are generally excluded. While special techniques used are sometimes mentioned in a publication, and the name of the craftsman or engineer may even be acknowledged, the actual discovery is normally credited solely to the astronomical observer.

Christian Huygens is known to have been aided by his brother Constantin in the instruments he built.[3] James Bradley's precision instruments were constructed for him by the instrument maker George Graham.[4] William Herschel's brother, Alexander, possessed great dexterity in the construction of mechanisms and worked with William on all the early telescopes.[5] Joseph Fraunhofer designed and built optical apparatus of unmatched quality that not only permitted him to discover the dark lines in the sun's spectrum, but also enabled Bessel to discover stellar parallax and the existence of Sirius B and Procyon B—both later shown to be white dwarfs. In more recent times, Martin Annis, president of American Science and Engineering Corporation, recalls the crucial role played by one of the technicians at the company who succeeded in the construction of large, thin-window counters essential for the sensitivity needed in the discovery of X-ray stars. Had this individual not been in the employ of the firm, would the project have failed? Would Herbert Friedmann's group at the Naval Research Laboratory have discovered X-ray stars first? Or would a replacement technician have readily been found to construct the required X-ray counters? In short, just how unique were these technical skills to the particular craftsman involved? It seems clear that we will need to gain a better understanding of the role of engineers and technical assistants before all the factors favoring discovery of cosmic phenomena can be fully understood.

Interpretation and Theory

There is a remarkable detachment between the observational recognition of a novel phenomenon and the discovery of the model that explains it. The theoretical basis for neutron stars was first outlined in the 1930s by Lev Landau in the Soviet Union, by the German-born astronomer Walter Baade and the Swiss astrophysicist Fritz Zwicky, working in the United States, and by J. Robert Oppenheimer and G. M. Volkoff at the University of California at Berkeley.[6] These men all agreed that neutron stars were logically and physically possible. But no one knew whether the universe produces them: Would massive stars contract to the neutron star state? Or would there be some intervening step that would abort the formation of a neutron star and lead to some other form of stellar death?

Thirty years after the conception of these theories, Anthony Hewish and Jocelyn Bell discovered pulsars. There was no doubt at all that here was a new phenomenon. No astronomer had ever before come across regular pulsations on a time scale of a second at radio wavelengths— or any other wavelengths for that matter. Thomas Gold at Cornell University soon was able to discern that the pulsars might well be the long-sought neutron stars. The early models of these stars had not figured on the the very rapid rotations that such a compact mass could exhibit and had not realized that the collapse of the star might also produce a magnetic field far stronger than any encountered elsewhere in the universe. These two features, rapid rotation and high magnetic field strength, were able to account for the radio pulses Hewish and Bell had observed. Even now, however, we have no theories that satisfactorily explain just how a massive star collapses to become a neutron star. We know that neutron stars are possible in our universe only because we see that they are there—not because we understand how they form.

The discovery of quasars shows a quite different sequence of events: Observational recognition clearly preceded theoretical explanations. Around 1960 the quasars were first recognized as highly compact, extremely bright radio sources. Their positions were identified accurately enough for optical observers to find the quasars with their own telescopes, and by 1963, highly red-shifted spectra had been discovered for two of the quasars. Here again the observational features corresponded to nothing we had ever seen before.

Everyone agreed that quasars represented a new phenomenon. But it took fifteen years even to reach somewhat of a concensus that quasars are very distant. There is now reasonably convincing proof that quasars are seen concentrated near galaxies that have red shifts close to those of the quasars. This observational result makes sense only if the quasars are distant. But as yet we have no unique theory and no single model that explains the nature of quasars, let alone their origin or source of energy.

Black holes have quite a different history, again. The idea that a body might be so massive and so compact as to devour itself gravitationally was first suggested two centuries ago, when Pierre-Simon Laplace conjectured about this possibility.[7] In a precise astronomical context, however, the concept is first found in a publication by Oppenheimer and H. Snyder, who considered the continued gravitational contraction of the neutron stars that Oppenheimer was then investigating with Volkoff.[8] Neutron stars have by now been found, but black holes remain controversial. Many observed sources in the sky have been proposed as candidate black holes, but there is no concensus that a black hole has really been found. The questions that used to be asked about neutron stars now apply to black holes: Will massive stars or clusters or galaxies contract to the black-hole state? Could there be some intervening step that will abort the formation of a black hole and lead to some other form of death for such massive objects? We do not know.

The point to emphasize is this: The astrophysical concepts that lead to an understanding of cosmic phenomena have a history that is all but decoupled from the actual discovery of the phenomena. The phenomena are recognized largely because they exhibit observational features bewilderingly different from anything noted before. This was true of Tycho Brahe's supernova, of Galileo's Jovian satellites, of Saturn's rings studied by Galileo and Huygens, of the earth's motion about the sun discovered by Bradley, of clusters of nebulae first seen by William Herschel, of the nebular emission lines seen by Huggins, of Hartwig's detection of an eruption in the Andromeda Nebula in 1885, of Hess's discovery of cosmic rays, of Jansky's discovery of the Galaxy's radio emission, of Giacconi's and Rossi's discovery of X-ray stars, of Byram, Chubb, and Friedman's discovery of X-ray galaxies. . . . Where was there any but the vaguest connection between theoretical prediction and actually observed events?

The theorist's dilemma is this: He cannot precisely predict the appearance of new phenomena because the laws of physics that he uses permit too broad a range of possible models. In turn, the observer has no time to pursue the many consequences of every model that might be proposed. And so a gap persists between theorists and observers. That is how Gamow, Alpher, and Herman's prediction of the microwave cosmic background radiation was able to lie dormant for fifteen years without any observer taking notice. It was lost in a profusion of other theoretical predictions concerning other conceivable phenomena, many of which may have had no merit at all.

At any rate, theory and observations pursue their own somewhat separate ways, and the major cosmic phenomena continue to be discovered mostly by chance. As we come to know more about our universe and the phenomena that characterize it, theoretical predictions may become increasingly accurate, and ultimately new phenomena may well be discovered through precise prediction. Right now, however, the observational base on which our theoretical predictions are founded still seems too shaky for that.

Military and Commercial Influence on the Progress of Astronomy

In his 1938 study on science, technology, and society in seventeenth-century England, Robert Merton demonstrated that even three hundred years ago the most influential scientists were very much concerned with practical problems. Hooke, Halley, and Newton, who all left their imprint on astronomy, actively engaged in applied research. Investigations on gunpowder, ballistics, transportation by land and by sea, as well as basic problems in mechanics, heat, metallurgy, and chemistry were handled by the members of Britain's Royal Society as a result of practical demands. Merton estimates that about 10 percent of the research in that era was directly or indirectly related to military problems, and about 59 percent to socioeconomic needs, including mining, marine transport, textiles, general and military technology, and husbandry.[9] Today we would call this applied science or technology. Only 41 percent of the work carried out at the time might be termed pure science.

Among the practical arts, navigation, in particular, has influenced astronomical observations throughout history—although superstition, and in particular astrology, may have had an even more important influence. David Clark and Richard Stephenson point out that the ancient Chinese, Japanese, and Korean records on supernova explosions, nova outbursts, comets, and notable alignments of planets, all were products of an uninterrupted astrological tradition. This tradition was maintained over millennia as a source of vital information by the Chinese imperial court and later also by Korean dynasties and to a less systematic extent by Japanese courtiers and scholars.[10]

Sir Bernard Lovell has written about progress in astronomy connected with military activities in the period between 1935 and 1957. In his view not only has all of the postwar effort in radio astronomy directly benefited from military investment in technology but also the space effort succeeded largely because military hardware was available for use. Lovell cites the fiasco of the Vanguard program which attempted to launch the first U.S. satellite through a purely civilian effort, and the subsequent rapid launch of an Army ballistic rocket that carried the first successful U.S. Explorer satellite into space.[11]

Military support for the improvement of detector sensitivity, for the development of rocket and spacecraft techniques, and for the application of new detection techniques provided enormous financial gains for astronomy in the 1960s. The development of space techniques by NASA also was invaluable.

None of this support would normally have been made available to astronomy for its own sake. Neither was most of this aid alloted to astronomy under a peer review system. Martin Annis of the American Science and Engineering Corporation, where much of the pioneering work on stellar X-ray work was carried out, attributes the technical advances to one factor:[12] In the late 1950s and early 1960s both the military and NASA were often willing to support research which at best could only claim to lie on what Annis calls "the fringes of legitimacy." This period of easy support ended in the late 1960s with the enactment of the Mansfield Amendment forbidding military support for basic research; with the budgetary cuts suffered by NASA after the completion of the Apollo landings on the moon; and with the increasing influence that centralized planning panels began to exert on the formulation of astronomical policy. By the early 1970s astronomical goals were being increasingly defined by prestigious panels of scientists, research supported by NASA was being subjected to ever-tightening peer review, and the support of interesting research on an individual contract officer's hunch became all but impossible.

I have not been able to find any comprehensive study relating military and astronomical activities throughout recent centuries. Such a study might, however, show extensive linkage. Astronomers, as well as military officers, have much to gain from efficient surveillance. Their efforts, of course, have quite different aims, but the technical means may frequently be the same; and that is the fundamental reason for a connection between advances in military and astronomical technology.

Contributions Made by Nonastronomers

Throughout the centuries about half the discoveries of new phenomena have been made by men who originally were trained for work in other areas. In the era following World War II the fraction was even higher—rising to two-thirds or perhaps even three-quarters. One question that naturally arises is whether a similarly high fraction of all working astronomers has been trained outside the field, or whether the outsiders in fact do contribute an unusually large share of the prime discoveries.

A thoroughly documented answer to this question would require further research. However, a 1975 National Academy of Sciences (NAS) study of employment patterns in astronomy gives the following statistics: Among astronomers working in the United States in 1973 there were 1,313 actively employed Ph.D.s. Of these 885 had been trained as astronomers, 390 had been trained outside astronomy, and 38 did not state their background training.[13] About one-third of the American Ph.D. astrono-mers active that year, therefore, had been trained outside their field. This figure may have been unusually high.

The number of active American Ph.D. astronomers more than dou-bled between 1970 and 1973, rising from 623 to 1,313 as the field rapidly expanded. To meet this increase of 690 Ph.D.'s, 380 new astronomers were trained and awarded doctorates during this time. But to account for the total increase, at least 310 positions must have been filled by other means. The NAS report surmises that the influx represents a flow of more mature Ph.D. physicists into the field. This suggests that a sub-stantial fraction of the 390 Ph.D. astronomers who had not been trained in astronomy entered the field between 1970 and 1973. By that time the bulk of the discoveries listed in table 5.1 had already been made, and the newcomers presumably were attracted by the successes of the field during the 1960s.

The conclusion we must reach then is that the fraction of astronomi-cal phenomena discovered by nonastronomers in the era following World War II—about 70 percent—is substantially larger than the fraction of nonastronomers working in astronomy—at most 33 percent, but quite possibly much lower. For earlier epochs the statistical data have not been assembled but are likely to lead to similar conclusions.

Many of these researchers were at first concerned primarily with the construction of instrumentation capable of searching the sky in ways that no one had tried before. Only after their instruments had become sufficiently sensitive to detect astronomical signals did these workers have to worry about the implications of what they had observed. At this stage they increasingly were drawn from the margins into the center of the astronomical community. Today, gamma-ray, X-ray, and infrared astronomy are considered central to modern astronomy. In time they too will be joined by even younger, more recent arrivals.

In summarizing the nature of radio astronomical innovation, the sociologist David Edge has provided us with a perspective that can be effectively used to characterize all of the major astronomical advances of the past three decades. Of radio astronomy he has written:

> . . . the innovation consisted of systematic, state-of-the-art exploitation of a set of observational *techniques* by those whose background and training equipped them with the necessary skills and competences, to produce results for an audience many of whom were not so equipped.[14]

Precisely the same words might have been written about men like Galileo, Huygens, Bradley, William Herschel, Fraunhofer, and Huggins in previous centuries. In many instances, the new technique was used by the innovator to gain entry to cosmic research and ultimate accep-tance by astronomers. A retrospective article written by William Hug-gins in the closing years of the nineteenth century speaks clearly of this.

> In 1856 I built a convenient observatory opening by a passage from the house, and raised so as to command an uninterrupted view of the sky except on the north side. . . . I commenced work on the usual lines, taking transits, observing and making drawings of planets.
> I soon become a little dissatisfied with the routine character of ordinary

astronomical work, and in a vague way sought about in my mind for the possibility of research upon the heavens in a new direction or by new methods. It was just at this time, when a vague longing after newer methods of observation for attacking many of the problems of the heavenly bodies filled my mind, that the news reached me of Kirchhoff's great discovery of the true nature and the chemical constitution of the sun from his interpretation of the Fraunhofer lines.

This news was to me like the coming upon a spring of water in a dry and thirsty land. Here at last presented itself the very order of work for which in an indefinite way I was looking—namely, to extend his novel methods of research upon the sun to the other heavenly bodies. A feeling as of inspiration seized me: I felt as if I had it now in my power to lift a veil which had never before been lifted; as if a key had been put into my hands which would unlock a door which had been regarded as for ever closed to man—the veil and door behind which lay the unknown mystery of the true nature of the heavenly bodies. This was especially work for which I was to a great extent prepared, from being already familiar with the chief methods of chemical and physical research.

It was just at this time that I happened to meet at a soirée of the Pharmaceutical Society, where spectroscopes were shown, my friend and neighbour, Dr. W. Allen Miller, Professor of Chemistry at King's College, who had already worked much on chemical spectroscopy. A sudden impulse seized me to suggest to him that we should return home together. On our way home I told him of what was in my mind, and asked him to join me in the attempt I was about to make, to apply Kirchhoff's methods to the stars.[15]

Huggins and Miller thus brought their expertise in chemical spectroscopy to bear on stellar and nebular spectroscopy. Within a handful of years Huggins then discovered the emission lines of gaseous nebulae and began radial velocity measurements on stars through observations of their Doppler-shifted spectral lines.

The Systematic Introduction of Novel Techniques

Table 5.1 gives strong support to a contention that the application of new instrumental techniques has directly produced a substantial number of the most striking astronomical discoveries throughout the past three or four centuries. We might ask ourselves to what extent the systematic implementation of new techniques could be encouraged in the future? Here we can learn from an illuminating chapter written by Philip Morrison in *Opportunities and Choices in Space Science, 1974,* a planning document prepared by the National Academy of Sciences's Space Science Board for the National Aeronautics and Space Administration. In listing the principal criteria that were used by the board in arriving at particular recommendations, Morrison illustrates the difficulties to be faced if we wish to infuse startling new techniques into an existing long-range research program. He presents the Space Science Board's views in these words:

The phrase "balance and continuity" may adequately describe the most used of all criteria. It serves to remind us that science is not a human activity without parallel, for in most human plans such a criterion must be employed. Momentum is a real property of social as well as physical systems. Plainly, this speaks to the danger of self-serving, but just as plainly it is inevitable that these considerations must play a role in any practical planning. For science, even in its search for novelty, is an ongoing enterprise. Results will never appear unless there are skilled and devoted people, adequate instruments, problems set by previous knowledge, and whole disciplines of study that are based on past results. It is no use to propose experiments if no one is interested in carrying them out or if no one is competent to make the measurements or use the results. The body of existing disciplines must play a role in plans for the future. Thus, the Solar Maximum Mission or the (1.2 meter imaging) x-ray telescope gain a place in our plans because there are powerful groups of experimenters who can use and will use well such opportunities and who cannot work without them. A gap of many years in such a series of experiments would waste a resource even more precious than the NASA budget. So much we share with every group that plans for the future of persons and organizations.

At the same time, exactly because of this strong parallel, we cannot find in this argument alone any justification for the future of any mission. If no problem exists or no means for solution is present or no novelty seems within reach, mere continuity would not serve for justification.[16]

Such an approach seems sensible enough, until we start to worry about how to construe the last sentence. In the late 1930s there were countless interesting problems left in visual astronomy. At the same time there were no apparent problems to be solved by radio or X-ray techniques, and equipment in existence was completely inadequate to measure the tiny flux of radio or X-ray emission that any theorist was willing to predict. Had radio, ultraviolet, X-ray, infrared, and gamma-ray astronomy not forced their way onto the scene by dint of spectacular observational results that could not be ignored, visual observations might still dominate astronomy to the exclusion of all else, and theorists might still persist in their timid predictions.

The Planning Wedge

If we are to take seriously the lessons learned from the discoveries of the past three decades, we must make a more deliberate effort to construct plans which systematically introduce radically new observational techniques, permitting a view of the universe through brand-new channels and subchannels of information. With some effort and considerable amounts of goodwill, it should become possible to overcome the human problems that such rapid change can entail and to make these plans compatible with the main social criteria that Morrison lists.

To see how bold new schemes could be introduced to astronomy, we must be aware of current commitments and of their decline in the future (figure 5.1). The further ahead we look, the smaller are our commitments. Usually few firm commitments are made more than ten years ahead of time. Because hard commitments decrease into the future, the budget available for planning always appears to increase as we look far enough ahead, and we therefore speak of a planning wedge. Our plans must start out modestly, but they can expand like a wedge.

As Philip Morrison points out, this planning wedge is not entirely free for use in any way we choose. The social momentum he mentions is a factor that cannot be ignored. Nevertheless, a certain fraction of the wedge is free and clear simply because momentum—just like the number of firm commitments—declines.

Social momentum declines because the inertia of research groups can be expected to diminish as we look ahead. The typical creative scientist works forty years before retiring. During the last years of his career, however, he frequently turns to administration. Research facilities and equipment typically are retired much sooner. Often they are outmoded in ten to fifteen years. Thus the radio interferometer, one of the finest instruments built at the National Radio Astronomy Observatory in the 1960s, has already been supplanted by the Very Large Array of radio antennas built by the observatory in the mid-1970s.

We can therefore count on virtually a total loss of inertia, both of men and machines over a span of twenty-five years. And if we are to

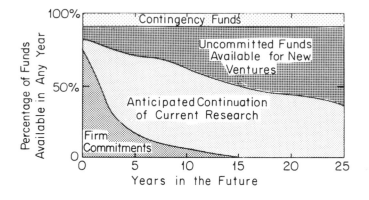

Figure 5.1 *Constraints on Research Support*

All research plans are subject to limitations imposed by competing demands for funds: Firm commitments to ongoing projects must be honored; current investments in long-term ventures that should be given an opportunity to mature generally are anticipated to continue—though at a slowly decreasing level; contingency funds must be kept in reserve to meet unanticipated crises that would otherwise threaten severe setbacks. When all these competing demands are subtracted from the anticipated budget for the present or for a future year, we are left with the uncommitted funds available to us for planning. For any current year, previously scheduled expenditures normally leave no more than a tiny fraction of the total budget available for new starts. But as we look into the future, the firm commitments rapidly decline, the anticipated expenditures also drop, and the funds available for new enterprises grow. This growth of uncommitted funds for future years gives rise to an expanding planning wedge available for one or more new ventures.

construct a plan that comes into effect five years from now, we should feel free to count on a drop in social inertia and momentum amounting to 20 percent during this interval. With this drop, those researchers who do not naturally retire are permitted to continue research in their field of interest, but the number of workers in the field is not replenished. If we follow this prescription, we need not fear the problems arising from social momentum and can count on 20 percent of the planning budget to be available for innovations leading to new discoveries. These are the funds that provide us with genuine means for creative planning.

We can, therefore, insist that astronomical plans which look five years into the future dedicate a substantial part of the funds in the planning wedge to a search for new cosmic phenomena. Longer term plans should set aside correspondingly larger portions. Ultimately, we might figure on a continuing budget amounting to 20 percent of research expenditures dedicated to the development of novel techniques. The precise figures may vary depending on the nature of existing commitments and budgetary increases or decreases, but a rational approach does exist to put into practice creative plans at a rate best geared to continuing searches.

New Directions

While this prescription may seem sensible, there are many obstacles to be overcome. One difficulty with the construction of long-range plans for astronomy is the virtually automatic self-perpetuation built into the planning process. A national plan for the development of astronomy is generally designed in successive stages. First, a number of subcommittees made up of experts is convened to list recommendations and priorities for individual disciplines. The reports of these subcommittees are then reviewed by a central group that seeks to establish a balanced program which includes at least the higher priority items recommended by each subcommittee. This group effectively assures the balance and continuity mentioned above. Individual observational disciplines like X-ray, visual, infrared, and radio astronomy all are assured some sense of stability by this process.

What this procedure lacks, however, is a means for introducing a desirable amount of imbalance and discontinuity. It needs an added mechanism for infusing the changes in direction periodically required by astronomy if it is to remain vigorous. To effect these changes a group of advisers is needed with the explicit charge of making recommendations on genuinely novel approaches in observational astronomy— schemes for arriving at new ways to probe the cosmos. The recommendations from this group could then be judged in the same spirit as those from established disciplines, and place would be made in long-range plans for the most promising new ideas.

Such a system of planning can, however, only succeed if it carries the approval of the whole community: A novel research enterprise has no firmly entrenched lobby and usually is an easy first victim for any budgetary cut. For this reason cutbacks are particularly serious in sciences, like astronomy, in which the more important discoveries seem clearly to result from daring ventures, from projects carried out on the remote frontiers, where established scientists have not yet trod, where groups with strong self-interests have not yet arisen.

Manpower

If astronomy is to stay active and healthy, we must find better ways to infuse new approaches into the astronomical search. This must include new ways of training our students, ways to permit the continuing growth of established astronomers, and ways to encourage the transfer of men with special skills into the field—all with the aim of increasing the technical expertise than can be brought to bear on observational astronomy.

Recommendation 1

As a first step toward attracting a younger generation of astronomers competent to make major discoveries, we should make sure that our best students of astronomy are trained in physics as thoroughly as the physicists themselves.

In the leading universities in the United States, this kind of training is now almost taken as a matter of course. With this preparation, young astronomers will be able to talk the language of the physicists and understand some of their more sophisticated techniques without difficulty. However, there will still remain a gap between the mere appreciation of, and the actual familiarity with, a technical innovation potentially useful to astronomy. It is this gap which is hardest to bridge under the present system of education.

There is a further, quite different type of difficulty. A young scientist just out of graduate school may well be trained in the most recent technical advances in astronomy, but to be at the forefront of innovation in twenty years, he may have to re-educate himself in radically new techniques emerging at that time. This may require an effort almost as comprehensive as that required to obtain a Ph.D. in the first place, involving an investment of two or three years concentrated on a new approach. Nothing less suffices if a scientist is to acquire sufficient competence in a newly chosen specialty and to win research grants in competition against established colleagues working on established projects.

The number of scientists who make a serious attempt to change fields, and the number of those who succeed, does not appear to be known. The sociologists Robert McGinnis and Vijai P. Singh have provided an analysis of mobility patterns for Ph.D. physicists and astronomers during the period from 1960 to 1966. Their study concerned itself with the mobility of physicists within sixteen subfields including astronomy, high-energy physics, solid-state physics, nuclear physics, the physics of fluids, acoustics, and so on. Approximately one-third of the physicists changed subfields during this time, younger physicists slightly more frequently than their more established colleagues.[17]

Many uncertainties, however, remain. The period between 1960 and 1966 was a time of unparalleled growth for American science. The increased funding available during this period may have made for greater mobility. Current tight funding policies may make a move into a new field of physics or astronomy considerably more difficult. That question needs to be investigated.

Recommendation 2

The community of astronomers, cognizant of the great advances that instrumental innovations can bring to the field, should deliberately establish ways for creative scientists to continue on their innovative searches by acquiring new technical skills in midcareer. The nation's investment in its research capacity can be bolstered by attractive long-term (three- or four-year) incentive grants to successful scientists seeking to strike new directions.

The implementation of such a plan would not be difficult under our present system of grants if we properly recognize that continual re-education is an important aspect of successful research. We should, however, also be aware that many astronomical tools will be devised by physicists, chemists, and engineers working in other fields. Some of these scientists will welcome an opportunity to apply their special skills to astronomy.

Recommendation 3

Astronomers should make special efforts to attract researchers with novel techniques from other fields: Manpower plans should take into account the enormous gains astronomy has inherited from the discoveries made by transplanted scientists.

This recommendation is particularly important in view of a finding general to many scientific disciplines: A technical innovation originating elsewhere is seldom introduced into a field by practitioners in the field. Rather, it is brought into the field by the originator himself. He appears to be the only one with the initial proficiency and interest to apply the technique successfully.[18]

This importation of new techniques by the original innovators characterized the beginnings of radio, X-ray, gamma-ray, infrared, and cosmic-ray astronomy. We should help this trend continue.

Rates of Progress and the Need for Vigorous Research Support

I have suggested that most of the major cosmic phenomena should be found by the year 2150, though a few rare phenomena difficult to discover could resist recognition for many centuries to come (figures 1.8 and 1.9). Astronomy, however, will progress at this predicted rate only if our present era of enthusiasm for science is not prematurely cut short. At the moment we do not understand enough about the historical patterns of intellectual endeavors to tell whether or not a serious threat to science is currently taking shape.

The National Aeronautics and Space Administration's Physical Sciences Committee (PSC), a group of distinguished astronomers and space scientists chaired by George Field, director of the Harvard University and Smithsonian Institution's Center for Astrophysics, has expressed dismay in its 1976 evaluation of the efforts of NASA's program in Supporting Research and Technology (S.R.&T.). This is the area of prime significance to future progress, since it is the program that supports the technological innovations which play a dominant role in astronomical discoveries. The Field Committee's comments are sufficiently serious to warrant extensive quotation:

> The Committee wishes to draw [attention to] a problem that is becoming more acute as the Shuttle era approaches.
>
> In 1971, the NASA Sustaining University Program, which provided core support at major university centers at a total annual level of $10 million, was drastically curtailed. (A modest amount of core support, amounting to only 10 percent of the previous program, was continued at Berkeley, Chicago, Harvard, MIT, and Texas.) Moreover, the use of step funding grants has progressively decreased (from 130 awards totaling $15.3 million in FY 1970, to 116 awards totaling $10.7 million in FY 1975), as NASA forced investigators to draw upon anticipated future funds to sustain current activities.
>
> These problems were compounded when young space scientists trained in the late 1960s began to form research groups of their own and to compete with already established groups for the diminishing funds available. This ever-fiercer competition for funds has led to a progressive decrease in the funding of any single project. . . . While such competition is of course necessary for the maintenance of high research quality, all investigators find that they must now devote an increasingly large fraction of their time to the preparation of NASA proposals and progress reports. The associated paperwork has already reached the level at which efficiency is being adversely affected, with the result that NASA is not getting optimum performance from the groups it supports. . . .
>
> We think it relevant to quote from the statement of 20 January 1976, by Dr. James C. Fletcher:
>
> . . . I continue to believe that an adequate continuing investment in the future—namely, in research and technology—is the key to increasing national productivity, efficiency, and well being. Continual squeezing of essential research and technology efforts, either directly by underfunding or indirectly by inflation, is bound to undercut the viability of the entire program and foreclose important opportunities for the nation's future. I trust that this will not be permitted to occur. . . .

It is the unanimous conclusion of the PSC that . . . continuing inaction will cause irreversible damage, very possibly removing the United States from a position of world leadership in space science.[19]

Dr. James Fletcher quoted by the PSC was NASA Administrator at the time of his statement. NASA's policies, however, do not seem to have changed substantially in response to the statement. A further recommendation may therefore be added:

Recommendation 4

NASA, the NSF, and other funding agencies supporting astronomy should vigorously sustain continuing development of new techniques for astronomy with the anticipation that these techniques will result in exciting new discoveries.

Continuity in the Support of Innovative Ventures— Accountability and the Peer Review

We should examine, with care, the Physical Science Committee's conclusions about the current need for frequent fund applications, progress reports, and final reports which burden researchers with paperwork well beyond normal needs for keeping a funding agency fully informed on progress and the proper disposal of funds. The difficulties arise in a number of ways.

Many government grants in support of research are only provided for spans of one year or two. The researcher cannot be sure of renewed support, and the projects he undertakes therefore must promise a return within a few months. Results are all-important if the grant is to be renewed. This type of support inspires timidity. Only small projects with limited scope can be realized with certainty in that short a time. To make matters worse, research grants often are awarded just weeks before the starting date for the project, making it difficult to hire the best engineers or technicians in time for the project's start.

To counter this trend, many scientists try to maintain at least a few projects that they consider safe. These projects may have been consistently funded in the past and may therefore be expected to receive support for another year. Safe projects tend not to be the most innovative, but they may permit the researcher to pay for technicians and a support staff, for maintenance of equipment, and for a variety of other costs that are incurred by every laboratory. The safe project makes it possible for the researcher to seek other support for innovative work on a small scale, without incurring the risks that he will be left with no research support at all. Innovation, after all, requires change, and change may bring with it the threat of loss of the grant.

It might be thought that well-established scientists with a reputation for skilled and imaginative work would almost be guaranteed support for new ventures because they have established credentials: For these scientists support from government agencies should virtually be automatic. In fact, quite the opposite seems to be true.

A recent study by Jonathan R. Cole and Stephen Cole describes the results of the peer review system within the National Science Foundation. Surprisingly, the two sociologists find that established scientists do not appear to be significantly favored by the review.

Under the rules governing research support through the National Science Foundation, a researcher first submits a written proposal to the foundation. The proposal is then routed to several reviewers who work in the same research area. Written judgments are returned to the NSF by each reviewer, and a research grant is then awarded or denied largely on the basis of the reviews.

The opinions of the grant reviewers, however, often exhibit wide disagreement. To the Coles this lack of consensus seems responsible for a low correlation between the status of a researcher, his prior success in the field, the quality of his research in the past—in other words, his scientific reputation—on the one hand, and his ability to attract the research support he needs, on the other.[20]

This feature of the peer review system is a matter of genuine concern. If the proposals for research support submitted by our most eminent scientists can be routinely judged to be no better than the proposals submitted by others, then a lack of continuity is almost built into our system of research funding from the start—at least at the level of small projects and individual grant applications.

An early study by the sociologist Harriet Zuckerman has shown that leading scientists, researchers with outstanding records in the past, maintain a consistently high level of competence over many years.[21] Despite this conclusion the grant system appears to treat each applicant as though he had no demonstrated abilities and judges the proposed research largely on the basis of the proposal's promises. Such a system does have the advantage of providing young researchers of unproven skills with an opportunity to compete against what might otherwise be termed the establishment, except that Jonathan and Stephen Cole do not find any truly established, entrenched group, when it comes to the awarding of individual grants.

Stuart Blume of the London School of Economics has applauded the findings of the brothers Cole as demonstrating the fairness of the selection procedure:

> Of course there are curiosities. A not infrequent one seems to be the eminent scientist whose application is rejected because it consists of little more than "I'm me—give me the money"! So the system seems to work as intended, and this study shows this to be so.[22]

An alternative explanation, however, is that the well-received proposal is not necessarily written by the best researcher. The ability to conduct first-class research may not necessarily imply the ability to write first-class proposals. As the Coles note:

> If . . . the correlations between the ratings and the characteristics of applicants are low to moderate, this would necessarily mean that the correlation between characteristics of applicants and the perceived quality of their proposals would be low to moderate. We are currently analysing a set of data which will allow us to determine the extent to which this hypothesis is correct and the meaning of these data for peer review and the distribution of federal funds to research scientists.[23]

Proposals frequently are written many months—sometimes years—before a project is actually funded. The work carried out with the funds is therefore often quite unrelated to the originally proposed work, either because new findings in a rapidly developing field have made the original idea unsound or because the proponent of the work feels he now has a better idea. Overly strict abidance to the original proposal would then be a mistake. The successful scientist knows that and may feel that a detailed proposal might shackle his work later. In a *Science* editorial, Raymond Orbach of the physics department at the University of California, Los Angeles, explains the problem:

> A scientists must be free to change research direction according to his instincts or hunches as new opportunities present themselves in the course of his inquiry. But the auditing practices of the federal government today, which are undoubtedly valuable for the procurement of hard goods, are increasingly being applied to the performance of basic research. A researcher will submit a proposal to the federal government outlining a program of activity sometimes 2 years in advance. If he is fortunate enough to obtain funding (few are in today's marketplace), he finds himself enmeshed in a web of controls and budget restrictions. Ostensibly these are to prevent fraud, but most research institutions already have safeguards to protect against this very rare abuse. In reality, the bureaucratic drive for uniformity seems a more likely explanation for the narrow auditing perspective imposed on the scientific researcher.
>
> However, the nature of research requires unfettered lateral movement. The scientist must be free to shift his momentum and move in a totally unexpected direction. . . . [He] constantly faces the possibility of failure. His ideas, when successful, are usually the breakthroughs that one hears about and that play such an important role in the development of science for mankind. This is where the payoff lies.[24]

Some of the drawbacks of the peer review system follow other lines. In another *Science* editorial, A. Carl Leopold, formerly aide to Guyford Stever, when he was director of the NSF, has deplored the massive waste of manpower involved in research proposals. His views reflect those of many others.

> In 1960, when U.S. science was beginning to depend heavily on grants and contracts for research support, Leo Szilard wrote a fanciful story. In it he suggested that if some person or some group should ever want to bring research progress to a standstill, they could do so by establishing a competitive grants system under which all researchers would be required to prepare written proposals describing what they wished to work on. The commitment

of time by the research community in writing, reviewing, and supervising such a universal grant system would effectively halt research progress.

With approximately 47,500 proposals annually submitted to the principal federal granting agencies, Leopold estimates that some 6,600 man-years are spent each year by academic researchers in preparing and judging the proposals. He concludes:

> It may be that now, in the late 1970's, we should ask ourselves whether the load of the competitive grant and contract system is becoming excessive and whether it is time to seek alternatives . . . to ask whether Szilard's fanciful story is turning into a serious matter. Should consideration be given to ways of providing research support without adding to the heavy burden of our present grants and contracts system?[25]

There is a further conservative influence that curbs innovation. In reporting on testimony given before the House Subcommittee on Science, Research and Technology, Luther J. Carter of *Science* quoted both Carl Leopold and Thomas J. Jones, Jr., vice president for research at the Massachusetts Institute of Technology, as expressing concern that grant proposals which are truly innovative and outside the mainstream of the subfield go unfunded. Carter writes:

> According to Leopold, NSF program directors are constrained to support "conservative proposals and proposals which are 'sure bets' in that they are most liable to provide some definable product in a short period of time." As for Jones, he said the peer review process discriminates against new interdisciplinary science and scientific thinking that is not "au courant" even though creative and ripe with "unusual possibilities for breakthroughs."
> Leopold attributed this undue conservatism in grant-making to the "imposition of increasingly bureaucratic regulation." NSF programs, he said, are under pressure "to show that they have supported maximal numbers of proposals which have paid off with evident successes." With available dollars used to support research projects for "very short intervals" and at insufficient levels, NSF programs are "under pressure not to take 'longer shots' on more imaginative or longer-term projects, especially if a reviewer has given the proposal poor marks."[26]

Adequate continuity of support not only has worried scientists but also has come to be recognized as a major problem by federal administrators. Elmer B. Staats, comptroller general of the United States has recently written:

> The federal government must continue to provide major support for basic research in both natural and social sciences and the engineering disciplines. Sponsors must recognize that the very nature of basic research is long-term and exploratory, with little or no assurance of predetermined positive results. While it is necessary to assure wise and accountable expenditure of public funds, we in the government should seek ways to fulfill this need without inhibiting freedom of intellectual inquiry and risk-taking.
> I believe that the government should establish a long-term plan for investment in basic research. In addition, I believe that it is important to provide a stable base for funding from year to year. As longer-range plans are developed, Congress should also consider greater use of multiyear and advanced funding methods for basic research and other selected R & D efforts

which require more than 1 year to complete. I stated these views in my testimony . . . before the House Committee on Science and Technology.[27]

There appears to be substantial agreement among scientists and administrators that federally funded research can be made innovative provided the researcher is given full freedom to pursue work along general directions outlined in advance, subject only to normal financial accountability. Putting that agreement into practice, however, will require special efforts. The fringe of legitimacy that straddles full accountability and full freedom of research is narrow indeed, and we must make major efforts to broaden this domain and return to scientists much of the flexibility they enjoyed in the highly productive 1960s.

Recommendation 5

Researchers should receive long-term support with grants having durations of the order of several years, replacing the currently predominant one- and two-year contracts. Short contracts prevent researchers from starting the kind of ambitious development program that leads to true innovation.

Recommendation 6

The current peer review system must be considerably loosened to permit individual researchers a greater measure of freedom to pursue novel ideas without excessive justification at each step. True innovation can seldom be fully justified in advance. Its proof lies in the results it achieves.

National Astronomy Centers

The institution of national centers serving a larger astronomical community dates back only twenty-five years. Prior to that era national observatories did, of course, exist in many countries, including the United States, but their prime functions were different. These earlier observatories usually were dedicated to timekeeping or satisfied other practical needs. The idea that astronomers all over a country could share in the use of equipment in the custody of a national center found its first realization in the late 1950s in the construction of the National Radio Astronomy Observatory (NRAO) at Green Bank, West Virginia, and the Kitt Peak National Observatory (KPNO) near Tucson, Arizona. Since then both NRAO and KPNO have become centers, respectively, of radio and optical astronomy that rank among the leading observatories anywhere in the world. In addition, there has been a drive to pattern other national astronomy centers after these first two, and we now also have the Cerro Tololo Interamerican Observatory located in Chile, the Sacramento Peak observatory near Sunspot, New Mexico, and the National Astronomy and Ionsphere Center at Arecibo, Puerto Rico, primarily serving American astronomers. The European Southern Observatory performs a similar communal service mainly for astronomers from a number of West European member nations.

Over the past decade, an appreciable portion of the funding assigned to astronomy has gone directly into these national centers, and it is natural to ask how the centers have helped in the search for new phenomena.

While it is important to recognize that the centers are only a quarter century old, it is worth noting that none of the phenomena listed in table 5.1 have been discovered through work carried on at a national center, though contributions were made by astronomers employed at the National Radio Astronomy Observatory to the rediscovery of several phenomena through radio techniques. None of the rediscoveries listed were made by visiting astronomers at a national center, even though visitors have a claim on the largest fraction of the observing time. These trends need to be watched in the future to see whether they persist.

There may be good reasons for a lack of major discoveries at national centers. The centers may be better suited to studying smaller analytical questions. The major instruments at a center are shared by two, three, or more different observers or groups in a single week. A given telescope may serve as many as a hundred different astronomers in a given year. During this time, the telescope's performance is continually being upgraded, new features are being added, and maintenance is going on; thus the telescope never is quite the same for any two separate observing runs that a given observer might have at half-year intervals. Therein lies a difficulty.

When Bessel first described the parallax observations he had made on the star 61 Cygni, he wrote in his paper that he had hoped to start these observations earlier but had had to postpone them in favor of pressing observations. He had wanted to wait until the telescope could be devoted to parallax measurements uninterruptedly for a long period. The exquisitely fine observations involved in this venture clearly required a dedicated instrument, in Bessel's view.[28] Later, in writing about further observations meant to follow up the first parallax measurements, Bessel described how he became aware that one of the adjustment screws had been wearing down on the telescope during the course of the observations. His publication of the results includes a page-long discussion of the effects of the wear and tear of this screw on the reliability of his measurements.[29] Had Bessell not been in sole control of his instrument, such an estimate might have been all but impossible.

Approximately half the discoveries listed in table 5.1 were made by observers who were in complete charge of the telescope they used. To judge the importance of individual control we would have to know just how large a fraction of all significant research instruments have in fact been under the control of one individual or of a small cohesive team. Rocket and balloon-borne pioneering instruments frequently are operated in this fashion, as are some of NASA's smaller—Explorer class—earth-orbiting stations. However, larger, more expensive orbiting or spacecraft instruments often involve sizable consortia of scientists.

Similarly, the NSF currently allocates roughly two-thirds of its astronomical budget to national centers and only some one-third to grants for individual researchers.[30]

These figures suggest that most of our current expenditure supports projects that are more or less communally controlled, even though discoveries of new phenomena tend to result from projects conducted by one or two individuals. More data will be needed to make this assessment reliable.

In the present mode of operation of national centers, no single telescope can be dedicated to any one observer for any length of time. Certain types of observations that are particularly promising for novel discovery therefore appear to be excluded at these centers.

The discovery of the microwave background radiation emphasizes this point. Penzias and Wilson, as well as Ohm and other workers at Bell Laboratories, had noted a slight excess of radio noise in their measurements of sky noise through the Bell Laboratory's horn antenna. Puzzled by this excess, the two researchers dismantled the antenna, taped up joints that might have accounted for the excess radiation, cleaned the insides of the horn, and resumed observations. The results then obtained proved to be almost identical with earlier measurements and convinced the two observers that the horn was above suspicion.

It is just this kind of careful calibration and cross-checking that lies at the root of many scientific discoveries. The observer must reach a point where he trusts his instrument more than he trusts current theories, his own prejudices and intuitions, or the weight of opinion of the scientific community. At a national center that trust in an instrument seldom has an opportunity to mature. The acquaintance between observer and instrument is too brief. The instrument itself is too much changed from one observing run to the next. The opportunity for careful checks and cross-checks is too limited.

If national centers are to become institutions at which new discoveries are to be made, it may become necessary to operate them in a different mode. A variety of ways come to mind:

1. A new instrument might be turned over to a single individual, possibly the instrument's architect, for a period of one or two years, before the instrument is made available to the astronomical community at large. This would provide an opportunity to fully extend the instrument's capabilities and would permit acquisition of full familiarity with the instrument's performance. It could also afford an opportunity to use the instrument to look at the sky in a variety of modes to search for unexpected peculiarities. Such a procedure might be followed for all instruments at a national center, from novel spectrographs to new giant telescopes. In practice such a routine is never followed. Most new telescopes and most new spectrographs are put into use almost as soon as completed, frequently before being fully tested. The pressure from potential observers is so great that a premature start is more the routine than exception.

2. A given instrument at a national center might be made available to just a limited group of observers who then would gain familiarity with the instrument. An extreme of this holds at Westerbork in The

Netherlands, where a permanent staff makes observations requested by outside observers. Under such circumstances sufficient expertise and familiarity with the equipment develops. Rather than having any competent observer with an interesting proposal handle his own observations, only a smaller number of observers have access to the instrument.

If some mode of limited access were generally implemented at national centers, certain visiting observers would be responsible for just one instrument. Other visitors might be allocated responsibility for a different telescope or piece of apparatus. Familiarity with equipment would then not be diluted so much at the centers.

Both these proposals have obvious drawbacks. They lack flexibility and favor overspecialization; and that is presumably why they have not been generally implemented in the past. However, the open access and continual change at national centers may also be the prime reason for the apparent lack of major discoveries. We may have to search for new ways to operate national centers to produce optimal results.

In the long run we will also need to search for ways to plan occasional changes of direction at centers: With the present trend for specialization of each center in one particular branch of astronomy—visual or radio— and a tradition for the construction of general-purpose telescopes that can appeal to a wide class of astronomers, there may ultimately be a danger of self-perpetuation of the centers, and a loss of vitality.

Recommendation 7

The most effective long-term use of national centers should be studied on the basis of past performance and in light of future needs of the field. The problems of self-perpetuation of disciplines around which large centers have been built should be faced, and the potential entry of existing national centers into new areas of research and research support should be considered.

Self-perpetuation

In the United States problems of self-perpetuation are seldom faced when a new organization is set up. The plans confine themselves to the building up and not to the closing down of the newly founded establishment. As a result, once a national laboratory or center has come into existence, it becomes difficult to close it down. Yet, termination of centers that have run the course of their useful existence is just as important to advances in science as the construction of new facilities. In an economy that cannot afford continually to expand its scientific budget, every new start must be accompanied by a corresponding termination of an outdated project.

Much can be learned in this respect from the organization of the

Max Planck Society in West Germany. The society runs more than fifty research institutes in Germany and one or two in other European countries. There are over eight thousand permanent staff members, one-quarter of whom are scientists and scholars. An added two to three thousand visiting scholars and students also participate in the society's activities each year. The annual budget currently is around half a billion U.S. dollars.[31] Leading research in astronomy, biology, chemistry, history, jurisprudence, medicine, physics, and many other fields is conducted in the individual institutes located all over the Federal Republic of Germany. These institutes are largely built around the country's leading researchers who act as directors of the institutes. The programs supported by the society carry an annual budget that runs in the hundreds of millions of dollars.

Professor Reimar Lüst, formerly an astrophysicist, now president of the society, has written an elucidating article "How the Max Planck Society Keeps Research Alive." The thrust of the article is that research vitality is maintained through timely replacement of institutes in fields that no longer are at the forefront of science.

> A particularly acute circumstance for considering a change in the direction of scientific work is the approaching retirement of a director of an institute or a section of an institute, or of a member of the leadership of the institute. In the Max Planck Society we have felt obliged to check, at each retirement, whether the area of study is worthy of continuation. We take this task very seriously, since herein lies a decided possibility to ensure that our limited means can be put to use in new enterprises. Our procedure is the following:
>
> Four years before a retirement is to take place, the responsible section of the Max Planck Society as a rule sets up a commission with the aid of scholars who are not members of the Society.
>
> . . . It is imaginable that an area of research twenty years ago seemed so pressing that a section dedicated to its pursuit was established. In the meantime the field may have lost currency. Even when continuation of a particular branch of research seems advisable, it may happen that there is no suitable scholar available to direct the work. To the Society, that too suggests discontinuation. Our well equipped institutes can be justified only through the quality of their research; that means that the leadership must lie in the hands of a particularly well qualified scientist or scholar.
>
> On this basis the Max Planck Society has, in the last decade, transferred leadership or terminated the existence of an institute or section on the average once a year.
>
> These numbers alone cannot make clear how difficult it was in each particular case, to reach such a necessary decision. In each case, the careers and personal fortunes of the [director's] coworkers are gravely affected. . . .
>
> . . . it goes without saying that the Max Planck Society feels itself responsible for each of these coworkers.[32]

The termination of an institute, according to Lüst, involves the help of the society's administration in placing former members of the institute into suitable new positions. With sufficient warning, through a review set into motion four years earlier, the closing of an institute or center can be met with greater responsibility to those directly affected.

New areas of concentration are centrally decided by a committee for research planning. Views of the German government, as well as of leading research organizations in England and France, are represented. The final decision concerning a new enterprise is made by the senate of the Max Planck Society whose forty-six members include sixteen scientists and scholars, with the remaining members coming from many other areas of society and public life.

Our methods of organizing science in the United States differ from those practiced in other countries. But it seems important to realize that reasonably successful methods have been worked out for terminating research in areas that have lost their immediacy and for using funds and manpower freed through this action in new ventures that carry greater promise. These procedures can be both humane and responsible, but only if confronted with resolution and with sufficient forewarning.

Currently Planned and Recently Completed National Facilities

It is worth examining a number of major astronomical facilities currently planned, or recently completed, in order to assess their potential for making major astronomical discoveries.

The Next Generation Telescope

A design study is currently underway at the Kitt Peak National Observatory for a Next Generation Telescope [NGT]—a telescope with a diameter of 25 meters. Its collecting area would be 25 times larger than that of existing telescopes. Were it not for the brightness of the night sky, it would see a galaxy at a distance of 500 million light-years with the same clarity with which the 5-meter telescope atop Mount Palomar detects a galaxy only 100 million light-years away.[33] The cost of the NGT has been estimated at several hundred million dollars.

The lure of increased telescope size has always been this: The angular resolving power promises to increase in direct proportion to the telescope's diameter. The spectral resolving power could increase in proportion to the square of the diameter in the absence of substantial night sky contributions or in proportion to the diameter if the night sky prevails. The time resolution ideally would increase as the diameter taken to the fourth power, but it increases only as the diameter squared if the night sky dominates. In practice, technical, as well as atmospheric problems, may reduce these gains well below optimum.

The Kitt Peak report on the Next Generation Telescope recognizes these limitations but points out that ground-based visual astronomy now can turn nowhere except to telescopes of increased size.[34] While this seems realistic within the framework of ground-based astronomy, a careful comparison will have to be made with the promise and cost of smaller but potentially more powerful telescopes placed in earth-orbit. Such space telescopes might have a smaller light gathering power but still show greater potential because of their freedom from atmospheric limitations.

The Space Telescope

The Space Telescope (ST), to be launched in the mid-1980s, will have an aperture of 2.4 meters in diameter (figure 5.2). Its angular resolving power is to be about $\frac{1}{10}$ of an arc second, 10 times better than for a telescope of the same size installed at a mountaintop observatory. Ultraviolet and infrared observations that cannot be carried out through the atmosphere will also be possible. Plans call for periodic maintenance and repair by astronauts, and the telescope may be retrieved from time to time by NASA's Space Shuttle for extensive refurbishing and subsequent relaunch into Earth orbit.[35]

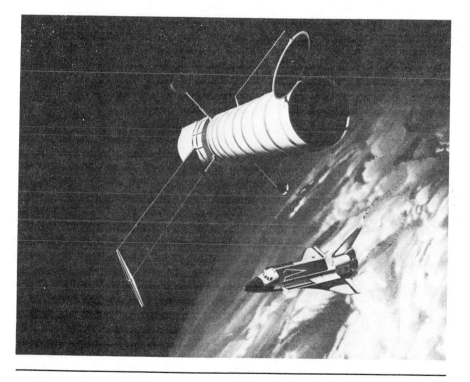

Figure 5.2 *The Space Telescope*

The Space Telescope is a 2.4-meter-aperture telescope to be launched from the Space Shuttle into earth-orbit in the mid-1980s. Because its operation will not be hampered by the earth's seething atmosphere, it will be able to focus radiation more sharply and attain an angular resolution of $\frac{1}{10}$ of a second of arc and a time resolution of a millisecond in observing stellar sources. The Space Telescope is most likely to serve well as a powerful analytic tool. Its fiftyfold increase in sensitivity and thousandfold increase in time resolution for variations in stellar sources may also make it an instrument capable of discovering new phenomena.

Photograph of artist's conception courtesy of NASA

At visual wavelengths the faintest observable stars should be fifty times fainter than any source we can observe from the ground. For extended visual sources many of these advantages in sensitivity admittedly are lost and only the improved angular resolving power remains. The increased sensitivity for stellar sources, however, also promises a thousandfold improvement in fast time resolution. Time variations that occur in a span of milliseconds should become accessible with the same reliability currently reserved for variations on a time scale of seconds. The spectral resolution that can be achieved should increase an order of magnitude over capabilities exhibited by ultraviolet instruments aboard the spacecraft *Copernicus* and *International Ultraviolet Explorer (IUE)* launched in the 1970s.

The Space Telescope will also offer gains at far-infrared and submillimeter wavelengths where the increase in sensitivity over ground-based astronomy will be enormous. The first launch of the ST will not be carrying aloft any instruments sensitive to these wavelengths, but infrared and submillimeter astronomers hope to have sensitive devices aboard with the next complement of instruments.

The Space Telescope is likely to prove itself a versatile analytic research tool at ultraviolet, visible, and infrared wavelengths. Its greatest potential for discovering new phenomena may lie in the visual regime where its fiftyfold increase in sensitivity and thousandfold increase in rapid time resolution for stellar sources afford the most striking advances in instrumental capability.

By the time it reaches completion, the Space Telescope will have cost the major part of a billion dollars.

The Very Large Array (VLA)

On New Mexico's Plains of San Augustin, the Very Large Array (VLA) of radio antennas has recently been completed (figure 5.3). The signals detected by the individual antennas are channeled to a central processing station by movable waveguides that stretch along tracks. At wavelengths of 6 centimeters, the VLA has a resolving power of 0.6 seconds of arc; at longer wavelengths the resolving power decreases in proportion to the wavelength.[36] The estimated cost of the full array is about $80 million.[37]

In light-gathering power, the combined aperture of the twenty-seven dishes rivals the capabilities of a 130-meter single dish. This is somewhat larger than the 100-meter telescope now operating at Effelsberg in West Germany but rather smaller than the 300-meter telescope at Arecibo in Puerto Rico. In angular resolving power the array is roughly comparable to the much smaller two- and four-element interferometers used by groups at the University of Manchester and at the U.S. National Radio Astronomy Observatory at Green Bank. This angular resolving power is larger than that of arrays in existence in Australia, England, and The Netherlands but 100 times less than the resolving power attained with transcontinental or intercontinentally operating very long baseline interferometers (VLBI). The VLBI technique has permitted us to resolve structures of the order of $\frac{1}{1,000}$ of an arc second.

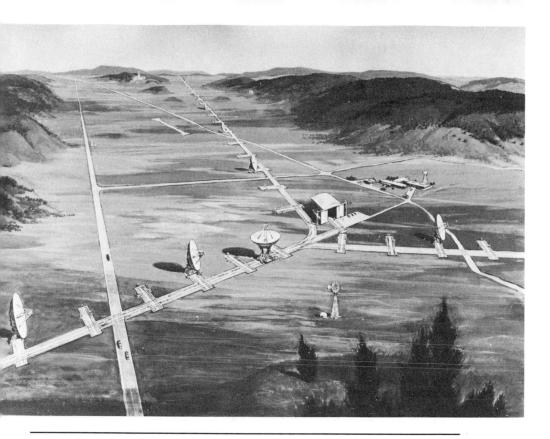

Figure 5.3 *The Very Large Array (VLA)*

The VLA is a collection of twenty-seven radio telescopes, each 82 feet in diameter, mounted on sets of railroad tracks that describe an enormous Y-shaped array on New Mexico's Plains of San Augustin. Each arm of the array is some 20 kilometers long. The spacings between antennas is varied by rolling the telescopes along tracks, and a sequence of interferometric observations yields radio pictures of celestial sources with a spatial resolution equivalent to that of a single radio telescope whose diameter would have to be roughly 30 kilometers.

Artist's conception courtesy of the National Radio Astronomy Observatory

At intermediate angular resolving power, the VLA should be capable of gathering roughly an order of magnitude more radiation than other existing arrays. This relative wealth of radiation might be used to improve spectral resolving power and to obtain detailed pictures of astronomical sources as observed in individual spectral lines. There is also the possibility of using the extra radiation to search for time variations in the structure of sources.

The VLA, therefore, appears to be a versatile, powerful research tool that will answer a host of questions we are able to pose right now. It will probably prove its worth largely in analytic research, since it appears to lack the prime traits of a discovery instrument. As shown in figures 5.4 to 5.6, it does not operate in a remote wavelength range never probed before, and it does not exhibit angular, spatial, or spectral resolving power several orders of magnitude more advanced than existing instruments.

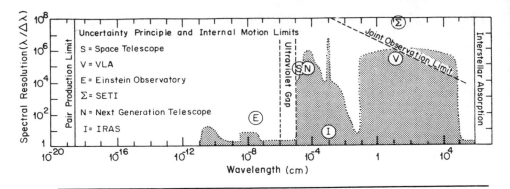

Figure 5.4 *Spectral Resolution Available in 1979 and in Planned Ventures*

Instrumental capabilities in existence at the beginning of 1979 are represented by shaded areas and are compared with the anticipated capabilities of recently completed facilities and of planned ventures. This particular figure presents these data for spectral resolving power. The following two plots present the same data for time resolution and angular resolution, and a final plot provides data on sensitivity. The new instruments for which comparisons are made are the Space Telescope; the Very Large Array of radio antennas (VLA); the Einstein Earth-orbiting X-ray Observatory, a radio instrument to search for extraterrestrial intelligence (SETI); the Next Generation twenty-five-meter equivalent aperture, ground-based, optical telescope (NGT); and the infrared astronomical satellite telescope (IRAS). Figures 5.4 to 5.6 use the same format as figures 3.6 to 3.8.

A Very Long Baseline Array (VLBA) of radio antennas stationed in Alaska, Hawaii, and across much of the continental United States is under consideration as a successor to the more informal VLBI network responsible for the discovery of superluminal sources. The array would consist of nine or ten identical antennas that could be operated more efficiently than the current network of varied instruments, particularly at the shortest wavelengths where the highest angular resolution can be attained. The array will increase sensitivity and angular resolution by modest factors and, like the VLA, may excel at analysis rather than discovery.

SETI—An Instrument in Search of Extraterrestrial Intelligence

A proposal to seriously search for radio messages beamed in our direction by transmitters set up by other civilizations has been receiving increasing attention in the newspapers and in Congress. This search for intelligent messages could also have an astronomical yield. Since we cannot tell in advance just which transmission frequency, if any, another civilization might select, we will have to search over a wide band of frequencies if we are to have any likelihood of success. This band will have to be sifted carefully to see whether unusual signals are reaching us in any one of millions of different narrow subchannels. And in this sifting we are quite likely to come across novel astronomical effects not noticed before.

Bernard M. Oliver of the Hewlett-Packard Corporation has described a variety of options for such a system. One of the less expensive ones would cost upward of $32 million for the antenna and processor, without buildings, auxiliary equipment, or operating costs. It would enable us to scan the entire sky in a wavelength range from about 17 to 21 centimeters, with a sensitivity of 10^{-28} Watts per square centimeter (figure 5.7), a spectral resolving power of 10^8, and an angular resolving power of roughly 10^3. The total dedicated time required for a complete coverage of the sky through the 30 million individual subchannels that comprise the entire system would be twelve days.[38] The large number of subchannels employed would lead to a rate of data collection unrivaled by any-

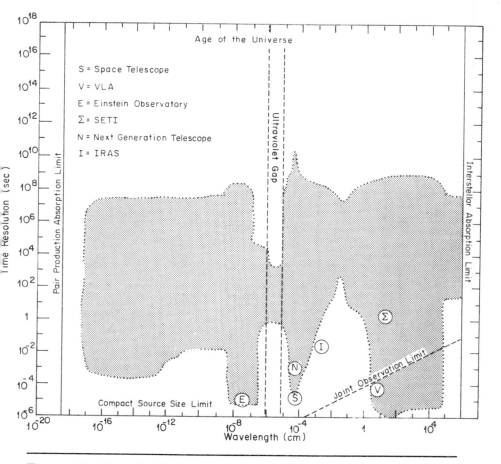

Figure 5.5 *Time Resolution Available in 1979 and in Planned Ventures*

As explained in the caption to figure 5.4, this figure is one of a series of four. The symbols used are explained in more detail in the caption to figure 5.4.

The instrumental capabilities shown for the new and planned ventures do not far outstrip current capabilities available from smaller, often less expensive apparatus. This indicates that many of the instruments are more likely to prove their worth in analytical work as contrasted to discovery. Discovery instruments normally are sensitive to carriers of information never detected before or provide large jumps in spectral, spatial, or temporal resolving power. Introduction of polarization capabilities sometimes also leads to new discoveries. Among the instruments analyzed here, the Einstein X-ray Observatory may prove the most likely to make new discoveries because it does operate consistently at or beyond the capabilities available before its launch. The infrared astronomy satellite also may prove to be a discovery instrument, primarily because it will be the first instrument to survey the sky at far infrared wavelengths in an unbiased manner—without preconceptions gained from radio, optical, or X-ray data. Such a survey can best be carried out with a satellite instrument, since atmospheric interference is too powerful at aircraft and balloon altitudes.

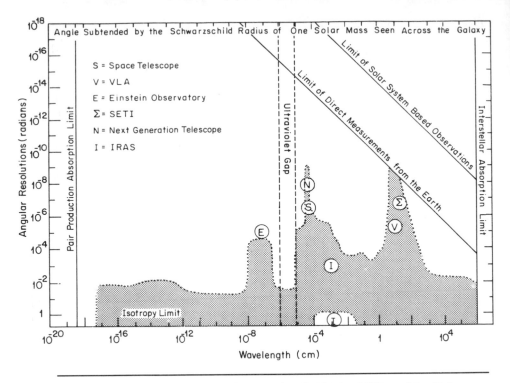

Figure 5.6 *Angular Resolution Available in 1979 and in Future Ventures*

This is the third in the series of figures whose format is described in the caption to figure 5.4.

Again, we see that the resolving power of which new instruments will be capable is not enormously different from that of existing equipment. The infrared astronomy satellite, however, in addition to presenting a capability for resolving structures of a few minutes of arc, will also be the first instrument capable of providing a measure of an isotropic background radiation bath, at wavelengths ranging between 10 and 100 μ—10^{-3} to 10^{-2} centimeters. Currently we only have upper limits on the power incident from the universe in this wavelength range. Cosmological data of significance might therefore be gained for the first time at far infrared wavelengths. The angular resolving power of the Einstein Observatory X-ray telescope also may lead to new discoveries. The instrument already has produced X-ray pictures of unprecedented quality.

thing apart from optical photography. A SETI installation might therefore lead to the discovery of astronomical phenomena through the same many-order-of-magnitude increase in the information rate that led to optical discoveries with the introduction of photographic plates.

The Einstein Observatory

The second high energy astronomical observatory, launched late in 1978, has been named after Albert Einstein. The array of X-ray detectors aboard the Einstein Observatory is illuminated through a 60-centimeter aperture, grazing incidence, reflecting X-ray telescope that forms images much in the manner of the reflecting telescopes used by visual astronomers. The cost of the spacecraft was $87 million.

The Einstein Observatory can detect X rays in the ¼- to 3-kilovolt range. These energies correspond to a wavelength range of 3×10^{-8} to 40×10^{-8} centimeters. The respective spectral resolving powers attained range from 50 to 1,000 across this wavelength span—roughly a factor-of-10 improvement over the resolving powers offered by earlier X-ray instruments. The time resolution available is 10 microseconds. The angular resolution amounts to 10 seconds of arc, and the sensitivity is 1,000 times higher than that of the spacecraft *Uhuru* which was launched a decade earlier and was responsible for determining the periodic nature of large numbers of Galactic X-ray binary stars, as well as for the discovery of a series of extragalactic sources.[39]

With its high sensitivity, the Einstein Observatory already has discovered X-ray emission from a variety of young main sequence stars. Its high angular resolving power and highly increased sensitivity should make this instrument capable of discovering a variety of novel phenomena. Its high spectral resolving power may also lead to new discoveries.

A potential successor to the Einstein Observatory is the Advanced X-Ray Astrophysics Facility. Plans call for a long-lived, Earth-orbiting X-ray telescope, sensitive in the 0.1 to 8 kilovolt range, with a hundredfold increase over the Einstein Observatory's sensitivity, a tenfold increase in angular resolution, enhanced spectral resolution, and capability to analyze for plane polarization. Such a jump in capability should favor discovery of new phenomena.

The Infrared Astronomy Satellite (IRAS)

Infrared astronomical observations can be carried out from the ground, from airplanes, and from balloons at selected wavelengths, provided we know precisely where to point our telescopes in order to make our observations. The infrared emission from the atmosphere even at balloon altitudes is so strong that astronomical sources are only detected if atmospheric sky-emission can be cancelled out by subtracting the emission incident from adjacent patches of the sky. This involves knowing, with great accuracy, the position of a source to be viewed. A large part of infrared astronomy, therefore, has involved the study of sources originally discovered by optical or radio astronomers. Sources actually discovered in the infrared are relatively rare. At near-infrared wavelengths, unbiased surveys of the sky have been carried out from the ground; and at wavelengths 20 times longer than visible light, a survey has been carried out from rockets. In the far-infrared, however, no systematic survey of the sky has been made, and there might well exist celestial sources that emit the bulk of their energy in this spectral range but are totally unknown.

There may also exist isotropic background radiation at infrared wavelengths, and again the instrumentation used in the past would have been incapable of detecting this flux, unless it had been far stronger

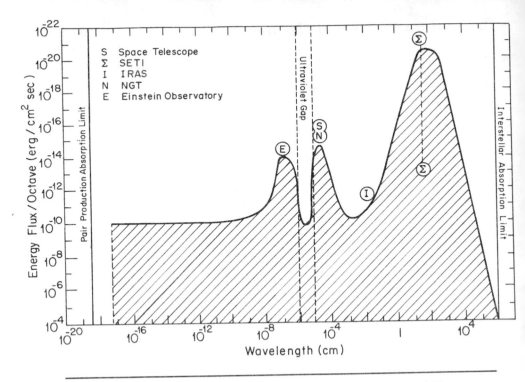

Figure 5.7   *Sensitivity Available in 1979 and in Planned Ventures*

This is the last of the series of four figures whose format is described in the caption to figure 5.4.

It presents data on the sensitivity of instrumentation available for astronomical observations. The shaded area represents sensitivity available in 1979. Sensitivity can be measured under a variety of different operating procedures, and the sensitivity of a given piece of apparatus may vary quite drastically, depending on the operating mode in which it is used. Here we assume that all of the instruments are to be used to observe a single point source in the sky, as distinct from a diffuse source whose picture we might like to obtain. We also assume that the source radiates over a broad range of wavelengths, and that the energy per octave—the energy residing between a wavelength range stretching from some wavelength, λ, to twice that wavelength, 2λ—is the quantity we are to measure. Under these conditions, the SETI instrument denoted by the symbol Σ does not appear to perform very well because the plan is to search for extraterrestrial intelligence through an enormous number of narrowly tuned filters, and each filter then receives energy from just the tiniest portion of an octave of wavelengths. A more appropriate sensitivity measure for SETI therefore is also shown by the upper circle marked Σ. This refers to the sensitivity that would be measured in units designated energy flux, rather than energy flux per octave, and would represent the actual energy, measured in erg/cm² sec, that could be detected by the instrument if the radiation were concentrated in a very narrow spectral line.

than the cosmic microwave background radiation. Finally, the corrections that have to be made to null out emission from the atmosphere, at balloon altitudes and below, are time-consuming and have not permitted the observation of rapidly varying infrared signals. We therefore do not know whether infrared sources radiate steadily or whether any of them exhibit rapid time variations.

The infrared astronomy satellite is designed to make a survey of the sky at wavelengths reaching out to ⅒ of a millimeter—200 times the wavelength of visible light. Since the sky has never been surveyed at these long wavelengths, there exists the potential for discovering new phenomena. With no atmospheric emission entering the IRAS telescope from above, there will also exist a possibility of determining whether radiation reaches us isotropically from the universe in any appreciable quantities at infrared wavelengths. Finally, time variations as short as $\frac{1}{10}$ of a second will be registered by detectors during the survey so that we could gain a better understanding of the variability of cosmic infrared sources.

The telescope to be flown is sophisticated, indeed. A telescope at room temperature would copiously emit infrared radiation, the normal heat radiated by any warm object. For the sensitive infrared detectors aboard IRAS not to see this confusing source of infrared radiation, the whole telescope is cooled to the temperature of liquid helium, a temperature just a couple of degrees above absolute zero. The liquid helium stored aboard the spacecraft for cooling must be capable of surviving at least six months in earth-orbit in order to allow the entire sky to be surveyed once. All this requires a special combination of thermal and optical engineering not encountered elsewhere. By means of this telescope, however, as many as a million sources are expected to be catalogued during the life of the spacecraft and their intensities measured at four infrared wavelengths.

The cost of the IRAS satellite will be comparable to the cost of erecting the VLA.

Recommendations for New Searches

We have seen how important it is for the discovery of new phenomena to open further channels of information through which the universe can be observed in novel ways. We might well ask ourselves which channels to open next? What are the choices? What is the likely yield?

We saw in chapter 3 the large fraction of all conceivable electromagnetic observations that still remains inaccessible. Eventually all of these channels of information should become available to astronomers, but we may have to decide on capabilities that might receive priority now. First, however, we should emphasize the importance of opening two new channels in which no astronomical signals whatsoever have been detected to date. These are the gravitational-wave and the neutrino channels, through which we might hope to recognize phenomena completely different from those viewed in the electromagnetic domain. Gravitational radiation might, for example, signal the collapse of the interior portions of a massive star before its outer shell explodes in a supernova eruption. To date we have never seen a star collapse. We merely have noted the existence of pulsars which we believe to be highly compact neutron stars. We conjecture that they are formed in a collapse that also triggers a supernova explosion. This hypothesis appears borne out by the existence of pulsars at the centers of the Crab Nebula and the Vela supernova remnants.

Neutrino Astronomy

The sun is so bright that we cannot long look at it without protecting our eyes. If we could see neutrinos as readily as we see light, we would find the sun almost as dazzling. Our star emits a small percent of its radiation in the form of neutrinos, and the solar neutrino flux incident on Earth is easily 10 thousand times brighter—more powerful—than the visible light from the full moon. Despite this enormous influx of solar neutrino energy, we have no observational evidence, as yet, that these neutrinos in fact are there. Our arguments, though theoretically sound, still need observational verification.

Neutrinos are elusive particles. They may have no rest mass; they have no charge; they can cross through a body as large as the sun with negligible probability of interacting with even a single atom. Though postulated as early as 1930 by the theoretical physicist Wolfgang Pauli, neutrinos were not detected experimentally until 1956, when Clyde Cowan and Frederick Reines succeeded in observing them at the Savannah River Plant nuclear reactor. That experiment firmly established the existence of the neutrino as a physical entity rather than a hypothetical particle.[40]

For many years the neutrino emission from the sun has been sought by Raymond Davis, Jr., who has used increasingly sensitive detectors for over a decade. The theoretically predicted neutrino flux should by now have been observed, but though the sensitivity of the detectors appears sufficiently high, no neutrino flux attributable to the sun has been reliably established.

This presents a serious problem to theoretical astrophysics. Detection of neutrinos emitted by the sun would provide the first direct confirmation that our theories concerning energy generation in the interior of stars are correct. In short, the neutrino observations should confirm our ideas of how stars manage to shine as brightly as they do for billions of years on end without significantly fading.

Neutrinos are expected not only from the sun. Supernovae and the attendant collapse of the core of a star, the collapse of supermassive objects at the center of a globular cluster or at the center of a galaxy, cosmic ray interactions with ambient hydrogen atoms, and the ubiquitous cosmic microwave background radiation, all can produce neutrinos at varying energies and in varying numbers. A cosmic background neutrino bath comparable to the 3 degree Kelvin microwave background radiation is also expected, though the neutrinos should be slightly cooler than the microwaves. The neutrino temperature might be as low as 2 degrees Kelvin.[41]

An extensive review of neutrino detection capabilities has been written by John N. Bahcall of the Institute for Advanced Study at Princeton. He estimates the capabilities of nine different types of detectors for the detection of neutrinos in different energy ranges. A neutrino detector generally consists of a particular isotope of an element, the chlorine isotope of mass 37, the gallium isotope of mass 71, and so forth. Absorption of the neutrino is accompanied by a nuclear reaction which changes chlorine into argon, or gallium into germanium. The presence of the tiny quantities of newly formed argon or germanium must then be determined. These are difficult measurements, but they appear to be entirely possible; and the cross sections for neutrino absorption, in at least a number of the detection schemes analyzed by Bahcall, are well established.[42]

Because the neutrinos interact so poorly with matter of any kind, these especially constructed detectors must contain many tons of material. To be capable of detecting one solar neutrino daily, a detector would have to contain ten thousand tons of chlorine 37, or fifty tons of gallium 71. Other detectors made of isotopes of various elements, such as lithium, vanadium, manganese, bromine, rubidium, indium, or thallium, also would have masses ranging from several tons to several thousands of tons.

The material for these detectors is bound to be very expensive, but it will not be used up; rather, it will be stored during the course of the experiment in vats that will permit isolation or detection of the atoms produced in the neutrino interactions. At the end of the measurement, the apparatus may be dismantled, and the detector material resold like any other substance unharmed by storage.

A massive chlorine detector has been in use by Raymond Davis, Jr. for his experiment. Bahcall, Davis, and a number of colleagues have suggested construction of the gallium detector. One estimate for the cost of this detector is 25 million dollars.[43] If this cost is verified on closer analysis, such an experiment would not be expensive by comparison to construction costs of other major astronomical facilities.

A quite different type of neutrino experiment has recently also been proposed. It is the Deep Underwater Muon and Neutrino Detector (DUMAND), a detector array that would be emplaced below 6 kilometers of water at the bottom of the ocean and would consist of thousands of photomultiplier tubes and hydrophones distributed over a volume of a cubic kilometer of water at those depths. This array would not interact with the neutrinos directly but rather with the particles produced when high energy neutrinos interact with matter. The array of detectors would have high directional sensitivity in contrast to the detectors mentioned before which will have no way of recording the incident direction of a neutrino. The neutrino energy ranges covered by DUMAND, however, are much higher than those to which the solar neutrino detectors would be mainly matched, and the phenomena to which these two classes of detectors would be sensitive would probably be quite distinct.[44]

Gravitational Waves

Efforts to detect gravitational waves have been going on for over a decade. A consequence of Einstein's general theory of relativity, these waves were long held to be on an unsound theoretical footing. By the late 1950s, however, sufficient theoretical work had been carried out to convince relativists that gravitational waves really must exist in nature. The intensity of waves that can be generated under astrophysical conditions has been estimated, and we know of varieties of sources that should produce an ultimately measurable flux of gravitons—gravitational waves.

Unlike the neutrinos, gravitational waves have not been observed in the laboratory. In fact, it is more likely that these waves will first be found emanating from some cosmic source than from a piece of laboratory apparatus. The coupling of these waves to matter is so poor that gravitons can only be generated by very massive bodies, such as stars collapsing onto themselves. Such stars are much more likely to generate a flux detectable even at the distance of Earth than any piece of massive, rapidly accelerating machinery we could currently construct in our laboratories.

Because the gravitons couple poorly to matter, they are both generated in very small numbers within most astronomical sources and are absorbed in very small numbers at the detector: Both the emission and the absorption of gravitons is generally inefficient. This contrasts to the situation for neutrinos which at least are generated with an efficiency of a few percent, when compared to the generation of light in the sun. For neutrinos, only the absorbing apparatus we construct is inefficient.

For all these reasons, we may expect that gravitational-wave astronomy will deal with the rapid acceleration—collapse, orbital motion, explosion—of highly massive objects such as stars, clusters of stars, nuclei of galaxies, or quasars. There may also exist an ambient graviton bath, analogous to the neutrino and microwave background radiation and, similarly, a remnant of the earliest stages of cosmic evolution.

Gravitational detectors built to date generally have been massive resonators that are designed to ring like a bell when hit by a graviton. The best chance of detecting gravitons will be with detectors that ring for a long time before their oscillations die out. Such detectors are said to have a high quality factor Q. The factor Q gives the number of oscillations the resonator experiences before the amplitude of the wave is appreciably damped. At very low temperatures the elasticity of material improves, and Q values as high as 10^{10} are now foreseeable. A detector with $Q = 10^{10}$, ringing with a frequency of 1,000 cycles a second, could go on ringing for many months. The best gravitational detectors on the drawing boards might be large, highly purified, single crystals of material cooled to temperatures well below those at which helium liquifies. The lower the temperature, the lower the expected detector noise, and the higher the anticipated Q value. Many theoretical problems still remain, and even the best detectors envisaged for the next decade will still fall many orders of magnitude shy of detecting any anticipated cosmic signal.[45]

A different class of gravitational-wave detectors would consist of two widely separated masses that could be accelerated toward and away from each other by an incident gravitational wave. Such masses might be emplaced in spacecraft orbiting the sun. By accurately monitoring the separation of the masses by means of lasers, or through Doppler-tracking procedures of the kind used by NASA's Deep Space Network, any sudden change in separation could be determined. While a variety of disturbances could lead to changes in the separation, the sequence of changes produced by a gravitational wave has its own signature, and gravitational-wave interaction could therefore be distinguished from other competing effects.[46]

A recent planning document published by the National Aeronautics and Space Administration suggests a third variant. It proposes placing a single structure weighing 16 tons into earth-orbit. This structure would consist of two girders, each 1 kilometer in length, joined at their centers to form a cross. The lengths of the girders would change slightly in response to a gravitational wave, and the changing separation between the ends of the girders would be determined by an interferometer. In earth-orbit, isolation from the usual vibrations that hinder sensitive laboratory experiments would be complete.[47]

This is not the place to compare the merits of the various proposed gravitational experiments, nor to judge the promise of different neutrino detection schemes.

It is not even the right forum for discussing the relative merits of neutrino and gravitational-wave detection—if indeed a choice between the two disciplines needs to be made. It is possible, however, to formulate a general recommendation:

Recommendation 8

Gravitational-wave and neutrino detection promise new information that may deal with phenomena not recognizable through the electromagnetic radiation channels. Facilities adequate to detect neutrinos and gravitons should therefore be constructed as a means for continuing our searches for novel phenomena.

The costs that have been cited for major neutrino and gravitational-wave detection facilities do not appear excessive when compared to the costs of erecting other major astronomical facilities. Serious priority should therefore be given to the start of at least one of these facilities during the next decade.

A careful analysis should be made both of the feasibility and costs of the various neutrino and graviton detectors before a decision is made on which is to receive priority for a first massive effort. The losing discipline in such a competition should, however, still be vigorously supported in anticipation of a substantial effort roughly a decade later.

Whatever detectors are first built, whether they turn out to be neutrino or gravitational-wave detectors, will almost certainly be highly specific—able to detect radiation only in one relatively confined energy band. Theoretical studies, however, predict a variety of astrophysical processes that would emit most of their flux in quite distinct energy ranges. We must therefore understand that both neutrino and gravitational-wave detection will require programs lasting many decades and will involve varieties of detectors as different from each other as radio antennas, visual telescopes and gamma-ray counters. We are not likely to exhaust neutrino and gravitational-wave astronomy quickly. A steady, protracted effort will be needed.

Technological Advances in the Electromagnetic Domain

Thousandfold increases in resolving power or thousandfold increases in wavelength or frequency of observations in the electromagnetic domain should be expected to systematically induce the discovery of new phenomena. Such technical advances are not always expensive, need not always rely on the support of industry or of the military, and should be resolutely pursued in future ventures. A number of currently feasible improvements in the electromagnetic domain show what can be accomplished with deliberate emphasis on instrumentation:

1. One capability that essentially is here now has been implemented at very low cost but should be applied over a much wider wavelength range. It is infrared heterodyne spectroscopy, a method pioneered by the group of Charles Townes at the University of California at Berkeley. With this technique the resolving power of some infrared astronomical observations was, in one recent year, abruptly increased by a factor of 1,000 beyond capabilities that had previously existed. Further extension of this technique will require the construction of powerful reference lasers that can be tuned over wide spectral ranges. Since such lasers are likely to have far-reaching applications in industry and in the military, it is quite likely that this technique will become generally available in infrared and optical astronomy in the next two decades.

2. Very little is known about the isotropic radiation reaching us from the universe in the ultraviolet, the visible, and throughout the infrared. At visual wavelengths, the zodiacal dust reflects sunlight; and Galactic starlight beyond makes the detection of extragalatic radiation difficult. In the infrared, the atmosphere and the zodiacal light provide difficulties. Nevertheless, gains can be made in these directions.

The Solar Polar Mission, which NASA and the European Space Agency (ESA) now plan to launch in the mid-1980s, will comprise a spacecraft that is to rise out of the ecliptic plane and go beyond the reaches of zodiacal dust. It is primarily designed to study the solar system. But because the spacecraft will travel beyond the realm of light-scattering interplanetary dust, it should also obtain a clear enough view of space beyond the solar system to detect extragalactic visual and near infrared radiation. With an instrument that has sufficient sensitivity and spectral coverage, we could obtain information about radiation that might have been emitted aeons ago when galaxies first formed. While the wavelength range covered by the photometer available on these flights may not be sufficiently wide to provide all this information, techniques do exist for implementing such observations, and no expensive development program would be required.

3. Currently there are no plans to construct infrared equipment with

improved time resolution. However, with equipment mounted behind cooled telescopes in spacecraft, the sensitivity of infrared equipment increases by many powers of 10, because noise due to foreground atmospheric and instrumental radiation is drastically cut. As sensitivity is increased, the capacity for short-time scale resolution increases rapidly. Much could therefore be learned if rapid data handling capabilities were to be installed on infrared astronomical spacecraft. Factors of a 1,000 in speed could be gained with existing techniques.

Spatial resolution may also be increased through a scheme that involves long baselines. Charles Townes's group has for some time been practicing interferometery at intermediate infrared wavelengths. Two ground-based telescopes are used in combination as an interferometer. At the moment the available telescopes are still rather close together. But with telescopes placed a few kilometers apart, the resolving power obtained through interferometry could be increased a thousandfold. Alternatively, lunar and planetary occultations could be used to study the structure of infrared sources. Occultations have been used to determine the structure of the bright infrared source IRC + 10216, and such methods can be extended. The use of spacecraft with rapid time resolution could again come into play. With these the disappearance of an astronomical source behind the limb of the moon or a planet could be timed with extreme precision, and fine structural features in the source would become defined on a scale that improves in proportion to improved time resolution.

I have selected these examples from just two spectral domains, the visible and the infrared. Thousandfold increases in instrumental capabilities can, however, be sought in all channels of the electromagnetic domain.

Recommendation 9

A variety of instrumental advances could increase current observing capabilities by factors of 1,000 at modest cost. Such potential advances should be systematically sought and pursued with full vigor for application in astronomical observations.

Astronomical Planning

When we look back at the early successes of radio astronomy, one main trend can be discerned, and this has largely been duplicated in the emergence of X-ray, gamma-ray and infrared observations. A pioneer of British radio astronomy summed it up somewhat brusquely for the history compiled by Edge and Mulkay.

... Radio astronomy was done with a peculiarly close relation between the actual lab work on the equipment, and the observations. It was done at the same place, and by the same people. One reason why optical astronomy has been so backward technically has been that telescopes were made by people at one place and then carted up some remote mountain and used by other people who, naturally, became interested in the work the telescope could do, and inevitably came down the mountain years later and said, "the most important problems in astronomy can only be solved by a larger telescope, just like the one we have, but much larger." No doubt radio astronomy approaches or has reached this stage. . . .[48]

Let us not ignore this remark simply because it may sound irreverent: The most important single planning document for ground-based astronomy in the 1960s was the Whitford committee report sponsored by the National Academy of Sciences. This committee, chaired by A. E. Whitford of the Lick Observatory, placed its highest priority on the construction of three telescopes comparable to the largest ground-based optical telescopes then in existence. At least one of these was to be erected in the Southern Hemisphere to make accessible portions of the sky at which the largest existing telescopes at Mount Palomar, Mount Wilson, and the Lick Observatory could not be pointed. The Whitford committee's second recommendation was even more concerned with size. It called for "an engineering study for the construction of the largest feasible optical telescope."[49]

The highest priority recommendation for radio astronomy was for "a very large high resolution pencil-beam array with low sidelobes to be constructed as a national facility." These recommendations have largely been put into practice. A 4-meter telescope was erected at the Kitt Peak National Observatory near Tucson, Arizona, and a second 4-meter telescope was emplaced on Cerro Tololo in Chile. An engineering study for a 25-meter optical telescope is in progress at the Kitt Peak National Observatory, and the Very Large Array of radio antennas on New Mexico's Plains of San Augustin conforms to the last cited recommendation.

A few years after the appearance of the Whitford report the National Academy sponsored a follow-on study, *Astronomy and Astrophysics for the 1970s*. It has become the most influential astronomical blueprint of the decade. In its preface, the survey committee's chairman, Jesse L. Greenstein, professor of astronomy at the California Institute of Technology, shows great insight into the needs of astronomy and contrasts his committee's approach to that of the Whitford report.

... The present Survey has a different emphasis. It reviews the present state and future need for facilities, flight programs, and ongoing support of all astronomy, including space science and solar physics; one of its main themes is the rapid progress of the field since the Whitford report. The effectiveness of ground-based facilities has increased extraordinarily as a result of new applications of sophisticated electronics, which have greatly enhanced the effectiveness of existing telescopes and extended their use far into new wavelength regions. The capability provided by the space-astronomy program resulted in observations at essentially all wavelengths unobservable from the ground. These advances led to the discovery of many new objects and phenomena and made it clear that the astronomical universe was in many ways still largely unexplored. New facilities are needed on the ground and in earth orbit to exploit fully the promising opportunities opened by advanced technologies. . . .[50]

These words are inspiring, and it is, therefore, particularly interesting to see how they relate to the survey committee's actual proposals. In order of importance the four programs of highest priority are listed as:

1. A very large radio array, designed to attain resolution equivalent to that of a single radio telescope 26 miles in diameter; this should be accompanied by increased support of smaller radio programs and facilities at the universities or other smaller research laboratories;
2. An optical program that will vastly increase the efficiency of existing telescopes by use of modern electronic auxiliaries and at the same time create the new large telescopes necessary for research at the limits of the known universe;
3. A significant increase in support and development of the new field of infrared astronomy, including construction of a large ground-based infrared telescope, high-altitude balloon surveys, and design studies for a very large stratospheric telescope;
4. A program for x-ray and gamma-ray astronomy from a series of large orbiting High Energy Astronomical Observatories, supported by construction of ground-based optical and infrared telescopes.[51]

The word *large* appears in each recommendation. Edge and Mulkay's anonymous radio astronomer seems to have been right. We might actually extend his last statement and give it greater generality: By now radio astronomy, optical astronomy, infrared, X-ray, and gamma-ray astronomy, all have reached the stage of asking, almost automatically, for ever larger telescopes. The early emphasis on sensitive detectors and increased resolving power, which led to the initial flourishing of these fields, has shifted toward a preoccupation with larger telescopes.

Professor Greenstein has pointed out that each of the four recommendations is double pronged and combines a request for new, large instruments with a recommendation for the funding of smaller complementing programs. He expresses disappointment that only one-half of many of these recommendations was implemented—mainly the half that dealt with a readily identified major instrument or the half that could be carried out as part of the efforts of a national research center. Thus the Very Large Array of radio telescopes has been completed in New Mexico, but there has been none of the increased support for smaller radio programs spelled out in the first recommendation. In fact, many universities had to close down their radio telescopes in the years following publication of the report.[52]

The Greenstein committee's remaining recommendations continue in much the same mood as the four highest priority items:

5. The construction of a very large millimeter-wavelength antenna to identify new complex molecules, to study their distribution in interstellar space, and to study quasars in their early, most explosive phases;

6. A doubling of support for astrophysical observations from aircraft, balloons, and rockets, at wavelengths ranging from the far infrared to gamma rays;
7. A continuation of the Orbiting Solar Observatories through OSO-L, -M, and -N, together with an updating of existing ground-based solar facilities;
8. A sizable increase of support for theoretical investigations, including an expansion of capability for numerical computation;
9. An expanded program of optical space astronomy, including high-resolution imagery and ultraviolet spectroscopy, leading to the launch of a large space telescope at the beginning of the next decade;
10. A large, steerable radio telescope designed to operate efficiently at wavelengths of 1 cm and longer to obtain observations with high angular resolution and record emission from more distant objects than is now possible;
11. Construction of several modern astrometric instruments at geographic locations chosen to permit systematic measurement of accurate positions, distances, and motions in both northern and southern hemispheres.[53]

Looking at this list, many astrophysicists would notice that there is no mention of neutrino astronomy or gravitational-wave detection. This is all the more curious because neutrinos are mentioned right on the first page of the Greenstein survey committee's report where recent astronomical discoveries are described in this fashion:

> As each new technology was applied to study light [photons] of different colors or energetic particles of different charge and mass [cosmic rays, neutrinos], new types of worlds were revealed.

With this kind of introduction, neutrino astronomy might have appeared to be slated for solid support. And indeed, at first it looks that way. On page 21 of the report the promise of neutrino observations is given a substantial buildup:

> Recently, an exciting new observational check of the theory of solar structure has been developed. Deep in a mine in South Dakota, physicists have been able to capture the elusive solar neutrino radiation with what must be the most remarkable telescope in existence, a 30,000-gallon tank of cleaning fluid. Such an exotic detector is needed because of the low interaction rate of neutrinos with matter; they pass almost unimpeded through the enormous mass of sun and earth. Occasionally, however, a neutrino reacts with a chlorine atom in the cleaning fluid and causes a measurable nuclear transmutation. The cleaning fluid is housed deep in a mine to avoid accidental transmutations from stray cosmic rays. Because the few neutrinos that are captured are generated as a consequence of the nuclear reactions that produce energy in the solar core, they give a direct measurement of the structure of the solar interior. This check shows that the sun produces six times fewer neutrinos than theory had predicted. The consequences of this experiment for solar physics are great. Either the central temperature is lower than expected, or the weak-interaction theory of particle physics is called into question.

With this paragraph, however, neutrino astronomy is ushered out of the plans. After a final note on page 54, neutrinos are never mentioned again in the survey committee's report—certainly nowhere in the high-priority recommendations.

Perhaps as a result, little has been done to increase neutrino observational work in the past few years. We are not much further than we were in the early 1970s. We are unable to detect solar neutrinos—the first direct verification we would ever obtain, by looking deep into the center of our nearest star, that the nuclear processes we hold responsible for stellar energy really go on as generally believed.

Gravitational-wave detectors fare even worse in the survey committee report. Here, it is true, there are still many unanswered questions to be settled before detectors 1,000 times more powerful than those we now possess could be built. Even then the detectors might be many orders of magnitude too insensitive to measure gravitational radiation at the levels theoretical astrophysicists currently predict. Nevertheless, a major long-term effort is worth launching along those lines, even though the steps taken now are only small, and the road ahead is long. But the survey committee report makes no provisions for gravitational-wave work.

When we ask how a major plan for astronomy, compiled by more than a hundred distinguished astronomers, could have omitted making recommendations on two such important fields of the future as neutrino and gravitational-wave observations, the answer is evident at once. The recommendations compiled in the Greenstein committee's survey report are largely those handed up by a series of panels that met individually to make recommendations about their own fields. These recommendations were then woven into one final report, apparently with few additions. The names of the panels and the corresponding numbered recommendations on the priority list are:

TABLE 5.2

Greenstein Committee Panels and Recommendations

Panel	Recommendations
Infrared	3
Optical	2
Radio	1,5,10
Solar	7
Space	9,6
Statistical	
Theoretical astrophysics	8
X-ray, gamma-ray	4
Working group on planetary astronomy	
Working group on dynamical astronomy	11
Astrophysics and relativity	

There is almost a one-to-one correspondence between the report's eleven recommendations and the eleven panels the survey committee had established. Had there been panels on neutrino and gravitational-wave astronomy the recommendations of the committee might well have included proposals for these two areas. But by selecting the particular panels it did, the Greenstein survey committee largely predetermined the kind of counsel it was going to receive and the types of recommendations it would be in a position to pass along.

According to Professor Greenstein[54] neutrino and gravitational-wave astronomy were excluded from his committee's report by agreement with Dr. D. Allan Bromley of Yale University, who, at the time, was chairman of the Physics Survey Committee of the National Academy of Sciences and was compiling a parallel study on the future of physics in the United States. This study appeared as a three-volume work *Physics in Perspective* and does include recommendations for neutrino and gravitational-wave observations.[55]

By common consent the Greenstein and Bromley committees had set up a panel on astrophysics and relativity to bridge any gaps that might otherwise be left open along the borders of physics and astronomy. This panel, headed by George B. Field, then at the University of California at Berkeley, submitted rather similar reports to both the Greenstein and Bromley survey committees.[56] But in the Greenstein survey committee's final report the Field panel's recommendations on neutrino and gravitational-wave astronomy have been deleted, while the Bromley report, which has a somewhat different structure, does retain these recommendations. By heeding the advice of Professor Field's cross-disciplinary committee, the Bromley survey committee appears to have arrived at a better balanced overall plan than the Greenstein committee.

Unfortunately, however, the Bromley Physics Survey Committee report is virtually unknown to astronomers who are wont to consider the Greenstein committee report as the prime astronomical plan of the 1970s. This is a pity because the Bromley committee provided a number of far-sighted recommendations. Sixty-nine subareas of physics are listed by priority from an overall viewpoint. Of special interest to astronomy are eleven items, given here in the order of their appearance.

12. Nuclear astrophysics
13. Theoretical relativistic astrophysics
16. Very Large Radio Array
17. X-ray and gamma-ray observatory
20. Infrared astronomy
21. General relativity tests
26. Digitized imaging devices for optical astronomy
28. Gravitational radiation
29. Aperture synthesis for infrared astronomy
34. Gamma-ray detectors in astronomy
42. Neutrino astronomy

Of peripheral interest to astronomy, also, are a number of other recommended programs. In particular:

1. Lasers and masers
18. Turbulence in fluid dynamics
30. Nonlinear optics
36. Holography and information storage
53. Optical systems and lens designs
55. Optical information processing[57]

All of these areas promise substantial advances in observational or theoretical technique. The list does omit the priority listings assigned by the Greenstein committee to millimeter astronomy, to solar observations, to optical and ultraviolet observations, to centimeter astronomy, and to astrometry. It therefore effectively leaves out Greenstein recommendations 5, for a very large millimeter wavelength antenna; 7, for the continuation of the Orbiting Solar observatories; 9, for a program on optical space astronomy; 10, for a large steerable radio telescope; and 11, for additional astrometric instruments. Omission of item 5 is serious. Some of the most important discoveries of the 1970s have been made in the area of interstellar chemistry, and these discoveries heavily depend on millimeter wavelength observations.

Astronomy is a small field compared to physics or chemistry. Whenever possible, astronomers therefore urge physicists and chemists to take an interest in astronomical problems they might be equipped to tackle. In planning councils at the highest levels, this sharing of tasks is often considered particularly important, because an astronomical program that physicists are willing to undertake is likely to be funded by a budget assigned to the field of physics; and that leaves the astronomy budget available for more traditional tasks of lesser interest to outsiders. The shifting of work on neutrino astronomy and gravitational waves to the Physics Survey Committee therefore may have appeared tactically sound to the Astronomy Survey Committee. However, in transferring responsibility for these areas, astronomers also lost control over the priority each area might be assigned.

In fact, the Physics Survey Committee assigned only intermediate levels of priority to these two fields. Had the Astronomy Survey Committee put these areas on its own final list of priorities, neutrino and gravitational-wave astronomy might have stood out as important, both to physicists and to astronomers. Omission from the Greenstein committee's list of priorities, however, may have implied a lack of adequate interest. At any rate, there was no rapid expansion of the number of research groups in either of these areas during the 1970s; and there was no attempt to build powerful new facilities requiring substantial investments.

The Greenstein committee's recommendations 7 and 9, respectively dealing with solar observations and optical space astronomy, are largely analytical in their promise. They are not likely to produce major new discoveries unless the resolving power attained in such observations were increased manyfold. Item 10 involving a large steerable radio telescope designed to operate efficiently at wavelengths of 1 centimeter and longer can well be deleted from the committee's list, since such a telescope already exists in the Federal Republic of Germany and supplies much of the most important information needed by astronomy. Item 11 involving modern astrometry is important, and the Physics Survey Committee might have given greater recognition to this area, particularly as carried out through the use of the highest resolution radio astronomical techniques. To some extent, however, this item may be covered by the Bromley committee's item 21, general relativity tests, which sometimes involve the high angular resolution techniques of modern astrometry.

All in all, the astronomical programs supported by the Bromley survey committee are remarkable in their promise to advance astronomical discoveries. They aim at providing the astronomer with new carriers of information and with improved resolving power in several ranges of electromagnetic frequencies. These new techniques should ultimately also prove their worth as analytical tools in future decades, as they join existing optical and radio astronomical instruments.

Recommendation 10

Future committees charting a plan for astronomy should decide to establish a panel of generalists who make recommendations on new areas that might be coming of age in astronomy. Such a panel should comprise astronomers and astrophysicists of unusual breadth of interest and far-ranging vision, and their advice should be carefully weighed against that of the more firmly established special interest groups representing traditional astronomical disciplines.

It is only a panel of this kind that can set new directions for astronomical searches. Without these new directions, astronomy must soon stagnate.

The Timing of Long-Range Plans

Many of the most striking discoveries take place within a few months or years following the introduction of new techniques into astronomy. These discoveries then change the directions of research for many investigators. The discoveries of quasars early in the 1960s completely revolutionized extragalactic research; the discovery of pulsars in 1968 led radio, optical, and X-ray astronomers into developing instrumentation capable of unprecedented fast response.

If astronomy is to remain flexible in its reactions to frequent revolutionary discoveries, then it cannot be saddled with a ten-year plan to which funding agencies feel tethered. But neither would most astronomers like to see the funding agencies disregard their recommendations for new searches entirely. Shorter intervals between major planning reviews, or at least periodic updating sessions should be contemplated. The present ten-year planning scheme endows the system with excessive inertia and must almost of necessity be counterproductive.

Recommendation 11

The timing between major policy reviews should be carefully examined and be more nearly matched to the rate of progress in the field. Flexibility permitting response to new discoveries should be an integral part of any planning document.

One final comment: There is no evidence right now that astronomical planning committees have in any way advanced the rate at which new astronomical phenomena are discovered. Figures 1.10 and 1.11, for example, do not show that the Greenstein or Whitford committee reports made the 1970s more prolific in discoveries of new phenomena than the 1960s had been. The available statistics, of course, are poor, but we should seriously analyze the impact of past planning on astronomy to learn to what extent and in what ways plans can be helpful or cause harm. If the current extent of planning were to be shown to stifle progress, and a greater degree of individual freedom and initiative were to appear desirable, considerable deregulation might be called for—even though that runs counter to trends toward centralized planning and control. Here enlightened planning may face its greatest challenge—striking a healthy balance between large-scale planning by national committees and spontaneity of action by individual scientists.

Recommendation 12

The benefits and harm accruing from astronomical plans should be studied with care to see the extent to which regulation or deregulation of astronomy might lead to greater progress.

My criticism of the Greenstein survey committee is intended mainly to point out weaknesses that should be avoided in future planning. It is in no way meant to be derogatory of Professor Greenstein or the members of his committee. The committee was attempting a pioneering task, the large-scale planning of astronomical research, a comprehensive effort that had never been tried before. In any initial attempt of this kind, difficulties are bound to arise, and mistakes are bound to be made. The members of the committee faced a huge task. The ground they were to cover had been inadequately prepared for them, and the time available for the preparation of their plan was inadequate for generating this required base. The committee had available neither adequate historical perspective on the nature of astronomical progress nor an appropriate study of sociological factors that have to enter any large-scale plan. The committee's consideration of manpower problems confined itself to standard education of Ph.D. candidates in graduate departments of astronomy. The documentation was lacking to show that outsiders are responsible for a substantial fraction of the major advances. Similarly, the importance of novel instrumentation of significantly increased resolving power was inadequately understood.

Recommendation 13

To facilitate future astronomical planning, a continuing effort must be made to improve our understanding of the factors that lead both to conceptual and observational advances in astronomy. That improvement can only come about through historical and sociological studies conducted at the highest levels of professionalism.

The stakes are too high for anything less.

Appendices

APPENDIX A

The Number of Undetected Species

To estimate how many more phenomena we might discover in further cosmic searches we must assume that past discoveries have in some sense been characteristic of those expected in the future. Let us examine our basic statistical scheme and point out the way in which some fundamental assumptions about the future are inserted.

We described a phase space of finite volume V each point of which corresponds to a particular cosmic observation that we may make. A small part, v, of this volume contains all the observations that have already been carried out to date. This subvolume v is likely to consist of several smaller domains v_1, v_2 . . . which are not connected to each other, and v is simply the sum of the volumes of these smaller domains.

To visualize our next few steps we may think of the phase space of observations as a box of volume V. A specific phenomenon, for example a supernova, can be represented by a set of red marbles. These marbles are placed in the box at locations corresponding to points in the phase space at which observations would lead to an independent recognition of the supernova phenomenon. There might be m such points, and we label each point with a red marble. At the moment we recognize supernovae in only one way—by the bright visual flash—but eventually we may expect to find other independent means of noting supernova bursts, perhaps through the emission of an intense pulse of gravitational waves or the emission of an intense pulse of neutrinos; then m would have to be set equal to 2 or 3 or higher.

We now consider these marbles to be distributed in subcompartments v_1, v_2, . . . , v_m in the box V. A new observation can be thought of as reaching blindly into the box and picking out one of these subcompartments. We open the compartment to see whether or not it contains a red marble. We repeat this process until we have picked out enough compartments to make up a volume v. If we really have searched through the box quite at random, then the number of marbles we expect to find in this way will have an average value of mv/V. This average value, which may not correspond to an integer number, is a mean value that should be obtained in successive independent attempts to repeat the experiment of picking compartments out of the box V. The actual number of marbles found in any given try would of course have to be an integer, and these integer values would fluctuate in a random way about the mean mv/V. We can label the ratio of volumes v/V with the symbol r. The ratio r tells us the probability of finding a number q of specifically selected red marbles inside v, and the remainder, namely $m - q$, distributed in the remaining fraction of the volume, $[1 - r]V$. That probability is $r^q[1 - r]^{m-q}$, an expression that arises because the relative probabilities for randomly finding a marble inside and outside

v, respectively, are r and $[1-r]$; and these probabilities for the individual marbles are independent of how any other marbles are distributed.

If we are interested merely in the probability P_q of finding any q red marbles—not a previously selected set—in volume v, then the probability is considerably higher, because there are so many different permutations of marbles which will allow us to obtain this less restrictive distribution. In fact, the probability for finding any set of q marbles in v is

$$P_q = \frac{m!}{q![m-q]!} r^q[1-r]^{m-q}$$

where $m! = m[m-1][m-2] \ldots 2$ is the factorial product. Zero factorial is unity, $0! = 1$.

The probability distribution P_q is called Bernoulli's binomial distribution. The sum of all probabilities P_q gives unity:

$$\sum_{q=0}^{m} P_q = \sum_{q=0}^{m} \frac{m!}{q!(m-q)!} r^q(1-r)^{m-q} = [r+(1-r)]^m = 1$$

The average value of q denoted by $<q>$ then can be shown to be

$$<q> = \sum_{q=0}^{m} qP_q = mr$$

If m is very large and r is very small, $m \gg 1 \gg r$, then for small values of q, $q \ll m$,

$$P_q \sim \frac{m^q r^q}{q!} (1-r)^m$$

Thus far the only assumption we have made is that the search in the volume V has been random in relation to the distribution of the red marbles. There is a necessary further complication which will require one additional assumption: The supernova phenomenon is not the only one that can be found in volume V; there are many others. Correspondingly we can consider a variety of different colored sets of marbles placed in our box, each set—that is each color—representing a different phenomenon. Each set also is randomly distributed within the box, relative to the chosen sequence of observations in the phase space V and will have its own multiplicity.

We previously said that there were m red marbles. Let us now say instead that there are m_r red marbles, m_b blue marbles, m_y yellow marbles and so on. The numbers m_r, m_b, m_y, . . . are unknown for the cosmic phenomena that the colored marbles represent, and we must now make an assumption concerning these numbers if we wish to make further headway. Let us therefore assume, since we do not know any better, that all the *multiplicities*, or *modalities* m_i, are the same

$$m_r = m_b = m_y = \cdots = m$$

We will first examine the results to which this assumption leads and will then look at some alternatives.

Let us now denote by A the number of colored marbles represented only once in the collection found in our volume v. The number of colors represented by precisely two marbles each in volume v can be written as B. The number of colors represented by exactly three marbles is C, and so on. Then if n is the total number of different colors found in the whole box of volume V,

$$A = nP_1 \sim nm\ r(1-r)^{m-1}$$

$$B = nP_2 \sim \frac{nm(m-1)}{2}\ r^2(1-r)^{m-2}$$

$$C = nP_3 \sim \frac{nm(m-1)(m-2)}{6}\ r^3(1-r)^{m-3}$$

$$\text{and}\ \frac{A^2}{2B} \sim \frac{AB}{3C}$$

which means that

$$C = \frac{2}{3}\frac{B^2}{A}$$

If we found that $A = 37$—that we had 37 marbles whose colors were represented only once in our sample—$B = 7$ and $C = 1$, then $A^2/2B \sim 98$ and $AB/3C \sim 86$. These are in fair agreement, but the ratio $A^2/2B$ is more reliable since it is based on a larger sample of marbles. We note that

$$2B/A \sim mr$$

and

$$n \sim (A + 2B)\frac{A}{2B}\left(\frac{m-1}{m}\right)\frac{1}{(1-r)}$$

Since m was assumed to be large, and r to be small,

$$n \sim \frac{A}{2B}(A + 2B)$$

With the values for A and B given above, we obtain $mr = 0.38$ and $n \sim 135$. Even though only $A + B + C = 45$ different colors are contained in our sample, we are able to estimate that the total set contains 135 different colors. The numbers A, B, and C selected here roughly correspond to the singly recognized, doubly recognized, and triply recognized cosmic phenomena, and n therefore corresponds to a rough estimate of the total number of phenomena characterizing our universe. We can still obtain a value for m if we take the fraction of the total volume examined to date to be $r \approx 0.01$, as estimated in chapter 3.* This yields $m \sim 40$, so that $m \gg 1 \gg r$ appears to hold true.

Let us now return to our assumption that all phenomena are encountered in our universe with equal frequency m. That assumption might equally well be replaced by another.

One of the earliest statistical studies of the kind described here was conducted by R. A. Fisher, A. S. Corbet, and C. B. Williams in 1943.[1] They were interested in determining the relation between the number of species in an animal population and the number of individuals found in a random sample. Fisher and his colleagues specified a parameter k which provides a measure of the range over which m can vary. At large values of k the various values m_i, m_j, m_k . . . corresponding to colors or phenomena $i, j, k, \ldots$ are nearly equal to some constant value m. That was the case we examined above. On the other hand, low values of k approaching $k = 0$ imply a very large range of m values, and this was the situation assumed for the collection of butterflies and moths studied by Fisher, Corbet, and Williams. We will return to one simple situation in which m values covering a wide range are considered, but first it might be useful to list a number of other studies in which similar problems of statistical estimation are encountered.

A mathematical paper published by L. A. Goodman in 1949 established a number of theorems on statistical estimates based on limited data. I. J. Good, in 1953, also considered a variety of measures for the heterogeneity of populations and in fact applied some of his results to the collections studied by Fisher, Corbet, and Williams.[2]

In 1968 Robbins provided a mathematical framework for estimating the total probability of the unobserved outcomes that are possible for a given experiment. And, in an application that exhibits similarities to the search for cosmic phenomena, H. W. Menard and G. Sharman, in 1975, mathematically analyzed histories of successes in oil exploration.[3] Again their methods show similarities to the statistical approach described above.

Word frequency distributions can be treated by similar means. In 1975 B. Efron and R. Thisted used a statistical approach similar to the one used here to estimate Shakespeare's vocabulary. Shakespeare left us 31,534 different words in his works.† Of these 14,376 appeared only once, 4,343 twice, and 2,292 three times. If we set $A = 14,376$ and $B = 4,343$, we can use our earlier derived formula to obtain $n \sim 38,200$ as an estimate. This is 6,600 words more than appear in Shakespeare's writing. Using far more detailed word counts based on the 884,647 words that appear in Shakespeare's known works, Efron and Thisted were able to provide a more informed estimate. There are 864 words that appear more than 100 times in the writings—compared to the 14,376 that appear only once. It is therefore clear that the distribution in frequency of occurrence of different words is extremely wide, in fact, generally a wide range of m values characterizes the vocabulary of all

* See page 191.
† Efron and Thisted considered each different form of a word, singular, plural, possessive, and so on, to be a *different word*.

classical literature. Making use of methods developed by Fisher, Efron and Thisted conclude that Shakespeare knew at least 35,000 words that he never used in his works, for a total of $n \sim 66,500$.[4] This number is 1.7 times higher than the 38,200 estimated on the basis of simply a constant value for m for all word types. But the disagreement between these two estimates is not huge. It suggests that an estimate assuming a constant value of m generally might err by no more than a factor of 2.

Since we have had absolutely no previous estimates of the number of cosmic phenomena, an error of only a factor of 2 would not appear excessive on a first try.

We can next demonstrate some bounds on our uncertainty of the number of cosmic phenomena by considering a wide distribution of the frequency of occurrence, m. A simple numerical example may suffice: Let us suppose that there are two distinct groups of phenomena. One group is distributed 16 times more frequently throughout the phase space volume V than the other, and there are a large number of different phenomena in each group. For the first group there will be a certain number of phenomena A_1 recognized only in one way and a certain number B_1 recognized in two ways; the symbols A_2 and B_2 have the same meanings for phenomena in the second group. For our cosmic counts the total then would be

$$A_t = A_1 + A_2 = 37$$
$$B_t = B_1 + B_2 = 7$$
$$C_t = C_1 + C_2 = 1$$

where A_t, B_t, and C_t respectively are the total number of singly, doubly, and triply recognized phenomena in our sample. The factor of 16 for the ratio of m_1/m_2 has been chosen mainly because it makes the arithmetic simpler when we use the above numbers. We then find that $A_1 = 5$, $B_1 = 5$, $A_2 = 32$, $B_2 = 2$ correspond to the numbers of singly and doubly recognized phenomena in the two groups. We can also obtain an estimate of the total number of expected phenomena belonging to each group separately using the equations

$$n_2 = A_2(A_2 + 2B_2)/2B_2$$
$$\frac{A_1}{B_1} = \frac{2(1-r)}{(m-1)r}, \frac{B_1}{C_1} = \frac{3(1-r)}{(m-2)r}, \cdots$$

with $n_1 = A_1 + B_1 + C_1 \cdots$

We then find the estimates $n_1 = 16$, $n_2 = 288$, and the total number of phenomena $n_t = n_1 + n_2 = 304$. This number is about twice as large as our previously estimated value $n \sim 135$. In other words, a blind substitution into our original formula which assumes identical frequency distributions m for all phenomena leads to an underestimate by a factor of 2 when compared to the total number of phenomena calculated with the understanding that there is a frequency spread of the order of $16:1$.

We still might ask whether a $16:1$ spread in the values of m is to be considered reasonable or unreasonable, since we know so little about how the cosmic phenomena are actually distributed in the space V. There is one added item of information that can help us here. It is the number of triply recognized phenomena C.

The values of C_1 and C_2 can be estimated from the relationship

$$C_1 = 2B_1{}^2/3A_1, \quad C_2 = 2B_2{}^2/3A_2$$

and these give

$$C_t = C_1 + C_2 = 3.4$$

for our spread of m values. The actually found value $C = 1$ is more consistent with an expectation obtained when all phenomena are distributed with a frequency distribution characterized by a single value of m; that leads to

$$C = 2B^2/3A = 0.88$$

The value of C is based on very weak statistics and cannot be used as much more than an indicator; but in this case it indicates, if anything, that the distribution of different groups of phenomena is not highly differentiated.

Incidentally, when we compute the value of C for Shakespeare's vocabulary, the formula just used would lead us to expect $C \sim 875$ while the value cited by Efron and Thisted is 2,292. This is just what we would expect, since the frequency of occurrence of different words in ordinary usage has a tremendously wide distribution, and we would therefore expect C_t to be high. A high value of C_t is a direct result of a widespread in m values. This follows directly from the relationship

$$C_t = C_1 + C_2 = \frac{2}{3}\left[\frac{B_1{}^2}{A_1} + \frac{B_2{}^2}{A_2}\right] \geq \frac{2}{3}\left[\frac{(B_1 + B_2)^2}{(A_1 + A_2)}\right] = C$$

This algebraic relationship shows equality only when m is the same for all groups of phenomena; and $C_t > C$ for any spread in m values.

We may still ask how the approach to ultimate completion of our search can be recognized? The answer again is given by our expressions for $A, B, C, \ldots$ the singly, doubly, triply, $\ldots$ recognized phenomena. If we once again choose a value of $m = 40$, then the ratios A/n, $B/n, \ldots$ can be readily evaluated for different values of r, the fraction of all possible observations permitted by current technology at any epoch in the development of astronomy. In figure 1.9 the factor r is the *competence ratio*. It represents our competence to perform astronomical observations. The ratios A/n, B/n and $[C + D + E + \ldots]/n$ are shown with different shadings, and those phenomena remaining unrecognized at a given state of competence, r, are left unshaded. The total fraction of all phenomena recognized by the time r has reached a given value is shown at the head of each column. We see that as the search nears completion, the number of singly recognized phenomena dwindles and is replaced by multiply recognized sources and events. We note from our formulae that the ratios A/n, B/n, and so on are independent of n and depend only on factors like m and r and on ratios such as A/B.

Since the possibility remains that some phenomena can only be singly recognized, no matter what the extent of our technical sophistication, we note that our formulae may lead to an underestimate of the total number of phenomena n. An upper limit can, however, be estimated by listing the number of phenomena already found that intrinsically appear capable of single recognition alone. Dividing that number by r gives an estimate of the total number of singly recognized phenomena that could ultimately be found. A slight upward revision might also be needed for phenomena that can only be doubly recognized, $m = 2$. If some phenomena had $m_1 = 40$ and some had $m_2 = 2$, the ratio of multiplicities would be $\sim 40/2 = 20$, and we just saw that a ratio of 16 might lead to a doubling of our earlier estimate for n. Beyond $m = 2$, therefore, the corrections should be small. The value of the competence ratio r shown in figure 1.9 is based on shown ratios A/B, B/C, . . . and on our best estimate for the multiplicity, $m = 40$.

APPENDIX B

Information, Capacity, and Information Rates

Information and Capacity

We have found a variety of restrictions governing the photons that can be transmitted across the universe and the information they can carry. We placed an upper limit on the rate at which information can be received in terms of the number of distinguishable photons that can be transmitted in a time T through a transmission line with bandwidth W. This upper bound, however, can only be approached when the transmission of information is undisturbed by extraneous noise and when the total number of available carriers of information—photons—is high.

While the actual rate of information transmission has not entered the principal arguments presented, it is a quantity that may play an important role in future discussions of the scope of conceivable cosmic searches. For that reason a few applicable results from information theory are presented in this appendix. They outline a relationship between information theory and the detectability of cosmic signals.

Suppose we constructed a message whose symbols consist entirely of the binary digits *1* and *0*. A symbol *1* is produced by transmitting a photon of a given frequency during a specified time interval and a *0* is produced by not transmitting the photon. Such a binary message is a sequence of *1*'s and *0*'s: *00011010100111. . . .* For this message the information transmitted per symbol is expressed in terms of the probabilities for the occurrence of each of the two symbols.

The quantum theory of radiation tells us that the maximum number of distinguishable photons that can reach a detector of area D^2 in time T and in the frequency range f to f_1 from one hemisphere in the sky, is

$$\frac{2\pi}{3c^2} D^2 T(f_1{}^3 - f^3)$$

If $f_1 - f$ represents a narrow bandwidth W, the maximum number of photons around frequency f distinguished by a detector of area D in time T is

$$W' T = \left[\frac{2\pi}{c^2} D^2 f^2\right] WT = \left[\frac{2\pi D^2 f^2}{c^2}\right] WT$$

As discussed below, $2W' T$ then represents the maximum bit rate we could expect at frequency f. In statistical mechanics we often picture distinguishable photons as occupying discrete cells in a fictitous phase space. A phase cell is considered occupied if it contains one or more photons.

Information theory as developed by Claude E. Shannon in 1948 and 1949 provides two concepts that are important to us. The first is the average information provided by each symbol—here each phase cell. The second is the channel capacity.

Let us first look at the information transmitted per symbol: Suppose that the probability of a phase cell being filled with 0, 1, 2, or more photons respectively is p_0', p_1', p_2', and so on. Then the average information $<I>$ carried by each phase cell is

$$<I> = -\sum_j p_j \log_2 p_j$$

and the logarithm is taken to the base 2 if the information is measured in bits. $<I> = \log_2 n$, for $j = 1, \ldots n$. However, the maximum information that can be transmitted is only $2W'T$, not $W'T \log n$, because with the bandwidth W', one cannot detect a symbol of 0 photons—a very low signal—immediately after a long string of symbols of n photons—a very high signal. The high bandwidth—fast time response—required to do that would also increase as log n and almost cancel the overall advantage. This was first shown by R. V. L. Hartley of Bell Laboratories in 1928.[1]

At high energies, for example in the visible, the ultraviolet, and the X-ray domains of astronomical observation, essentially all cells may be taken to be either empty or else filled by just a single photon; and to a sufficiently good approximation, we can then write a probability p for a cell to be filled and a probability $(1 - p)$ for it to be unfilled. The average information per symbol then becomes

$$<I> = -p \log_2 p - (1 - p) \log_2 (1 - p)$$

We know that p is a very small number. If it were not, everything on the surface of Earth would be instantaneously vaporized by the photons bearing down from space. And for small values of p the value of $-(1 - p) \log_2 (1 - p)$ approaches p which is small compared to $-p \log_2 p$.

Because p is small, $p \log_2 p$ is small too. If the information rate had to be maximized, p and $(1 - p)$ would have to be chosen equal, with values of $1/2$, and in that case the value of $<I>$ would rise to a peak of 1 bit per symbol. This leads to the second concept mentioned above, the *channel capacity*—the highest information rate that can be achieved for any given channel. For a bandwidth W' we can sample W' phase cells in 1 second and the channel capacity therefore is W' bits per second. For a binary channel with p very low, the bit rate becomes $-W' p \log_2 p$ bits per second.

Observations Against a Noisy Background

We often need to understand the limitations on observing radio signals detected against a luminous cosmic background. This is readily done. Consider a signal with power S observed in the presence of noise power N. For this case each phase cell may be filled with more than one photon, and in fact the occupation number often can be quite large. Shannon noted that in this type of transmission there are $[1 + S/N]$ distinguishable power levels between 0 and S, namely $0, N, 2N . . . S$. Suppose that signals are transmitted with equal probability at each power level. Under these conditions the maximum amount of information can be transmitted and the information per transmitted symbol is

$$< I> = - \sum_j p_j \log_2 p_j = \log_2 (1 + S/N)$$

If the bandwidth over which observations are to be carried out is W', we obtain a channel capacity of $W'/2 \log_2[1 + S/N]$ bits per second, since the antenna will be sensitive to one particular polarization and there are $W'/2$ distinguishable phase cells arriving at the receiver each second. This is the channel capacity for a noisy communication channel. It is higher than the capacity of the noise-free binary channel considered above by a factor of $\frac{1}{2}\log_2[1 + S/N]$, and at first this might seem puzzling. But each symbol in a noisy signal consists of a larger number of photons, and the information transmitted per quantum of energy, or photon, is still optimum for a noise-free channel.

For the specific search for information in the presence of the cosmic 3 degree Kelvin microwave radiation bath, the energy required for each transmitted bit of information is readily evaluated. A signal lasting some time interval T at a power level S uses up an energy ST. During this time CT bits of information are transmitted and the energy per bit, E, is therefore

$$E = \frac{S}{C} = \frac{2S}{W' \log_2 [1 + S/N]}$$

This energy is a minimum when the transmission takes place at very low signal levels. In that case the channel capacity also drops, but if we have the choice of making S small and increasing the transmission bandwidth W' to keep the capacity useful, we can minimize the energy cost per transmitted bit of information. For very small signal-to-noise ratios this becomes equal to $1.386N/W'$. At long wavelengths and at temperature T the noise is $N = kTW'/2$, where $W'/2$ is the number of phase cells sampled by the detector. The energy per bit of information is $0.693\ kT$ per bit. For the 3 degree Kelvin temperature of the cosmic background this amounts to 3×10^{-16} erg per bit.[2] The energy per bit of transmitted information is shown as a function of S/N in figure B.1.

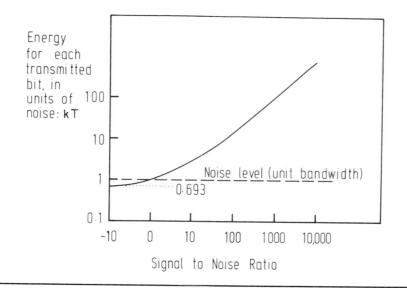

Figure B.1 *The Energy Expended for Each Transmitted Bit of Information for Signals of Different Strengths*

The energy that must be expended to transmit one bit of information across the universe decreases as the signal strength is made smaller in comparison to the noise at the transmitted wavelength. In the spectral range around centimeter wavelengths, that noise is largely due to the 3 degrees Kelvin microwave background radiation. At first sight, we might think that the most informative signal might be one that stands out strongly above the noise; and certainly that kind of signal might well attract an observer's attention. The most economical transmission of information, however, can be effected by keeping the signal power small. The minimum energy for each transmitted bit of information then approaches 0.693 kT, where T is the effective background radiation temperature—3 degrees Kelvin at centimeter wavelengths—and k is Boltzmann's constant, $k = 1.38 \times 10^{-16}$ erg per degree Kelvin. Messages we might expect from intelligent civilizations may therefore be buried in cosmic noise. Any other messages would most probably be uneconomical to transmit.

REFERENCES

Preface

1. F. Zwicky, *Morphological Astronomy* (Berlin: Springer, 1957). This book made a serious attempt to approach astronomical discovery in a systametic way. Zwicky realized, however, that his effort still was premature. On page 29 he writes, "We cannot . . . hope to explore here morphologically the whole field of astronomy. We shall have to be satisfied with a demonstration of the new method as applied to limited objectives."

2. Jurgen Schmandt, "Scientific Research and Policy Analysis," editorial in *Science* 201, no. 4359 (1978).

3. M. Granger Morgan, "Bad Science and Good Policy Analysis," editorial in *Science* 201, no. 4360 (1978).

4. Albert A. Michelson, *Light Waves and Their Uses* (Chicago: University of Chicago Press, 1903), pp. 23–24.

5. Walter Meissner, "Max Planck, the Man and His Work" (translated from German) *Science* 113, no. 75 (1951).

6. Stephen G. Brush, Letter to the Editor in *Physics Today,* January 1969, p. 9; Lawrence Badash, "The Completeness of 19th Century Science," *ISIS* 63 (1972): 48–58.

7. Silvanus P. Thompson, *The Life of William Thomson, Baron Kelvin of Largs,* 2 vols. (London: Macmillan & Co., 1910). The author provides many insights into the nineteenth century physicist's thoughts and working habits.

8. David Edge, "The Sociology of Innovation in Modern Astronomy," *Quarterly Journal of the Royal Astronomical Society* 18 (1977): p. 327.

9. David Edge, "The Sociology of Innovation," p. 337.

Chapter 1

1. Albert Van Helden, "The Invention of the Telescope," *Transactions of the American Philosophical Society* 67 pt. 4 (1977).

2. Stillman Drake, *Discoveries and Opinions of Galileo* (New York: Doubleday & Co., Anchor Books, 1957), p. 21; Arthur Koestler, *The Watershed—A Biography of Johannes Kepler* (New York: Doubleday & Co., Anchor Books, 1960), pp. 185–86.

3. Viktor F. Hess, "Über Beobachtungen der durchdringenden Strahlung bei sieben Freiballonfahrten," *Physikalische Zeitschrift* 13 (1912): 1084–91.

4. Karl G. Jansky, "Directional Studies of Atmospherics at High Frequencies," *Proceedings of the Institute of Radio Engineers* 20 (1932): 1920–32.

5. Karl G. Jansky, "Electrical Disturbances Apparently of Extraterrestrial Origin," *Proceedings of the Institute of Radio Engineers* 21 (1933): 1387–98.

6. T. R. Burnight, "Soft X-Radiation in the Upper Atmosphere," *Physical Review* 76 (1949): 165.

7. J. D. Purcell, R. Tousey, and K. Watanabe, "Observations at High Altitudes of Extreme Ultraviolet and X-rays from the Sun," *Physical Review* 76 (1949): 165.

8. H. Friedman, S. W. Lichtman, and E. T. Byram, "Photon Counter Measurements of Solar X-Rays and Extreme Ultraviolet Light," *Physical Review* 83 (1951): 1025–30.

9. R. Giacconi, H. Gursky, F. Paolini, and B. Rossi, "Evidence for X Rays From Sources Outside the Solar System," *Physical Review Letters* 9 (1962): 439–43.

10. R. W. Klebesadel, I. B. Strong, and R. A. Olson, "Observations of Gamma-Ray Bursts of Cosmic Origin," *Astrophysical Journal* 182 (1973): L85–88.

11. A Hewish, S. J. Bell, J. D. H. Pilkington, P. F. Scott, and R. A. Collins, "Observations of a Rapidly Pulsating Radio Source," *Nature* 217 (1968): 709–13.

12. David O. Edge and Michael J. Mulkay, *Astronomy Transformed—The Emergence of Radio Astronomy in Britain* (New York: John Wiley & Sons, 1976). Pages 204–13 provide an informed description of the discovery of quasars and early work on them.

13. David O. Edge and Michael J. Mulkay, *Astronomy Transformed,* chap. 10; David Edge, "The Sociology of Innovation in Modern Astronomy," *Quarterly Journal of the Royal Astronomical Society* 18 (1977): 330.

14. Edge and Mulkay, *Astronomy Transformed*, p. 362.

15. Edge and Mulkay, *Astronomy Transformed*, pp. 12–38.

16. J. S. Hey, "Solar Radiations in the 4-6 Metre Radio Wavelength Band," *Nature* 157 (1946): 47–48. German radar antennas in Denmark also detected the sun in 1939–40, but their findings were kept secret as well.

17. J. S. Hey, *The Radio Universe* (New York: Pergamon Press, 1971), p. 4. The author gives an account of these early observations which arose from spurious radar echoes received in tracking German V-2 rockets toward the end of the war.

18. J. S. Hey, S. J. Parsons, and J. W. Phillips, "Fluctuations in Cosmic Radiation at Radio-Frequencies," *Nature* 158 (1946): 234; J. S. Hey, "The First Discovery of Point Sources," in *Paris Symposium on Radio Astronomy*, edited by R. N. Bracewell (Stanford: Stanford University Press, 1959), pp. 295–96.

19. Van Helden, "The Invention of the Telescope," in Koestler, *The Watershed*, pp. 185–86.

20. Robert K. Merton, "Science, Technology and Society in Seventeenth Century England," *Osiris* 4 (1938): 360–632. Chapter 9 (pp. 543–57) deals with science and military technique. (New York: H. Fertig, 1970).

21. I. B. Strong and R. W. Klebesadel, "Cosmic Gamma-Ray Bursts," *Scientific American*, October 1976, pp. 66–79A.

22. R. Giacconi in *X-Ray Astronomy*, edited by R. Giacconi and H. Gursky (Dordrecht, The Netherlands: Reidel Publishing Co., 1974), p. 7.

23. A. A. Penzias and R. W. Wilson, "A Measurement of Excess Antenna Temperature at 4080 Mc/s," *Astrophysical Journal* 142 (1965): 419–21; Hewish et al., "Observations of a Rapidly Pulsating Radio Source."

24. H. Arp, "A Very Small, Condensed Galaxy," *Astrophysical Journal* 142 (1965): 402–6.

25. Harlan J. Smith and Dorrit Hoffleit, "Light Variations in the Superluminous Radio Galaxy 3C 273," *Nature* 198 (1963): 650–51.

26. Hilary Putnam, "How Not to Talk About Meaning—Comments of J. J. C. Smart," in *Boston Studies in the Philosphy of Science*, vol. 2, edited by R. S. Cohen and M. W. Wartofsky, (New York: Humanities Press, 1965), p. 217.

27. Michael Polanyi, *Personal Knowledge—Toward a Post-Critical Philosophy* (Chicago: University of Chicago Press, 1958), pp. 110–11.

28. Michael Polanyi, *Personal Knowledge*, p. 351.

29. Edge and Mulkay, *Astronomy Transformed*, p. 362.

30. Military Procurement Authorization Act of 1970, Public Law 91–121, Section 203, The Mansfield Amendment, directed that "none of the funds authorized to be appropriated by this act may be used to carry out any research project or study unless such project or study has a direct and apparent relationship to a specific military function or operation"; National Academy of Sciences, *Astronomy and Astrophysics for the 1970's—Report of the Astronomy Survey Committee,* 2 vols. chaired by Jesse L. Greenstein (Washington, D.C., 1972).

31. Bruce Margon, Holland C. Ford, Jonathan I. Katz, Karen B. Kwitter, Roger K. Ulrich, Remington P. S. Stone, and Arnold Klemola, "The Bizarre Spectrum of SS 433," *Astrophysical Journal* 230 (1979): L41–45.

32. National Science Board, *Science Indicators 1976,* National Science Foundation, (Washington, D.C.: U.S. Government Printing Office, 1977), pp. 206, 226, 229, and letter from Dr. William E. Howard III of the National Science Foundation, dated May 14, 1980 (author's file).

Chapter 2

1. A. Hewish, S. J. Bell, J. D. H. Pilkington, P. F. Scott, and R. A. Collins, "Observations of a Rapidly Pulsating Radio Source," *Nature* 217 (1968): 709–13.

2. T. Gold, "Rotating Neutron Stars as the Origin of the Pulsating Radio Sources," *Nature* 218 (1968): 731–32.

3. Thomas S. Kuhn, *The Structure of Scientific Revolutions* (Chicago: University of Chicago Press, 1962).

4. For an account of this work, see Bessel, "Ueber Sternschnuppen," *Astronomische Nachrichten* 16 (1839): 321.

5. Fred L. Whipple and Gerald S. Hawkins in *Handbuch der Physik*, vol. 52, edited by S. Flügge, (Berlin: Springer, 1959), p. 559.

6. W. Grotrian, "Über das Fraunhofersche Spektrum der Sonnenkorona," *Zeitschrift für Astrophysik* 8 (1934): 124–46.

7. J. S. Hey, *The Evolution of Radio Astronomy* (London: Elek Science, 1973), pp. 19–23; J. S. Hey and G. S. Stewart, "Derivation of Meteor Stream Radiants by Radio Reflexion Methods," *Nature* 158 (1946): 481–82.

8. J. P. Schafer and W. M. Goodall, "Observations of Kennelly-Heavyside Layer Heights During the Leonid Meteor Shower: November 1931," *Proceedings of the Institute of Radio Engineers* 20 (1932): 1941–45.

9. A. M. Skellett, "The Ionizing Effect of Meteors in Relation to Radio Propagation," *Proceedings of the Institute of Radio Engineers* 20 (1932): 1933–40.

10. Karl G. Jansky, "Electrical Disturbances Apparently of Extraterrestrial Origin," *Proceedings of the Institute of Radio Engineers* 21 (1933): 1398.

11. Woodruff T. Sullivan III, "A New Look at Karl Jansky's Original Data," *Sky and Telescope* 56, August 1978, pp. 102–3; Harald T. Friis, "Karl Jansky: His Career at Bell Telephone Laboratories," *Science* 149 (1965): 841–42.

12. Arthur Koestler, *The Watershed—A Biography of Johannes Kepler* (New York: Doubleday & Co., Anchor Books, 1959), chap. 6.

13. Nicolaus Copernicus, *On the Revolutions of the Heavenly Spheres,* translated by A. M. Duncan, bk. 1 (New York: Barnes & Noble, 1976), chap. 10.

14. "A Letter from the Reverend Mr. James Bradley, Savilian Professor of Astronomy at Oxford, and F. R. S. to Dr. Edmund Halley Astronomer Royal & c. giving an Account of a new discovered Motion of the Fix'd Stars," *Philosophical Transactions of the Royal Society of London* 35 (1728): 637–61. Stephen Peter Rigaud, *Miscellaneous Works and Correspondence of James Bradley,* 1832 (New York: Johnson Reprint Corp., 1972).

15. Thomas Thomson, *History of the Royal Society from Its Institution to the End of the Eighteenth Century* (London: Robert Baldwin, 1812), pp. 345–47.

16. A. Pannekoek, *A History of Astronomy,* translated from the 1951 Dutch version (London; George Allen & Unwin, 1951), p. 259.

17. William Herschel, "Account of a Comet," *Philosophical Transactions of the Royal Society of London* 71 (1781): 492–501.

18. Mrs. John Herschel, *Memoir and Correspondence of Caroline Herschel* (New York: D. Appleton & Co. 1876). This book has many observations by Caroline Herschel on her brother William's working habits.

19. Agnes M. Clerke, *A Popular History of Astronomy During the Nineteenth Century,* 4th ed. (London: Adam & Charles Black, 1902), p. 78; Koestler, *The Watershed,* p. 184.

20. K. L. Franklin, "An Account of the Discovery of Jupiter as a Radio Source," *Astronomical Journal* 64 (1959): 37–39.

21. Harlow Shapley and Helen E. Howarth, *A Source Book in Astronomy* (New York: McGraw-Hill, 1929), pp. 180–82. This gives a translation of Bode's 1802 account of the discovery of Ceres.

22. Agnes M. Clerke, *A Popular History of Astronomy,* pp. 73–75.

23. Ewan A. Whitaker, "Galileo's Lunar Observations, and the Dating of the Composition of Sidereus Nuncius," *Journal for the History of Astronomy* 9 (1978): 155–69; Rigaud, *Miscellaneous Works of James Bradley.*

24. A. Pannekoek, *A History of Astronomy,* pp. 246, 255–56, 278.

25. A. Van Helden, "Saturn and his Anses," *Journal for the History of Astronomy* 5 (1974): 105–21.

26. A. Van Helden, " 'Annulo Cingitur,' The Solution of the Problem of Saturn," *Journal for the History of Astronomy* 5 (1974): 155–74.

27. Rolf Riekher, *Fernrohre und Ihre Meister* (Berlin: VEB Verlag Technik, 1957), pp. 50–53.

28. James L. Elliot, Edward Dunham, and Robert L. Millis, "Discovering the Rings of Uranus," *Sky and Telescope* 53 (June, 1977): 412–16, 430.

29. International Astronomical Union Circular no. 3338 (Cambridge, Mass., 1979).

30. E. E. Becklin and C. G. Wynn-Williams, "Detection of Jupiter's ring at 2.2 μm," *Nature* 279 (1979): 400–401.

31. See, for example, David H. Clark and F. Richard Stephenson, *The Historical Supernovae* (New York: Pergamon Press, 1977), pp. 21–27.

32. J. L. E. Dreyer, *A History of Astronomy from Thales to Kepler,* 2nd ed. (New York: Dover Publications, 1953), p. 366.

33. Y. C. Chang, "Halley's Comet: Tendencies in its Orbital Evolution and its Ancient History," (translated from *Acta Astronomica Sinica* 19 (1978): 109–18. *Chinese Astronomy,* 3 (1979): 120–31.

34. George Sarton, *Six Wings—Men of Science in the Renaissance* (Bloomington: Indiana University Press, 1957), pp. 72–74.

35. *Sir Isaac Newton's Mathematical Principles of Natural Philosophy and His System of the World* (trans. Andrew Motte, 1729), revised translation by Florian Cajori, (Berkeley: University of California Press, 1943), p. 596.

36. F. W. Bessel, "Bestimmung der Entfernung des 61sten Sterns des Schwans," *Astronomische Nachrichten* 16 nos. 365–66 (1838): cols. 65–96; F. W. Bessel, "Fernere Nachricht von der Bestimmung der Entfernung des 61sten Sterns des Schwans," *Astronomische Nachrichten* 17, no. 401 (1840): cols. 257–72; F. W. Bessel, "Fernere Nachricht von der Bestimmung der Entfernung von 61 Cygni," *Astronomische Nachrichten* 17, no. 402 (1840):cols. 273–76.

37. Thomas Henderson, "On the Parallax of α Centauri," *Memoirs of the Royal Astronomi-

cal Society London 11 (1840): 61–68. (Read January 3, 1839.) F. G. W. Struve, "Uber die Parallaxe des Sterns α Lyrae nach Micrometermessungen am grossen Refractor der Dorpater Sternwarte," *Astronomische Nachrichten* 17 no. 396 (1840): cols. 177–80.

38. Edmund Halley, "Considerations on the Change of the Latitudes of some of the principal fixt Stars," *Philosophical Transactions of the Royal Society of London* 30 (1718): 736–38.

39. W. Herschel, "On the Proper Motion of the Sun and Solar System; With an Account of Several Changes that have Happened Among the fixed Stars Since the Time of Mr. Flamsteed," *Philosophical Transactions of the Royal Society of London* 73 (1783): 247–83. See also "The Scientific Papers of Sir William Herschel," vol. 1 (London: Royal Society and Royal Astronomical Society, 1912), pp. 108–30.

40. C. W. Allen, *Astrophysical Quantities,* 3rd ed. (London: University of London, Athlone Press, 1973), pp. 234–41.

41. Henderson, "On the Parallax of α Centauri," in Struve, "Uber die Parallaxe."

42. David H. DeVorkin, "Steps Toward the Hertzsprung-Russell Diagram," *Physics Today,* March 1978, pp. 32–39.

43. Agnes M. Clerke, *A Popular History of Astronomy,* p. 385.

44. Allan Sandage, "Observational Approach to Evolution II. A Computed Luminosity Function for K0–K2 Stars from $M_v = +5$ to $M_v = -4.5$," *Astrophysical Journal* 125 (1957): 436.

45. H. Bethe, "Energy Production in Stars," *Physical Review* 55 (1939): 434–56.

46. John Goodricke, "A Series of Observations on, and a Discovery of the Period of the Variation of the Light of the Star Marked δ by Bayer near the Head of Cepheus," *Philosophical Transactions of the Royal Society of London* 76 (1786): 48–61.

47. Agnes M. Clerke, *A Popular History of Astronomy,* p. 10.

48. Henrietta S. Leavitt, *Periods of 25 Variable Stars in the Small Magellanic Cloud,* Harvard Circular no. 173 (1912). The circular actually appears under the name of Edward C. Pickering, at the time director of the Harvard College Observatory, but the opening sentence is "The following statement regarding the periods of 25 variable stars in the Magellanic Cloud has been prepared by Miss Leavitt."

49. Harlow Shapley, "On the Nature and Cause of Cepheid Variation," *Astrophysical Journal* 40 (1914): 448–65.

50. A. S. Eddington, "On the Pulsations of Gaseous Stars and the Problem of the Cepheid Variables," *Monthly Notices of the Royal Astronomical Society* 79 (1918): 2–22; (1919): 177–188.

51. F. W. Bessel, Letter to Sir J. F. W. Herschel, "On the Variations of the Proper Motions of Procyon and Sirius," August 10, 1844, translated and reprinted in part in *Monthly Notices of the Royal Astronomical Society* 6 (1844): 136–41.

52. Agnes M. Clerke, *A Popular History of Astronomy,* pp. 41–42.

53. Agnes M. Clerke, *A Popular History of Astronomy,* p. 465. The author attributes the first sizable reflecting telescope to H. Draper. It was finished in 1870. Its aperture was 28 inches.

54. Walter S. Adams, "The Spectrum of the Companion of Sirius," *Publications of the Astronomical Society of the Pacific* 27 (1915): 236–37.

55. Owen Gingerich, "Abbé Lacaille's List of Clusters and Nebulae," *Sky and Telescope* 19 (February 1960): 207–208.

56. S. I. Bailey, "The Periods of the Variable Stars in the Cluster Messier 5," *Astrophysical Journal* 10 (1899): 255–65.

57. Harlow Shapley, *Through Rugged Ways to the Stars* (New York: Charles Scribner's Sons, 1969), chap. 4.

58. William Herschel, "On Nebulous Stars, properly so called," *Philosophical Transactions of the Royal Society of London* 81 (1791): 71–88. See also "The Scientific Papers of Sir William Herschel," vol. 1 p. 421–22.

59. Gustav Kirchhoff and Robert Bunsen, "Chemical Analysis by Spectrum-observations," *Philosophical Magazine* 4th Series 20 (1860): 89–109.

60. William Huggins, "On the Spectra of Some of the Chemical Elements," *Philosophical Transactions of the Royal Society of London* 154 (1864): 139–60.

61. William Huggins and W. A. Miller, "On the Spectra of Some of the Fixed Stars," *Philosophical Transactions of the Royal Society of London* 154 (1864): 413–35.

62. William Huggins, " 'On the Spectra of Some of the Nebulae,' " A Supplement to the Paper " 'On the Spectra of Some of the Fixed Stars' " by William Huggins and W. A. Miller, *Philosophical Transactions of the Royal Society of London* 154 (1864): 437–44.

63. C. R. Lynds, "Observations of Planetary Nebulae at Centimeter Wavelengths," *Publications of the National Radioastronomy Observatory* 1 (1961): 85–97.

64. Syuzo Isobe, "H-Beta Observation of the Barnard Loop," *Publications of the Astronomical Society of Japan* 30 (1978): 499–506.

65. William Huggins, "On the Spectrum of the Great Nebula in the Sword-Handle of Orion," *Proceedings of the Royal Society of London,* 14 (1865): 39–42.

66. N. S. Kardashev, "On the Possibility of Detection of Allowed Lines of Atomic Hydrogen in the Radio-Frequency Spectrum," *Soviet Astronomy, A. J.* 3 (1960): 813–20. This is a translation from the original article in Russian in *Astronomicheskii Zhurnal* 36 (1959): 838–44.

67. F. T. Haddock, C. H. Mayer, and R. M. Sloanaker," Radio Emission from the Orion Nebula and Other Sources at λ9.4 cm," *Astrophysical Journal* 119 (1954): 456–59.

68. J. Hartmann, "Investigations of the Spectrum and Orbit of δ Orionis," *Astrophysical Journal* 19 (1904): 268–86.

69. A. Pannekoek, *A History of Astronomy*, p. 451.

70. Agnes M. Clerke, *A Popular History of Astronomy*, p. 406.

71. H. I. Ewen and E. M. Purcell, "Radiation from Galactic Hydrogen at 1,420 Mc./sec.," *Nature* 168 (1951): 356.

72. Conversation with E. M. Purcell, in Ithaca, New York, April 9, 1979.

73. H. C. van de Hulst, "Radiogolven Uit Het Wereldruim II. Herkomst der radiogolven," *Nederlands Tijdschrift voor Natuurkunde* 11 (1945): 210–21.

74. C. A. Muller and J. H. Oort, "The Interstellar Hydrogen Line at 1,420 Mc./sec., and an Estimate of Galactic Rotation," *Nature* 168 (1951): 357–58.

75. Andrew McKellar, "Evidence for the Molecular Origin of Some Hitherto Unidentified Interstellar Lines," *Publications of the Astronomical Society of the Pacific* 52 (1940): 187–92.

76. D. M. Rank, C. H. Townes, and W. J. Welch, "Interstellar Molecules and Dense Clouds," *Science* 174 (1971): 1083–1101.

77. C. H. Townes, "Microwave and Radio-Frequency Resonance Lines of Interest to Radio Astronomy" in *Radio Astronomy*, International Astronomical Union Symposium no. 4, edited by H. C. van de Hulst (Cambridge, England, 1957), pp. 92–103.

78. E. E. Barnard, "On the Dark Markings of the Sky—With a Catalogue of 182 Such Objects," *Astrophysical Journal* 49 (1919): 1–23, 360.

79. Robert J. Trumpler, "Absorption of Light in the Galactic System," *Publications of the Astronomical Society of the Pacific* 42 (1930): 214–27.

80. W. A. Hiltner, "Polarization of Light From Distant Stars by Interstellar Medium," *Science* 109 (1949): 165; John S. Hall, "Observations of the Polarized Light from Stars," *Science* 109 (1949): 166–67.

81. V. M. Slipher, "The Spectrum of NGC 7023," *Publications of the Astronomical Society of the Pacific* 30 (1918): 63–64.

82. Tycho Brahe, *Astronomiae Instauratae Progymnasmata*, in *The Historical Supernovae*, chap. 3, translated and quoted by David H. Clark and F. Richard Stephenson (New York: Pergamon Press, 1977), p. 174.

83. David H. Clark and F. Richard Stephenson, *The Historical Supernovae*, p. 83.

84. Clark and Stephenson, *The Historical Supernovae*, pp. 25–26.

85. W. Baade and F. Zwicky, "On Super-Novae," *Proceedings of the National Academy of Sciences* 20 (1934): 254–59.

86. Kenneth Glyn Jones, "S Andromedae, 1885: An Analysis of Contemporary Reports and a Reconstruction," *Journal for the History of Astronomy* 7 (1976): 27–40.

87. M. A. Hoskin, "Ritchey, Curtis and the Discovery of Novae in Spiral Nebulae," *Journal for the History of Astronomy* 7 (1976): 47–53; G. W. Ritchey, "Novae in Spiral Nebulae," *Publications of the Astronomical Society of the Pacific* 29 (1917): 210–12.

88. H. Shapley, "Note on the Magnitudes of Novae in Spiral Nebulae," *Publications of the Astronomical Society of the Pacific* 29 (1917): 213–17.

89. R. Berendzen and R. Hart, "Adriaan van Maanen's Influence on the Island Universe Theory," *Journal for the History of Astronomy* 4 (1973): 46–56, 73–98. The authors have discussed these observations by the astronomer van Maanen.

90. Harlow Shapley, *Through Rugged Ways*.

91. Walter Baade, *Evolution of Stars and Galaxies*, edited by Cecilia Payne-Gaposchkin, (Cambridge, Mass.: Harvard University Press, 1963), pp. 30–31.

92. Clark and Stephenson, *The Historical Supernovae*, p. 53.

93. W. Strohmeier, *Variable Stars* (New York: Pergamon Press, 1972).

94. Alfred H. Joy, "T Tauri Variable Stars," *Astrophysical Journal* 102 (1945): 168–95.

95. George H. Herbig, "The Spectra of Two Nebulous Objects Near NGC 1999," *Astrophysical Journal* 113 (1951): 697–99; Guillermo Haro, "Herbig's Nebulous Objects Near NGC 1999," *Astrophysical Journal* 115 (1952): 572.

96. Agnes M. Clerke, *A Popular History of Astronomy*, p. 403.

97. C. O. Lampland, "Observed Changes in the Structure of the 'Crab' Nebula (NGC 1952)," *Publications of the Astronomical Society of the Pacific* 33 (1921): 79.

98. G. Neugebauer and R. B. Leighton, "Two-Micron Sky Survey," National Aeronautics and Space Administration, NASA SP-3047 (Washington, D.C., 1969).

99. Neugebauer and Leighton, "Two-Micron Sky Survey"; G. Neugebauer, D. E. Martz, and R. B. Leighton, "Observations of Extremely Cool Stars," *Astrophysical Journal* 142 (1965): 399–401.

100. W. J. Luyten, "A New Star of Large Proper Motion, (L726-8)," *Astrophysical Journal* 109 (1949): 532–37.

101. In a letter to the author from W. J. Luyten, April 1, 1979.

102. A. C. B. Lovell, F. L. Whipple, and L. H. Solomon, "Radio Emission From Flare Stars," *Nature* 198 (1963): 228–30.

103. J. S. Hey, "Solar Radiations in the 4-6 Meter Radio Wavelength Band," *Nature* 157 (1946): 47–48; Grote Reber, "Cosmic Static," *Astrophysical Journal* 100 (1944): 279–87; G. C. Southworth, "Microwave Radiation from the Sun," *Journal of the Franklin Institute* 239 (1945): 285–97.

104. J. S. Hey, "Solar Radiations."

105. George E. Hale, "The Earth and Sun as Magnets," Smithsonian Report for 1913, pp. 145–58, reprinted in part in H. Shapley, *Source Book in Astronomy 1900–1950* (Cambridge, Mass.: Harvard University Press, 1960), pp. 32–35.

106. A telephone conversation with H. W. Babcock, April 16, 1979.

107. H. W. Babcock, "Zeeman Effect in Stellar Spectra," *Astrophysical Journal* 105 (1947): 105–19.

108. H. Weaver, D. R. W. Williams, N. H. Dieter, and W. T. Lum, "Observations of a Strong Unidentified Microwave Line and of Emission from the OH Molecule," *Nature* 208 (1965): 29–31.

109. S. Weinreb, A. H. Barrett, M. L. Meeks, and J. C. Henry, "Radio Observations of OH in the Interstellar Medium," *Nature* 200 (1963): 829–31; S. Weinreb, M. L. Meeks, J. C. Carter, A. H. Barrett, and A. E. E. Rogers, "Observations of Polarized OH Emission," *Nature* 208 (1965): 440–41; R. D. Davies, G. de Jager, and G. L. Verschuur, "Detection of Circular and Linear Polarization in the OH Emission of Sources Near W3 and W49," *Nature* 209 (1966): 974–77; R. D. Davies, B. Rowson, R. S. Booth, A. J. Cooper, H. Gent, R. L. Adgie, and J. H. Crowther, "Measurements of OH Emission Sources With an Interferometer of High Resolution," *Nature* 213 (1967): 1109–110.

110. A. H. Barrett and A. E. Lilley, "A Search for the 18-cm Line of *OH* in the Interstellar Medium," *Astronomical Journal* 62 (1957): 5–6.

111. A telephone conversation with Allan H. Barrett, May 11, 1979.

112. A variety of accounts, not all of them consistent, have been given of this discovery. Anthony Hewish has given one account in his Nobel Prize lecture "Pulsars and High Density Physics" reprinted in *Science* 188 (1975): 1079–83. Another report in the same journal is by Nicholas Wade, "Discovery of Pulsars: A Graduate Student's Story," *Science* 189 (1975): 358–64. Much of the controversy has surrounded the credit awarded to Bell or Hewish. There does not, however, seem to be any doubt about Bell's persistence in following up a hunch that her "bit of scruff" was something that required explaining. She has written her own good-natured account: S. Jocelyn Bell Burnell, "Little Green Men, White Dwarfs or What?" *Sky and Telescope* 55 (March 1978): 218–21.

113. Bruno Rossi in *X-ray Astronomy,* edited by R. Giacconi and H. Gursky (Dordrecht-Holland: Reidel Publishing Co., 1974), p. vii.

114. Two internal publications, both dated January 15, 1960, from American Science and Engineering Corporation (Cambridge, Mass.) record these activities. ASE-TN-49 "A Brief Review of Experimental and Theoretical Progess in X-Ray Astronomy" by R. Giacconi, G. W. Clark, and B. B. Rossi; and ASE-TN-50 "Instrumentation for X-Ray Astronomy" by R. Giacconi and G. W. Clark. The quotation comes from the first of these.

115. R. Giacconi, H. Gursky, F. R. Paolini, and Bruno B. Rossi, "Evidence for X Rays from Sources Outside the Solar System," *Physical Review Letters* 9 (1962): 439–43.

116. Riccardo Giacconi in *X Ray Astronomy,* p. 10

117. Herbert Friedman, "Rocket Astronomy," *Scientific American* 200, June 1959, pp. 57, 59.

118. Bruno Rossi, "X-ray Astronomy," *Daedalus* 106 (Fall 1977), p. 40.

119. Bruno Rossi, "X-ray Astronomy," pp. 40–41.

120. A telephone conversation with Herbert Friedman, May 14, 1979.

121. Giacconi, Clark, Rossi, "A Brief Review of Experimental and Theoretical Progress"; Giacconi, and Clark, "Instrumentation for X-ray Astronomy."

122. P. C. Gregory, P. P. Kronberg, E. R. Seaquist, V. A. Hughes, A. Woodsworth, M. R. Viner, D. Retallack, R. M. Hjellming, and B. Balick, "The Nature of the First Cygnus X-3 Radio Outburst," *Nature, Physical Science* 239 (1972): 114–17; E. E. Becklin, G. Neugebauer, F. J. Hawkins, K. O. Mason, P. W. Sanford, K. Matthews, and C. G. Wynn-Williams, "Infrared and X-ray Variability of Cyg X-3," *Nature* 245 (1973): 302–4.

123. J. J. L. Duyvendak, "Further Data Bearing on the Identification of the Crab Nebula with the Supernova of 1054 A.D., Part 1. The Ancient Oriental Chronicles," *Publications of the Astronomical Society of the Pacific* 54 (1942): 91–94; N. U. Mayall and J. H. Oort, "Further Data Bearing on the Identification of the Crab Nebula with the Supernova of 1054 A.D., Part 2. The Astronomical Aspects," *Publications of the Astronomical Society of the Pacific* 54 (1942): 95–104.

124. John C. Duncan, "Changes Observed in the Crab Nebula in Taurus," *Proceedings of the National Academy of Sciences* 7 (1921): 179–80; "Second Report on the Expansion of the Crab Nebula," *Astrophysical Journal* 89 (1939): 482–85.

125. Lampland, "Observed Changes in the Crab Nebula."

126. N. U. Mayall, "The Spectrum of the Crab Nebula in Taurus," *Publications of the Astronomical Society of the Pacific* 49 (1937): 101–5.

127. N. U. Mayall, "The Story of the Crab Nebula," *Science* 137 (1962): 91–102.

128. J. G. Bolton and G. J. Stanley, "The Position and Probable Identification of the Source of Galactic Radio-Frequency Radiation Taurus-A," *Australian Journal of Scientific Research Ser. A.,* 2 (1949): 139–48; J. H. Oort and T. Walraven, "Polarization and Composition of the Crab Nebula," *Bulletin of the Astronomical Institutes of the Netherlands* 12 (1956): 285–308; W. Baade, "The Polarization of the Crab Nebula on Plates Taken with the 200-Inch Telescope," *Bulletin of the Astronomical Institutes of the Netherlands* 12 (1956): 312. The work by Oort and Walraven also refers to the earlier observations by Vashakidze and Dombrovsky.

129. S. Bowyer, E. Byram, T. Chubb, and H. Friedman, "Lunar Occultation of X-ray Emission from the Crab Nebula," *Science* 146 (1964): 912–17.

130. Enrico Fermi, "On the Origin of the Cosmic Radiation," *Physical Review* 75 (1949): 1169–174.

131. L. Davis, Jr. and J. L. Greenstein, "The Polarization of Starlight by Aligned Dust Grains," *Astrophysical Journal* 114 (1951): 206–40.

132. H. Alfvén and N. Herlofson, "Cosmic Radiation and Radio Stars," *Physical Reveiw* 78 (1950): 616; K. O. Kiepenheuer, "Cosmic Rays as the Source of General Galactic Radio Emission," *Physical Review* 79 (1950): 738–39.

133. I. S. Shklovsky, *Cosmic Radio Waves,* trans. R. B. Rodman and C. M. Varsavsky (Cambridge, Mass.: Harvard University Press, 1960), pp. 292–316.

134. C. H. Mayer, T. P. McCullough, and R. M. Sloanaker, "Evidence for Polarized Radio Radiation from the Crab Nebula," *Astrophysical Journal* 126 (1957): 468–70.

135. G. L. Verschuur, "Further Measurements of Magnetic Fields in Interstellar Clouds of Neutral Hydrogen," *Nature* 223 (1969): 140–42.

136. R. N. Manchester, "Pulsar Rotation and Dispersion Measures and the Galactic Magnetic Field," *Astrophysical Journal* 172 (1972): 43–52.

137. William Parsons, 3rd earl of Rosse, "Observations on the Nebulae," *Philosophical Transactions of the Royal Society of London* 140 (1850): 499–514.

138. John C. Duncan, "Three Variable Stars and a Suspected Nova in the Spiral Nebula M 33 Trianguli," *Publications of the Astronomical Society of the Pacific* 34 (1922): 290–91.

139. E. P. Hubble, "Cepheids in Spiral Nebulae," *Observatory* 48 (1925): 139–42.

140. Walter Baade, *Evolution of Stars and Galaxies,* chap. 8.

141. J. H. Oort, "Observational Evidence Confirming Lindblad's Hypothesis of a Rotation of the Galactic System," *Bulletin of the Astronomical Institute of The Netherlands* 3 (1927): 275–82.

142. N. U. Mayall, "The Occurrence of λ 3727 [OII] in the Spectra Extragalactic Nebulae," *Publications of the Society of the Pacific* 51 (1939): 282–86.

143. F. J. Kerr, J. V. Hindman, and B. J. Robinson, "Observations of the 21-cm Line from the Magellanic Clouds," *Australian Journal of Physics* 7 (1954): 297–314.

144. Edwin Hubble, "The Distribution of Extra-Galactic Nebulae," *Astrophysical Journal* 79 (1934): 8–76.

145. John Herschel, "Catalogue of Nebulae and Clusters of Stars; Part 1," *Philosophical Transactions of the Royal Society of London* 154 (1864): 1–137.

146. Günther Buttman, *The Shadow of the Telescope: A Biography of John Herschel,* trans. B. E. J. Pagel (New York: Charles Scribner's Sons, 1970), pp. 96–97.

147. William Herschel, "On the Construction of the Heavens," *Philosophical Transactions of the Royal Society of London* 75 (1785): 213–66. *The Scientific Papers of Sir William Herschel,* The Royal Society, London (1912), Vol. I, p. 253.

148. Karl G. Jansky, "Electrical Disturbances," pp. 1387–98.

149. J. S. Hey, S. J. Parsons, and J. W. Phillips, "Fluctuations in Cosmic Radiation at Radio-Frequencies," *Nature* 158 (1946): 234.

150. F. G. Smith, C. G. Little, and A. C. B. Lovell, "Origin of the Fluctuations in the Intensity of Radio Waves from Galactic Sources," *Nature* 165 (1950): 422–24.

151. W. Baade and R. Minkowski, "Identification of the Radio Sources in Cassiopeia, Cygnus A and Puppis A.," *Astrophysical Journal* 119 (1954): 206–14.

152. K. I. Kellermann, "Radio Emission from Compact Objects," in *External Galaxies and Quasi-Stellar Objects,* International Astronomical Union Symposium no. 44, edited by David S. Evans (Dordrecht-Holland: Reidel Publishing Co., 1972), pp. 190–213.

153. J. Kristian, A. Sandage, and B. Katem, "On the Systematic Optical Identification of the Remaining 3C Radio Sources. I. A Search in 47 Fields," *Astrophysical Journal* 191 (1974): 43–50.

154. B. J. Harris, "QSQs and Radio Galaxies—Their Spectra and Time Variations at Radio Frequencies," in *External Galaxies and Quasi-Stellar Objects,* International Astronomical Union Symposium no. 44, edited by David S. Evans (Dordrecht-Holland: Reidel Publishing Co., 1972), pp. 232–48.

155. The list of forty-one galaxies is published in a paper by G. Strömberg, "Analysis of Radial Velocities of Globular Clusters and Non-Galactic Nebulae," *Astrophysical Journal* 61 (1925): 353–62.

156. Edwin Hubble, *The Realm of the Nebulae* (New York: Dover Publications, 1958), pp. 105–6.

157. Edwin Hubble, "A Relation Between Distance and Radial Velocity Among Extra-Galactic Nebulae," *Proceedings of the National Academy of Sciences* 15 (1929): 168–73.

158. C. Wirtz, "De Sitters Kosmologie und die Radialbewegungen der Spiralnebel," *Astronomische Nachrichten* 222 (1924): 21–26; W. de Sitter, "On Einstein's Theory of Gravitation, and its Astronomical Consequences," *Monthly Notices of the Royal Astronomical Society* 78 (1917): 3–28; in particular the discussion on pp. 27–28.

159. M. L. Humason, "The Apparent Radial Velocities of 100 Extra-Galactic Nebulae," *Astrophysical Journal* 83 (1936): 10–22.

160. C. Hazard, M. B. Mackey, and A. J. Shimmins, "Investigation of the Radio Source 3C 273 by the Method of Lunar Occultations," *Nature* 197 (1963): 1037–39.

161. David O. Edge and Michael J. Mulkay, *Astronomy Transformed—The Emergence of Radio Astronomy in Britain* (New York: John Wiley & Sons, 1976), pp. 204–208.

162. David O. Edge and Michael J. Mulkay, *Astronomy Transformed,* p. 206.

163. M. Schmidt, "3C 273: A Star-Like Object with a Large Red-Shift," *Nature* 197 (1963): 1040.

164. Jesse L. Greenstein and Thomas A. Matthews, "Red-Shift of the Unusual Radio Source: 3C 48," *Nature* 197 (1963): 1041–42.

165. A report on this meeting appeared in *Sky and Telescope* 21 (March 1961): 148. A copy of the original draft paper was kindly sent to me by Dr. Matthews.

166. From a personal interview with Cyril Hazard, August 13, 1979.

167. L. R. Allen, H. P. Palmer, and B. Rowson, "New Limits to the Diameters of Some Radio Sources," *Nature* 188 (1960): 731–32.

168. Allan Sandage, "The Existence of a Major New Constituent of the Universe: The Quasi-Stellar Galaxies," *Astrophysical Journal* 141 (1965): 1560–78.

169. M. J. Rees, "Appearance of Relativistically Expanding Radio Sources," *Nature* 211 (1966): 468–70; M. J. Rees, "Studies in Radio Source Structure—I. A Relativistically Expanding Model for Variable Quasi-Stellar Radio Sources," *Monthly Notices of the Royal Astronomical Society* 135 (1967): 345–60.

170. C. A. Knight, D. S. Robertson, A. E. E. Rogers, I. I. Shapiro, A. R. Whitney, T. A. Clark, R. M. Goldstein, G. E. Marandino, and N. R. Vandenberg, "Quasars: Millisecond-of-Arc Structure Revealed by Very-Long-Baseline Interferometry," *Science* 172 (1971): 52–54.

171. A telephone conversation with Irvin I. Shapiro, July 19, 1979.

172. W. D. Cotton, C. C. Counselman III, R. B. Geller, I. I. Shapiro, J. J. Wittels, H. F. Hinteregger, C. A. Knight, A. E. E. Rogers, A. R. Whitney, and T. A. Clark, "3C 279: The Case for 'Superluminal' Expansion," *Astrophysical Journal* 229 (1979): L115–17; G. A. Seielstad, M. H. Cohen, R. P. Linfield, A. T. Moffet, J. D. Romney, R. T. Schilizzi, and D. B. Shaffer, "Further Monitoring of the Structure of Superluminal Radio Sources," *Astrophysical Journal* 229 (1979): 53–72.

173. J. Gubbay, A. J. Legg, D. S. Robertson, A. T. Moffet, R. D. Ekers, and B. Seidel, "Variations of Small Quasar Components at 2,300 MHz," *Nature* 224 (1967): 1094–95. The D. S. Robertson cited here is David S. Robertson, and the D. S. Robertson mentioned in reference 170 is Douglas S. Robertson. I am indebted to Prof. I. I. Shapiro for pointing that out to me.

174. A. T. Moffet, J. Gubbay, D. S. Robertson, and A. J. Legg, "High Resolution Observations of Variable Radio Sources," in *External Galaxies and Quasi-Stellar Objects,* International Astronomical Symposium no. 44, edited by David S. Evans (Dordrecht-Holland: D. Reidel Publishing Co. and New York: Springer-Verlag, 1972), pp. 228–29.

175. A telephone conversation with A. T. Moffet, September 4, 1979.

176. Shapiro conversation, July 19, 1979.

177. E. T. Byram, T. A. Chubb, H. Friedman, "Cosmic X-Ray Sources, Galactic and Extragalactic," *Science* 152 (1966): 66–71.

178. Harold L. Johnson, "Infrared Photometry of Galaxies," *Astrophysical Journal* 143 (1966): 187–91.

179. E. E. Becklin and G. Neugebauer, "Infrared Observations of the Galactic Center," *Astrophysical Journal* 151 (1968): 145–61.

180. William F. Hoffmann and Carl L. Frederick, "Far-Infrared Observation of the Galactic-Center Region at 100 Microns," *Astrophysical Journal* 155 (1969): L9–L13.

181. D. E. Kleinmann and F. J. Low, "Observations of Infrared Galaxies," *Astrophysical Journal* 159 (1970): L165–72; G. H. Rieke and F. J. Low, "Infrared Photometry of Extragalactic Sources," *Astrophysical Journal* 176 (1972): L95–100.

182. Ray W. Klebesadel, Ian B. Strong, and Roy A. Olson, "Observations of Gamma-Ray Bursts of Cosmic Origin," *Astrophysical Journal* 182 (1973): L85–88.

183. Letter from R. W. Klebesadel to author.

184. Andrew McKellar, "Evidence for the Molecular Origin of Some Hitherto Unidentified Interstellar Lines," *Publications of the Astronomical Society of the Pacific* 52 (1940): 187–92.

185. Walter S. Adams, "Some Results with the Coudé Spectrograph of the Mount Wilson Observatory," *Astrophysical Journal* 93 (1941): 11–23.

186. Gerhard Herzberg, *Molecular Spectra and Molecular Structure I. Spectra of Diatomic Molecules,* 2nd ed. (New York: Van Nostrand, 1950), p. 496.

187. A. A. Penzias and R. W. Wilson, "A Measurement of Excess Antenna Temperature at 4080 Mc/s," *Astrophysical Journal* 142 (1965): 419–21.

188. G. Gamow, "Expanding Universe and the Origin of Elements," *Physical Review* 70 (1946): 572–73; Ralph A. Alpher and Robert C. Herman, "Remarks on the Evolution of the Expanding Universe," *Physical Review* 75 (1949): 1089–95.

189. R. H. Dicke, P. J. E. Peebles, P. G. Roll, and D. T. Wilkinson, "Cosmic Black-Body Radiation," *Astrophysical Journal* 142 (1965): 414–19.

190. Penzias and Wilson, "A Measurement of Excess Antenna Temperature."

191. Arno A. Penzias in *Cosmology, Fusion and Other Matters—George Gamow Memorial Volume,* edited by Frederick Reines, (New York: Colorado Associated University Press, 1972), pp. 32–34.

192. Steven Weinberg, *The First Three Minutes* (New York: Basic Books, 1977). See chapter 6 for a discussion of this question. Robert H. Dicke, Robert Beringer, Robert L. Kyhl, and A. B. Vane, "Atmospheric Absorption Measurements with a Microwave Radiometer," *Physical Review* 70 (1946): 340–48.

194. J. S. Hey, *The Evolution of Radio Astronomy* (London: Elek Science, 1973), p. 174; Gamow, "Expanding Universe."

195. Alpher and Herman, "Remarks on the Expanding Universe."

196. E. A. Ohm, "Project Echo Receiving System," *Bell System Technical Journal* 40 (1961): 1065–94.

197. W. C. Jakes, Jr., "Participation of the Holmdel Station in the *Telstar* Project," *Bell System Technical Journal* 42 (1963): 1424. I am indebted to Professor J. R. Pierce for forwarding to me a copy of a letter to him by David C. Hogg, dated January 3, 1979. Hogg, a participant of this project carried out at a 4GHz frequency, calls attention to the roughly 2.5 degree-Kelvin discrepancy between the measured noise of 17 degrees Kelvin and all other calculated sources of noise, roughly 14.5 degrees Kelvin, which included the by-then-known atmospheric contribution, as well as instrumental noise contributions.

198. A conversation in Ithaca, New York, with Robert H. Dicke, November 13, 1978, and a letter, dated November 20, 1978.

199. Giacconi et al., "Evidence for X Rays."

200. Herbert Friedman has recounted these events in a set of personal recollections, "Rocket Astronomy," *New York Academy of Sciences, Annals* 198 (1972): 271–72. There he also records that he and James E. Kupperian, Jr. reported on this radiation at the 1958 International Astronomical Union Assembly in Moscow.

201. G. W. Clark, G. P. Garmire, and W. L. Kraushaar, "Observation of High-Energy Cosmic Gamma Rays," *Astrophysical Journal* 153 (1968): L203–7.

202. The Caravane Collaboration for the COS-B Satellite consists of the Cosmic-Ray Working Group, Huygens Laboratorium, Leiden, The Netherlands; Laboratorio di Fisica Cosmica e Tecnologie Relative del CNR, Instituto di Scienze Fisische dell'Università di Milano, Italy; Instituto Fisica, Università di Palermo, Italy; Max Planck Insistut für Physik und Astrophysik, Institut für Extraterrestrische Physik, Garching bei München, Germany; Service d'Electronique Physique, Centre d'Etudes Nucléaires de Saclay, Gif-sur-Yvette, France; Space Science Department of the European Space Agency, ESTEC, Noordwijk, The Netherlands.

203. C. T. R. Wilson, "On the Ionisation of Atmospheric Air," *Proceedings of the Royal Society of London* 68 (1900): 151–61.

Chapter 3

1. John R. Carson, "Notes on the Theory of Modulation," *Proceedings of the Institute of Radio Engineers* 10 (1922): 57–64.

2. Harry Nyquist, "Certain Factors Affecting Telegraph Speed," *Bell System Technical Journal* 3 (1924): 324–46.

3. D. Gabor, "Theory of Communication," *Journal of the Institute of Electrical Engineers* 93, Part III (1946): 429–57.

4. J. Willard Gibbs, *Elementary Principles in Statistical Mechanics* (New Haven: Yale University Press, 1902; New York: Dover Publications, 1960), p. 5.

5. R. Cruddace, F. Paresce, S. Bowyer, and M. Lampton, "On the Opacity of the Interstellar Medium to Ultrasoft X-rays and Extreme-Ultraviolet Radiation," *Astrophysical Journal* 187 (1974): 497–504.

6. Robert J. Gould and Yoel Rephaeli, "The Effective Penetration Distance of Ultrahigh-Energy Electrons and Photons Traversing a Cosmic Blackbody Photon Gas," *Astrophysical Journal* 225 (1978): 318–24.

7. Gérard de Vaucouleurs, *Astronomical Photography* (New York: Macmillan Co., 1961), pp. 34, 72; J. C. Marchant and A. G. Millikan, "Photographic Detection of Faint Stellar Ob-

jects," *Journal of the Optical Society of America* 55 (1965): 907–11; Allan G. Millikan, "Image Detection at the Telescope," *American Scientist* 62 (1974): 324–33; Rolf Riekher, *Fernrohre und ihre Meister* (Berlin: VEB Verlag, 1957); Jesse L. Greenstein, "Astronomical Implications of Future Very Large Telescopes," in *ESO Conference on Optical Telescopes of the Future,* Geneva, December 12–15, 1977, edited by F. Pacini, W. Richter and R. N. Wilson (February 1978), p. 534; William Kitchiner, M. D., *Of Telescopes* (London: Whittaker, 1825).

8. Bessel, "Fernere Nachricht von der Bestimmung der Enterfernung von 61 Cygni," *Astronomische Nachrichten* 17 no. 402 (1840) cols. 273–76; C. W. Allen, *Astrophysical Quantities,* 3rd ed. (London: University of London, Athlone Press, 1973), p. 237.

9. Claude E. Shannon, "A Mathematical Theory of Communication," *Bell System Technical Journal* 27 (1948) 379–423, 623–56.

Chapter 4

1. Martin Harwit, "The Number of Class A Phenomena Characterizing the Universe," *Quarterly Journal of the Royal Astronomical Society* 16 (1975): 378–409.

2. Stirling A. Colgate, "Early Gamma Rays From Supernovae," *Astrophysical Journal* 187 (1974): 333–35.

3. A. H. Rots and W. W. Shane, "Distribution and Kinematics of Neutral Hydrogen in the Spiral Galaxy M81," *Astronomy and Astrophysics* 45 (1975): 25–42.

4. B. M. Oliver, "The Rationale for a Preferred Frequency Band: The Water Hole" in *The Search for Extraterrestrial Intelligence—SETI,* edited by Philip Morrison, John Billingham, and John Wolfe, National Aeronautics and Space Administration, NASA SP–419 (Washington, D.C.: U. S. Government Printing Office, 1977), pp. 63–74.

5. Philip Morrison, John Billingham, and John Wolfe, *The Search for Extraterrestrial Intelligence—SETI.*

6. Claude Bernard, *An Introduction to the Study of Experimental Medicine,* 1865, reprint, trans. Henry Copley Greene (New York: Henry Schuman, Inc., 1949).

Chapter 5

1. A conversation in Tucson Arizona, with Frank J. Low, March 27, 1980; and a telephone conversation April 9, 1980.

2. Sir Bernard Lovell, "The Effects of Defense Science on the Advance of Astronomy," *Journal for the History of Science* 8 (1977): 151–73.

3. A. Pannekoek, *A History of Astronomy* (London: George Allen & Unwin, 1951), p. 246.

4. "A Letter from the Reverend Mr. James Bradley, Savilian Professor of Astronomy at Oxford, and F. R. S. to Dr. Edmund Halley, Astronomer Royal & c. giving an Account of a new discovered Motion of the Fix'd Stars," *Philosophical Transactions of the Royal Society of London* 35 (1728): 637–61.

5. Mrs. John Herschel, *Memoir and Correspondence of Caroline Herschel* (New York: D. Appleton & Co., 1876), pp. 36, 53.

6. L. D. Landau, "On the Theory of Stars," *Physikalische Zeitschrift der Sowjetunion, 1,* 285–288 (1932); W. Baade and F. Zwicky, "Cosmic Rays from Super-Novae," *Proceedings of the National Academy of Sciences* 20 (1934): 259–63; J. R. Oppenheimer and G. M. Volkoff, "On Massive Neutron Cores," *Physical Review* 55 (1939): 374–81.

7. Eric Sheldon, "A Champion to the Rescue of Laplace," *Observatory* 99 (1979): 91–93. Sheldon's letter gives a modern evaluation of Pierre-Simon Laplace's contribution which originally appeared in *Exposition du Système du Monde,* vol. 2 (Paris: l'Imprimerie du Cercle-Social, 1796), p. 305.

8. J. R. Oppenheimer and H. Snyder, "On Continued Gravitational Contraction," *Physical Review* 56 (1939): 455–59.

9. Robert K. Merton, "Science, Technology and Society in Seventeenth Century England," *Osiris* 4 (1938): 360–632. Chapter 9 (pp. 543–57) deals with science and military technique. (New York: H. Fertig, 1970.)

10. David H. Clark and F. Richard Stephenson, *The Historical Supernovae* (New York: Pergamon Press, 1977), chap. 2.

11. Lovell, "The Effects of Defense Science on the Advance of Astronomy."

12. Martin Annis, telephone conversation September 8, 1978.

13. National Academy of Sciences, Committee on Science and Public Policy, *Employment Problems in Astronomy—Report of the Astronomy Manpower Committee* (Washington, D.C., March 1975). This study, chaired by Leo Goldberg, director of the Kitt Peak National Observatory, drew heavily on data compiled by the American Institute of Physics and the National Science Foundation and provides further references to these and other studies of manpower problems of astronomy.

14. David Edge, "The Sociology of Innovation in Modern Astronomy," *Quarterly Journal of the Royal Astronomical Society* 18 (1977): 329.

15. William Huggins and Lady Huggins, eds., *The Scientific Papers of Sir William Huggins* (London: William Wesley & Son, 1909), pp. 5–6. Excerpted from *The Nineteenth Century Review* (June 1897).

16. Philip Morrison in National Academy of Sciences, National Research Council, Space Science Board, "Opportunities and Choices in Space Science, 1974" (Washington, D.C.: NASA, 1975), pp. 21–22.

17. Robert McGinnis and Vijai P. Singh, "Three Types of Mobility and their Covariation among Physicists" (Paper presented to the American Sociological Association, August 1972).

18. David O. Edge and Michael J. Mulkay, *Astronomy Transformed—The Emergence of Radio Astronomy in Britain* (New York: John Wiley & Sons, 1976), pp. 359–64.

19. National Aeronautics and Space Administration, Physical Science Committee, "Report on NASA's Office of Space Science, Supporting Research and Technology and Data Analysis Programs," chaired by George B. Field, (Washington, D.C.: U.S. Government Printing Office, May 1976).

20. Jonathan R. Cole and Stephen Cole, "Which Researcher Will Get the Grant?" *Nature* 279 (1979): 575–76.

21. Harriet Zuckerman, *Scientific Elite: Nobel Laureates in the United States* (New York: The Free Press, 1977).

22. Stuart S. Blume, "Peer Review in the NSF," *Nature* 278 (1979): 807.

23. Cole and Cole, "Which Researcher Will Get the Grant?"

24. Raymond Orbach, "Basic Research: The Need for Lateral Movement" editorial in *Science* 205, no. 4402 (1979).

25. A. Carl Leopold, "The Burden of Competitive Grants," editorial in *Science* 203, no. 4381 (1979).

26. Luther J. Carter, "A New and Searching Look at NSF," *Science* 204 (1979): 1064–65.

27. Elmer B. Staats, "Federal Research Grants—Maintaining Public Accountability without Inhibiting Creative Research," *Science* 205 (1979): 18–20.

28. F. W. Bessel, "Bestimmung der Entfernung des 61sten Sterns des Schwans," *Astronomische Nachrichten* 16 nos. 365–66 (1838): cols. 65–96.

29. F. W. Bessel, "Fernere Nachricht von der Bestimmung der Entfernung des 61sten Sterns des Schwans," *Astronomische Nachrichten* 17, no. 401 (1840): cols. 257–72.

30. William E. Howard III, Memorandum to the Astronomy Advisory Committee of the National Science Foundation and Directors of the National Astronomy Centers, April 13, 1978, p. AST: 4/3/78 LRP (3). I thank Dr. Howard for providing me with a copy of this memorandum. (See also *Physics Today*, March 1978, pp. 103–4).

31. Max-Planck-Gesellschaft, Jahrbuch, 1979 (Munich), p. 119.

32. Reimar Lüst, "Wie die MPG Forschung lebendig hält," *MPG Spiegel*, nos. 3, 4 (1979), pp. 56–57, (a publication of the Max Planck Society of West Germany). The translation given here is my own—M. H.

33. Kitt Peak National Observatory, *Next Generation Telescope*. A design study conducted at the Kitt Peak National Observatory, published by KPNO as a series of reports (1977–79).

34. D. N. Hall, R. C. M. Learner, and L. D. Barr, *Next Generation Telescope, Astronomical Potential and Scientific Uses for a Large Aperture Optical Telescope*. Report no. 1 (Tucson, Arizona: Kitt Peak National Observatory, 1977), p. 3.

35. Lyman Spitzer, Jr., "History of the Space Telescope," *Quarterly Journal of the Royal Astronomical Society*, London, 20 (1979): 29–36; M. S. Longair, "The Space Telescope and Its Opportunities," *Quarterly Journal of the Royal Astronomical Society*, London, 20 (1979): 5–28.

36. "The VLA Takes Shape," *Sky and Telescope*, 52 (November 1976).

37. Ralph Kazarian, "World's Most Powerful Radio Telescope Ready for Use by Scientists," National Science Foundation News Release, NSF PR 78–15, February 20, 1978.

38. B. M. Oliver in *The Search for Extraterrestrial Intelligence—SETI*, edited by Philip Morrison, John Billingham and John Wolfe, National Aeronautics and Space Administration, NASA SP–419 (Washington, D.C., 1977), pp. 136–37.

39. R. Giacconi et al. "The Einstein (HEAO-2) X-Ray Observatory," *Astrophysical Journal* 230 (1979): 540–50.

40. Frederick Reines, "The Early Days of Experimental Neutrino Physics," *Science* 203 (1979): 11–16.

41. Steven Weinberg, *The First Three Minutes* (New York: Basic Books, 1977), p. 118.

42. John N. Bahcall, "Solar Neutrino Experiments," *Reviews of Modern Physics* 50 (1978): 881–903.

43. J. N. Bahcall, B. T. Cleveland, R. Davis, Jr., L. Dostrovsky, J. C. Evans, Jr., W. Frati, G. Friedlander, K. Lande, J. K. Rowley, R. W. Stoenner, and J. Weneser, "Proposed Solar-Neutrino Experiment Using ⁷¹Ga," *Physical Review Letters* 40 (1978): 1351–54; "Solar-Neutrino Hunters Still Seek Explanation," *Physics Today,* December 1978, p. 20.

44. David Eichler, "Deep-Sea Neutrinos," *Nature* 276 (1978): 15.

45. J. Anthony Tyson and R. P. Giffard, "Gravitation-Wave Astronomy," *Annual Review of Astronomy and Astrophysics* 16 (1978): 521–54.

46. Bahram Mashhoon, "On the Detection of Gravitational Radiation by the Doppler Tracking of Spacecraft," *Astrophysical Journal,* 227 (1979): 1019–36.

47. National Aeronautics and Space Administration, "Astrophysics Project Concept Summary: Gravity Wave Interferometer," 1978–261–371:11 (Washington, D.C.: U.S. Government Printing Office, March 1978).

48. David O. Edge and Michael J. Mulkay, *Astronomy Transformed,* p. 437.

49. National Academy of Sciences, National Research Council, *Ground Based Astronomy, A Ten Year Program—A Report Prepared by the Panel on Astronomical Facilities,* Committee on Science and Public Policy of the National Academy of Sciences, chaired by A. E. Whitford (Washington, D.C., 1964), p. 74.

50. National Academy of Sciences: *Astronomy and Astrophysics for the 1970's—Volume 1, Report of the Astronomy Survey Committee,* 2 vols., chaired by Jesse L. Greenstein (Washington, D.C., 1972), 1:xiii–xiv.

51. *Astronomy and Astrophysics for the 1970's,* 1:8–9.

52. A private discussion with Professor Jesse Greenstein, December 11, 1978, at Cal Tech.

53. *Astronomy and Astrophysics for the 1970's.*

54. Greenstein conversation, December 11, 1978.

55. National Academy of Sciences, National Research Council, *Physics in Perspective,* 3 vols., chaired by D. Allan Bromley, vol. 2, pt. B, "The Interfaces," chap. 8 (Washington, D.C., 1972–73), pp. 749–848.

56. National Academy of Sciences: *Astronomy and Astrophysics for the 1970's.—Report of the Panels,* 2 vols., chaired by Jesse L. Greenstein, (Washington, D.C., 1973), 2:282–83; National Academy of Sciences, National Research Council, *Physics in Perspective,* 3 vols., chaired by D. Allan Bromley, vol. 1, vol. 2A, "The Core Subfields of Physics"; vol. 2B, "The Interfaces." (Washington, D.C., 1972–1973).

57. National Academy of Sciences, National Research Council, *Physics in Perspective,* 1:381–453.

Appendix A

1. R. A. Fisher, A. S. Corbet, and C. B. Williams, "The Relation Between the Number of Species and the Number of Individuals in a Random Sample of an Animal Population," *Journal of Animal Ecology* 12 (1943): 42–58.

2. L. A. Goodman, "On the Estimation of the Number of Classes in a Population," *Annals of Mathematical Statistics* 20 (1949): 572–79; I. J. Good, "The Population Frequencies of Species and the Estimation of Population Parameters," *Biometrika* 40 (1953): 237–64.

3. H. E. Robbins, "Estimating the Total Probability of the Unobserved Outcomes of an Experiment," *Annals of Mathematical Statistics* 39 (1968): 256–57; H. W. Menard and G. Sharman, "Scientific Uses of Random Drilling Models," *Science* 190 (1975): 337–43.

4. B. Efron and R. Thisted, "Estimating the Number of Unseen Species: How Many Words Did Shakespeare Know?" *Biometrika* 63 (1976): 435–47.

Appendix B

1. Claude E. Shannon, "A Mathematical Theory of Communication," *Bell System Technical Journal* 27 (1948): 379–423, 623–56. Reprint ed. C. E. Shannon and W. Weaver (Urbana, Illinois: University of Illinois Press, 1949); Gordon Raisbeck, *Information Theory—An Introduction for Scientists and Engineers,* (Cambridge, Massachusetts: MIT Press, 1964). This gives clear explanations for several theoretical concepts discussed in this appendix. R. V. L. Hartley, "Transmission of Information," *Bell System Technical Journal* 7 (1928): 535–63.

2. Raisbeck, *Information Theory.*

Glossary/Index

ABERRATION, 63. The apparent displacement of the position of a star in the sky resulting from the earth's orbital velocity around the sun. The angular displacement is of the order of the orbital speed divided by the speed of light.

ABSOLUTE MAGNITUDE, 79; see MAGNITUDE OF A STAR.

ACHROMATIC REFRACTOR, 76, 175. A telescope with lenses chosen to produce images that are sharply focused in one plane for all wavelengths of light. This avoids color-distortion of the image.

Adams, Walter S., 86, 147, 235.

ADVANCED X-RAY ASTROPHYSICS FACILITY, 271. A planned X-ray observatory.

ALGORITHM, 217. A recipe that provides a step-by-step procedure for handling a mathematical or a sorting problem.

ALMAGEST, 76. A compendium on ancient astronomy compiled by Ptolemy (Claudius Ptolemaeus of Alexandria) in the second century A.D. The name of the work is derived from the Arabian Al-majisti, a corruption of the original Greek title.

Alpher, Ralph, 148, 245.

Al Sufi, 82.

ANALYTICAL WORK, 24, 46, 264.

ANGSTROM UNIT (Å), 97, 203, 205. A measure of length equaling 10^{-8} cm or 10^{-10} m (table G.1). The distance between atoms in a solid is typically one Angstrom unit.

TABLE G.1
Relations Between Units of Length

Unit	Length in Centimeters	Length in Meters
Angstrom unit	10^{-8} cm	10^{-10} m
Micron	10^{-4}	10^{-6}
Millimeter	10^{-1}	10^{-3}
Centimeter	1	10^{-2}
Meter	10^{2}	1
Kilometer	10^{5}	10^{3}
Light-year	9×10^{17}	9×10^{15}
Parsec	3×10^{18}	3×10^{16}

ANGULAR DIAMETER, 139. The angle subtended by the diameter of an object as viewed from the distance of the observer.

ANGULAR RESOLUTION, 29, 141, 160, 172, 173, 184, 270. The angular separation at which two points of equal brightness can barely be seen to be apart.

ANNIHILATION OF MATTER, 26; see ANTIMATTER and table G.3. The destruction of matter on encountering antimatter, with an accompanying liberation of energy and the formation of pairs of particles and their antiparticles.

Annis, Martin, 119, 243, 246.

ANTIMATTER, 26, 177. Matter consisting of antiparticles.

ANTINUCLEON, 177. Antiparticle of a nucleon.

ANTIPARTICLE, 26; see ANNIHILATION OF MATTER. Matter consists of atoms that contain neutrons, protons, and electrons. Corresponding to each of these three particles there exists an antiparticle with identical mass, but opposite charge, if any. Neutrinos, which are neutral massless particles, are distinguished from antineutrinos by the direction of their spin. Particles and antiparticles annihilate on encounter.

TABLE G.2

Relations Between Units of Energy and Corresponding Wave-
lengths of Photons (Quanta of Electromagnetic Radiation)

Unit		Energy in Ergs*	Corresponding Photon Wavelength and Frequency†	
1 electron Volt (eV)	=	1.6×10^{-12} erg	1.2×10^{-4} cm	2.5×10^{14} Hz
1 keV $= 10^3$ eV	=	1.6×10^{-9}	1.2×10^{-7}	2.5×10^{17}
1 MeV $= 10^6$ eV	=	1.6×10^{-6}	1.2×10^{-10}	2.5×10^{20}
1 GeV $= 10^9$ eV	=	1.6×10^{-3}	1.2×10^{-13}	2.5×10^{23}

* The conversion between wavelength λ, measured in centimeters (cm), and frequency f, measured in Hertz (Hz), is

frequency × wavelength = speed of light
$$f\lambda = 3 \times 10^{10}$$

† The conversion between energy E, measured in ergs, and frequency, measured in Hz, is

energy = frequency × Planck's constant
$$E = 6.6 \times 10^{-27} f$$

DECLINATION, 216. The position of a star in the sky is defined by two sets of coordinates. One gives its east-west position and is known as the right ascension; the other coordinate is the position angle north or south of the equator.
de Sitter, W., 135.
Dicke, Robert H., 148.
DIFFERENTIAL ROTATION, 129. Planets at larger distances from the sun complete their orbits more slowly than the period of a year required by Earth. Similarly, stars at large distances from the center of a galaxy orbit the gravitationally attracting core more slowly than stars at lesser distances. Both systems of masses exhibit differential rotation about a massive center.
DIFFRACTION, 85, 87, 182. The spreading of a light beam around a body that blocks part of the beam.
DIFFUSE SOURCE, 86. An extended tenuous source of radiation.
DISCOVERY, 13, 17, 34, 43, 48, 234.
DISK, 225; see GALACTIC PLANE, RINGS.
Dollond, John, 175.
Dollond, Peter, 175.
Dombrovsky, V. A., 125, 126, 238.
DOPPLER BROADENING, 98, 169. A broadening of a spectral feature produced when different components of a source move at different velocities with respect to an observer and are Doppler shifted by different amounts.
DOPPLER SHIFT, 61, 80, 88, 124, 129, 136, 138, 222. The systematic shift of an entire spectrum of radiation toward longer wavelengths—lower frequencies—when the source of radiation rapidly recedes from the observer, and toward shorter wavelengths—higher frequencies—when the source approaches.
Draper, Henry, 175.
Duncan, J. C., 122, 127, 237.
Dunham, Edward, 72.
Dunham, Theodore, 147.
DUST, 61, 93, 106, 109, 146. Fine grains of solid matter.
DUST CLOUD, 98. In interstellar space dust appears aggregated in dark irregular clouds.
Duyvendak, J. J. L., 122, 237.
DWARF CEPHEID, 81. A variable star, somewhat brighter than the sun, pulsing with a period in the range of one to eight hours.
ECLIPSING BINARY, 82. A pair of orbiting stars in which one star passes in front of the other and blocks its light.
ECLIPTIC, 278. The plane in which the planets orbit the sun.
Eddington, Arthur Stanley, 81.
Edge, David O., 8, 20, 137, 247, 279.
EDUCATION OF ASTRONOMERS, 20, 21, 47, 234, 241, 242, 246, 252.
Efron, B., 294.
Einstein, Albert, 8, 135.
EINSTEIN OBSERVATORY, 270. An X-ray astronomical observatory launched into earth orbit in late 1978, containing a special X-ray reflecting telescope with high sensitivity and angular resolving power.

TABLE G.3

Energy Content of a Thimbleful of Water under a Variety of Conditions

Condition	Energy Content*	
	Measured in Ergs	Measured in Electron Volts
Heat gained in absorbing one quantum of sunlight	10^{-12} erg	1 eV
Heat gained if all the energy from the most energetic known cosmic ray (nuclear) particle were absorbed	10^8	10^{20}
Energy due to motion in an airplane travelling at the speed of sound	10^9	10^{21}
Thermal energy at rest at room temperature	10^{10}	10^{22}
Thermal energy at rest at sun's surface temperature	10^{11}	10^{23}
Energy of motion due to Earth's orbital velocity around the sun	10^{13}	10^{25}
Thermal energy at rest at sun's central temperature	10^{14}	10^{26}
Thermonuclear conversion to derive the nuclear energy of water†	10^{18}	10^{30}
Annihilation of total mass to liberate energy	10^{21}	10^{33}

* These values are only approximate and are rounded off. A unit of 1 electron volt (eV) should actually equal 1.6×10^{-12} erg.

† A thimbleful of solar material is nearly ten times richer in hydrogen than the same mass of water and could liberate another factor of ten more nuclear energy.

TABLE G.4
*Relations Between Units
of Mass*

Unit	Mass in Grams
Microgram	10^{-6} g
Milligram	10^{-3}
Gram	1
Kilogram	10^3
Metric ton	10^6
Solar mass	2×10^{33}

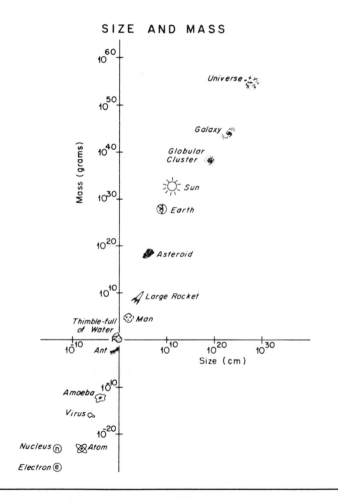

SIZE AND MASS

Figure G.1 *Size and Mass*

The masses and sizes of different bodies are shown on a logarithmic scale, expressed in the powers of 10 by which they exceed that of a cubic centimeter, roughly a thimbleful, of water. The range stretches from the electron, whose mass is $\sim 10^{-27}$ g and whose radius is $\sim 10^{-13}$ cm, to the universe whose mass is $\sim 10^{55}$ g and whose size is $\sim 10^{28}$ c.

$$10^n = N$$

TABLE G.5
Energy Liberated Each Second by Various Bodies

Source	Emission Rate*	
	In Ergs/Second	In Solar Luminosities, $L_\odot$
Ant (metabolism)	10^2 erg/sec	$10^{-32}L_\odot$
Man (metabolism)	10^9	10^{-25}
Large rocket during launch	10^{16}	10^{-18}
Earth's re-emission of absorbed sunlight	10^{24}	10^{-10}
White dwarf star	10^{31}	10^{-3}
Sun's emission of light	10^{34}	1
Supernova at peak brightness	10^{44}	10^{10}
Light emitted by Milky Way	10^{45}	10^{11}
All cosmic sources (total)	10^{56}	10^{22}

* These quantities are approximate and rounded off. The solar luminosity, the sun's emission of light, actually amounts to 4×10^{33} erg/sec.

SPEED OF LIGHT, 25, 136, 140. The speed of light is 3×10^{10} cm per second or, equivalently, 300,000 kilometers a second.

SPIN, 167; see PLANCK'S CONSTANT. Every fundamental particle is characterized by a spin. For electrons, protons, neutrons, and neutrinos that spin is ½, measured in a unit related to Planck's constant. For light quanta the spin is 1, in the same units. For gravitational waves the spin is 2.

SPIRAL GALAXY, 39, 127, 222. A galaxy that exhibits stars, gas, and dust arranged in lanes or segments of lanes that stretch outward from the galaxy's center in a spiral pattern. Barred spirals are galaxies in which spiral arms appear at the ends of an elongated bar-shaped aggregate of stars at the galaxy's center.

SPYGLASS, 13, 20, 65, 68, 127. The name used for a telescope until about 1610.

Staats, Elmer B., 258.

STABILITY, 224. The ability to withstand small disturbances and return to equilibrium.

Stanley, Gordon, 125.

STAR, 75; see BINARY STAR, EVOLVED STAR, FLARE STAR, GIANT STAR, MAGNETIC STAR, MAIN SEQUENCE STAR, OLD STAR, RED GIANT, SUBGIANT STAR, SYMBIOTIC STARS, VARIABLE STAR, YOUNG STAR. A gravitationally bound compact mass containing between 10^{32} and 10^{35} grams of matter. It can keep shining as long as nuclear or gravitational energy keeps being released by activity in the star's highly compressed central regions. Young stars are those formed within the last ten to a hundred million years. Old stars are those whose appearance suggests an age of 10^8 to 10^{10} years. Unusually luminous stars are classified as giants and supergiants.

STATISTICAL RANDOMNESS, 30, 181, 220. Random behavior constrained only by the physical makeup of a system.

STATISTICAL REASONING, 219. Reasoning based on the probabilities of randomly occurring events in a system of known structure.

STELLAR INTERFEROMETRY, 161. A method that makes use of interference of electromagnetic waves to measure the angular diameter of a source.

Stephenson, F. Richard, 100, 104, 245.

Stever, Guyford, 257.

Strong, Ian B., 16, 146, 239.

Struve, Wilhelm, 75, 175.

SUBGIANT STAR, 79, 80; see SPECTRAL TYPE. A star whose brightness, on a Hertzsprung-Russell diagram, lies part way between that of a main sequence star and a normal giant of the same spectral type.

SUN, 15, 20, 75, 112, 113, 120.

SUPERLUMINAL SOURCE, 140. A source whose components are expanding at a rate apparently faster than the speed of light.

SUPERLUMINAL VELOCITIES, 140. Speeds that appear to be greater than the speed of light.

SUPERNOVA, 100; see table G.5. A star whose brightness can increase by a factor of ~10^8 over a period of hours or days as the star explodes. Supernovae are the brightest individual stars known. Their bright phase declines over a period of months.

SUPERNOVA REMNANT, 122. Nebulosity surrounding the site of an earlier supernova explosion.

SURFACE BRIGHTNESS, 202. The energy emanating from each unit of area in a second.

SYMBIOTIC STARS, 104. Close binary stars that exchange mass.

TEMPERATURES IN THE UNIVERSE

	Degrees Kelvin
Central Temperature Inside Exploding Supernova	10^9
	10^8
Temperature at Center of Sun	10^7
	10^6
	10^5
Surface of a Hot Young Giant Star	
Sun's Surface Temperature	10^4
Temperature Above Which All Solids Vaporize	10^3
Man, and Earth's Surface Temperature	
	10^2
	10
Cosmic Microwave Background Radiation	
	1
	10^{-1}
	10^{-2}
	10^{-3}
	10^{-4}
Lowest Temperature Achieved in Laboratory	10^{-5}